case studies
in immunology

A CLINICAL COMPANION

FIFTH EDITION

case studies in immunology

A CLINICAL COMPANION

FIFTH EDITION

Raif Geha • **Fred Rosen**

Harvard Medical School

Garland Science
Taylor & Francis Group

NEW YORK AND LONDON

Vice President: Denise Schanck
Senior Editor: Janet Foltin
Text Editor: Eleanor Lawrence
Assistant Editor: Sigrid Masson
Editorial Assistant: Katherine Ghezzi
Senior Production Editor: Simon Hill
Copyeditor: Bruce Goatly
Indexer: Merrall-Ross International Ltd.
Illustration: Blink Studio
Layout: Georgina Lucas

10-digit ISBN 0-8153-4145-8 (paperback)
13-digit ISBN 978-0-8153-4145-1 (paperback)

Library of Congress Cataloging-in-Publication Data

Geha, Raif S.
  Case studies in immunology : a clinical companion / Raif Geha, Fred Rosen. -- 5th ed.
    p. ; cm.
  Rosen's name appears first on the earlier edition.
  Includes index.
  ISBN 978-0-8153-4145-1
  1. Clinical immunology--Case studies. I. Rosen, Fred S. II. Title.
  [DNLM: 1. Immune System Diseases--Case Reports. 2. Allergy and Immunology--
Case Reports. 3. Immunity--genetics--Case Reports. WD 300 G311c 2007]
  RC582.R67 2007
  616.07'9--dc22
                              2007002977

Published by Garland Science, Taylor & Francis Group, LLC, an informa business
270 Madison Avenue, New York, NY 10016, USA, and
2 Park Square, Milton Park, Abingdon, OX14 4RN, UK.

Printed in the United States of America.

15 14 13 12 11 10 9 8 7 6 5 4 3 2 1

Taylor & Francis Group, an informa business

Visit our web site at http://www.garlandscience.com

Preface

The science of immunology started as a case study. On May 15, 1796 Edward Jenner inoculated a neighbor's son, James Phipps, with vaccinia (cowpox) virus. Six weeks later, on July 1, 1796, Jenner challenged the boy with live smallpox and found that he was protected against this infection. During its 208 year history the basic science of immunology has been closely related to clinical observations and has shed light on the pathogenesis of disease.

The study of immunology provides a rare opportunity in medicine to relate the findings of basic scientific investigations to clinical problems. The case histories in this book are chosen for two purposes: to illustrate in a clinical context essential points about the mechanisms of immunity; and to describe and explain some of the immunological problems often seen in the clinic. For this fifth edition, we have added five completely new cases that illustrate both recently discovered genetic immunodeficiencies and some more familiar and common diseases with interesting immunology. We have revised other cases to add newly acquired information about these diseases. Fundamental mechanisms of immunity are illustrated by cases of genetic defects in the immune system, immune complex diseases, immune mediated hypersensitivity reactions and autoimmune and alloimmune diseases. These cases describe real events from case histories, largely but not solely drawn from the records of the Boston Children's Hospital and the Brigham and Women's Hospital. Names, places, and time have been altered to obscure the identity of the patients described; all other details are faithfully reproduced. The cases are intended to help medical students and pre-medical students to learn and understand the importance of basic immunological mechanisms, and particularly to serve as a review aid; but we hope and believe they will be useful and interesting to any student of immunology.

Each case is presented in the same format. The case history is preceded by basic scientific facts that are needed to understand the case history. The case history is followed by a brief summary of the disease under study. Finally there are several questions and discussion points that highlight the lessons learned from the case. These are not intended to be a quiz but rather to shed further light on the details of the case.

The Garland Science website (www.garlandscience.com) now provides instructors who adopt *Case Studies* with a link to Garland Science Classwire, where the textbook art can be found in a downloadable, web-ready format, as well as in PowerPoint-ready format.

We are grateful to Dr. Peter Densen of the University of Iowa for C8 deficiency case material, Dr. Sanjiv Chopra of Harvard Medical School for the case on mixed essential cryoglobulinemia and Dr. Peter Schur of the Brigham and Women's Hospital for the rheumatoid arthritis case. We also thank Dr. Jane Newburger of the Boston Children's Hospital for the case on rheumatic fever and Dr. Eric Rosenberg of the Massachusetts General Hospital for the AIDS case. We are also greatly indebted to our colleagues Drs. David Dawson, Susan Berman, Lawrence Shulman and David Hafler of the Brigham and Women's Hospital, to Dr. Razzaque Ahmed of the Harvard School of Dental Medicine, to Drs. Ernesto Gonzalez and Scott Snapper of the Massachusetts General Hospital and to Drs. Peter Newburger and Jamie Ferrara of the Departments of Pediatrics of the University of Massachusetts and the University of Michigan and Dr. Robertson Parkman of the Los Angeles Children's Hospital as well as Henri de la Salle of the Centre régional de Transfusion sanguine in Strasbourg and Professor Michael

Levin of St. Mary's Hospital, London for supplying case materials. Our colleagues in the Immunology Division of the Children's Hospital have provided invaluable service by extracting summaries of long and complicated case histories; we are particularly indebted to Drs. Lynda Schneider, Leonard Bacharier, Francisco Antonio Bonilla, Hans Oettgen, Jonathan Spergel, Rima Rachid, Scott Turvey, Jordan Orange, Eamanuela Castigli, Andrew McGinnitie, Marybeth Son, Melissa Hazen, Douglas McDonald and John Lee, and to Lilit Garibyan, third year medical student at Harvard Medical School, in constructing several case histories. In the course of developing these cases, we have been indebted for expert and pedagogic advice to Fred Alt, Mark Anderson, John Atkinson, Hugh Auchincloss, Stephen Baird, Zuhair K. Ballas, Leslie Berg, Corrado Betterle, Kurt Bloch, Jean-Laurent Casanova, John J. Cohen, Michael I. Colston, Anthony DeFranco, Peter Densen, Ten Feizi, Alain Fischer, Christopher Goodnow, Edward Kaplan, George Miller, Luigi Notarangelo, Peter Parham, Jaakko Perheentupa, Jennifer Puck, Westley Reeves, Patrick Revy, Peter Schur, Anthony Segal, Lisa Steiner, Stuart Tangye, Cox Terhorst, Emil Unanue, André Veillette, Jan Vilcek, Mark Walport, Fenella Woznarowska, and John Zabriskie.

Eleanor Lawrence has spent many hours honing the prose as well as the content of the cases and we are grateful to her for this. We would also like to acknowledge the Garland Science team for their work on the fifth edition.

A note to the reader

The cases presented in this book have been ordered so that the main topics addressed in each case follow as far as possible the order in which these topics are presented in the seventh edition of *Janeway's Immunobiology* by Kenneth Murphy, Paul Travers, and Mark Walport. However, inevitably many of the early cases raise important issues that are not addressed until the later chapters of *Immunobiology*. To indicate which sections of *Immunobiology* contain material relevant to each case, we have listed on the first page of each case the topics covered in it. The color code follows the code used for the five main sections of *Immunobiology*: yellow for the introductory chapter and innate immunity, blue for the section on recognition of antigen, red for the development of lymphocytes, green for the adaptive immune response, purple for the response to infection and clinical topics, and orange for methods.

Dedication

This fifth edition is dedicated to Fred Rosen (1935-2005). Fred dedicated his career of more than 50 years to the investigation and care of patients with primary immunodeficiency disease. Above all, he loved to teach and he did so superbly, aided by an encyclopedic knowledge of immunology, an incisive intelligence, an incredible memory, and charisma combined with an aura of authority. Fred had an enormous influence on many generations of both basic and clinical immunologists. This book is his brainchild and his contribution to it will be sorely missed.

CONTENTS

CASE 1 Congenital Asplenia

The role of the spleen in immunity.

Topics bearing on this case:

Circulation of lymphocytes through secondary lymphoid tissues

Toxoid vaccines

Hemagglutination tests

T-cell help in antibody response

The adaptive immune response occurs mainly in the secondary lymphoid tissue—the lymph nodes, the gut-associated lymphoid tissue, and the spleen (Fig. 1.1). Pathogens and their secreted antigens are trapped in these tissues, and presented to the naive lymphocytes that constantly pass through. Microorganisms that enter the body through the skin or the lungs drain to regional lymph nodes where they stimulate an immune response. Microorganisms and food antigens that enter the gastrointestinal tract are collected in the gut-associated lymphoid tissue. Microbes that enter the bloodstream stimulate an immune response in the spleen.

Fig. 1.1 The distribution of lymphoid tissues in the body. Lymphocytes arise from stem cells in bone marrow, and differentiate in the central lymphoid organs (yellow)—B cells in bone marrow and T cells in the thymus. They migrate from these tissues through the bloodstream to the peripheral lymphoid tissues (blue)—the lymph nodes, spleen, and mucosal-associated lymphoid tissues such as tonsils, Peyer's patches, and appendix. These are the sites of lymphocyte activation by antigen. Lymphatics drain extracellular fluid as lymph through the lymph nodes and into the thoracic duct, which returns the lymph to the bloodstream by emptying into the left sub-clavian vein. Lymphocytes that circulate in the bloodstream enter the peripheral lymphoid organs, and are eventually carried by lymph to the thoracic duct where they reenter the bloodstream.

The spleen

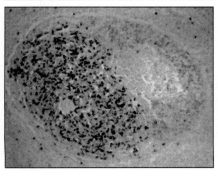

The spleen is organized to accomplish two functions (Fig. 1.2). In addition to being a secondary lymphoid organ, it acts as a filter of the blood to remove aged or abnormal red cells and other extraneous particles that may enter the blood-stream, including microorganisms. The lymphoid function of the spleen is carried out in the white pulp and the filtration function by the red pulp. Many microorganisms are recognized directly and engulfed by the phagocytes of the red pulp. Others are not removed efficiently until they are coated by antibodies generated in the white pulp. In experimental animals, an immune response (as measured by antibody formation) can be detected in the white pulp of the spleen approximately 4 days after the intravenous injection of a dose of microorganisms. The clearance of antibody- and complement-coated bacteria or viruses by the phagocytic cells of the red pulp of the spleen is very rapid. Rapid clearance from the blood is important as it prevents these bacteria from disseminating and causing infections of the meninges (meningitis), the kidney (pyelonephritis), the lung (pneumonia), or other distant anatomical sites.

Fig. 1.2 Schematic views and light micrograph of a section of spleen. The spleen consists of red pulp (pink areas), which is a site of red blood cell destruction, interspersed with lymphoid white pulp. The center panel shows an enlargement of a small section of the spleen showing the arrangement of discrete areas of white pulp around central arterioles. The white pulp is shown in transverse section. Although the organization of the spleen is similar to that of a lymph node, antigen enters the spleen from the blood rather than from the lymph. Photograph courtesy of J.C. Howard.

Bacteria enter the bloodstream all the time, such as when we brush our teeth or when we have a local infection, for example of the skin or middle ear. Normally these bacteria are disposed of efficiently by the spleen. When, for one reason or another, the spleen is not present, serious, even fatal, infections occur.

The case of Susan Vanderveer: a fatality because of an absent spleen.

Mr and Mrs Vanderveer owned a farm in the Hudson Valley in lower New York State. They were both descended from Dutch settlers who came to the Hudson Valley in the mid 17th century. There were multiple consanguineous marriages among their ancestors, and Mr and Mrs Vanderveer were distantly related to each other. At the time of this case, they had five children—three girls and two boys. Their youngest daughter, Susan, was 10 months old when she developed a cold, which lasted for 2 weeks. On the 14th day of her upper respiratory infection, she became sleepy and felt very hot. Her mother found that her temperature was 41.7°C. When Susan developed convulsive movements of her extremities, she was rushed to the emergency room but she died on the way to the hospital. Post-mortem cultures of blood were obtained, and also from her throat and cerebrospinal fluid. All the cultures grew *Haemophilus influenzae*, type b. At autopsy Susan was found to have no spleen.

Susan Vanderveer, age 10 months, dead on arrival in Emergency.

At the time of Susan's death her 3-year-old sister, Betsy, also had a fever of 38.9°C. She complained of an earache and her eardrums were found to be red. She had no other complaints and no other abnormalities were detected on physical examination. Her white blood count was 28,500 cells μl^{-1} (very elevated). Cultures from her nose, throat, and blood grew out *Haemophilus influenzae*, type b. She was given ampicillin intravenously for 10 days in the hospital and was then sent home in good health. Her cultures were negative at the time of discharge from the hospital. She was seen by a pediatrician on three occasions during the following year for otitis media (inflammation of the middle ear), pneumonia, and mastoiditis (inflammation of the mastoid bone behind the ear).

Betsy Vanderveer, age 3 years, presents with severe H. influenzae infection.

David, Susan's 5-year-old brother, had been admitted to the hospital at 21 months of age with meningitis caused by *Streptococcus pneumoniae*. He had responded well to antibiotic therapy and had been discharged. Another occurrence of pneumococcal meningitis at 27 months of age had also been followed by an uneventful recovery after antibiotics. He had had pneumonia at age 3½ years. At the time of Susan's death he was well.

The two other children of the Vanderveers, a girl aged 8 years and a newborn male, were in good health.

All the Vanderveer children had received routine immunization at ages 3, 4, and 5 months with tetanus and diphtheria toxoids and killed *Bordetella pertussis* to protect against tetanus, diptheria, and whooping cough, which are three potentially fatal diseases caused by bacterial toxins (Fig. 1.3). Serum agglutination tests were used to test their antibody responses to these and other immunogens. Samples of serum from both Betsy and David caused hemagglutination (the clumping of red blood cells) when added to red blood cells (type O) coated with tetanus toxoid. Hemagglutinating antibodies to tetanus toxoid were seen at serum dilutions of 1:32 for both Betsy and David, and were found at a similar titer in their 8-year-old sister. All three children were given typhoid vaccine subcutaneously and 4 weeks later

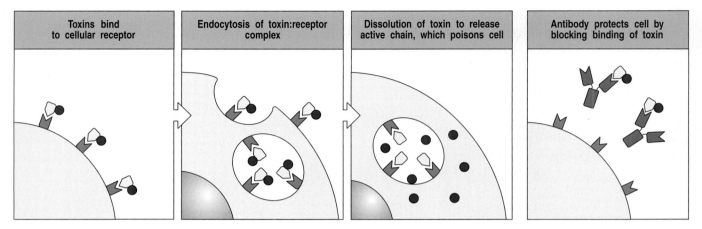

Toxins bind to cellular receptor	Endocytosis of toxin:receptor complex	Dissolution of toxin to release active chain, which poisons cell	Antibody protects cell by blocking binding of toxin

Fig. 1.3 Neutralization by antibodies protects cells from toxin action. Secreted bacterial toxins usually contain several distinct moieties. One piece of the toxin must bind a cellular receptor, which allows the molecule to be internalized. A second part of the toxin molecule then enters the cytoplasm and poisons the cell. In some cases, a single molecule of toxin can kill a cell. Antibodies that inhibit toxin binding can prevent, or neutralize, these effects. Protective antibodies can be generated by immunizing subcutaneously with toxoids. Toxoids are toxins rendered harmless by treating with denaturing agents, such as formalin, which destroy their toxicity but not their ability to generate neutralizing antibodies. In the case of the DPT vaccine, the killed *Bordetella pertussis* cells act as an adjuvant, which enhances the immune response to all components of the vaccine by delivering activating signals to antigen-presenting cells.

Agglutination tests to diphtheria, tetanus and pertussis toxins normal.

samples of their sera were tested for the ability to agglutinate killed *Salmonella typhosa*. The results indicated a normal immune response. David had an agglutination titer of 1:16, Betsy 1:32, and their normal 8-year-old sister 1:32. All three children were given 1 ml of a 25% suspension of sheep red cells intravenously. David had a titer of 1:4 for hemagglutinating antibodies against sheep red blood cells prior to the injection. He was tested again 2 and 4 weeks later and there was no increase in titer. Betsy had an initial titer of 1:32 and her titer did not rise either. The 8-year-old normal sister had a preimmunization titer of 1:32. She was tested 2 and 4 weeks after the immunization, when she was found to have a hemagglutinating titer of 1:256 against sheep red blood cells.

All the children and their parents were injected intravenously with radioactive colloidal gold (^{198}Au), which is taken up by the reticuloendothelial cells of the liver and spleen within 15 minutes after the injection. A scintillation counter then scans the abdomen for radioactive gold. The pattern of scintillation reveals that Betsy and David have no spleens (Fig. 1.4).

Asplenia and splenectomy.

The genetic defect causing asplenia has not yet been identified. The Vanderveer family is unusual in that three of their first four children were born without spleens. After the events described in this case, the Vanderveers had three more children. One of the boys and the girl were also born without spleens; the other boy had a normal spleen. This family provides us with an uncomplicated circumstance in which to examine the role of the spleen. The major consequence of its absence is a susceptibility to bacteremia, usually caused by the encapsulated bacteria *Streptococcus pneumoniae* or *Haemophilus influenzae*. This susceptibility is caused by a failure of the immune response to these common extracellular bacteria when they enter the bloodstream.

Surgical removal of the spleen is quite common. The capsule of the spleen may rupture from trauma, for example in an automobile accident. In such cases, the spleen has to be surgically removed very quickly because of blood loss into the abdominal cavity. The spleen may also be removed surgically for therapeutic reasons in certain autoimmune diseases, or because of a malignancy in the spleen. After splenectomy, patients, particularly children, are susceptible to bloodstream infections by microorganisms to which they have no antibodies. Microorganisms to which the host has antibodies are removed quickly from the bloodstream by the liver, where the Kupffer cells complement the role of the red pulp of the spleen. Antibodies to the encapsulated bacteria that commonly cause trouble with bloodstream infections persist for a very long time in the bloodstream of exposed individuals, even in the absence of a spleen (for reasons that are not fully understood). Adults who already have antibodies to these microorganisms are therefore much less vulnerable to problems of bacteremia than children who have not yet developed antibodies to these germs.

Questions.

1   *Nicholas Biddleboy, a 5-year-old boy, had his spleen removed following a sledding accident, during which both he and his sled struck a tree trunk. In the emergency room of a nearby hospital, it was determined that his spleen had ruptured. The surgeon, following removal of a spleen that had indeed ruptured, calls you for an immunology consultation. What do you advise?*

2   *Why did David and Betsy have normal responses to the typhoid vaccine but not to the sheep red blood cells?*

3   *The Vanderveer family is unique in the medical literature. The parents, who were distantly related, were normal and had normal spleens. Five of their eight children were born without spleens. Of these, only Betsy subsequently had children—four boys and one girl. They are all normal and have spleens. What is the inheritance pattern of congenital asplenia in this family? According to Mendelian laws how many of the eight Vanderveer children would be expected to have no spleen?*

Fig. 1.4 A scintillation scan of the abdomen after intravenous injection with radioactive colloidal gold (^{198}Au) reveals that Betsy and David Vanderveer have no spleens. The top panel shows an abdominal scan of Betsy's mother. The large mass on the left is the liver and the small mass on the right is the spleen. The reticuloendothelial cells of both liver and spleen take up the labeled gold within 15 minutes after the injection. No spleen is seen in either Betsy (middle panel) or David (lower panel).

CASE 2 | Chronic Granulomatous Disease

A specific failure of phagocytes to produce H_2O_2 and superoxide.

One of the most important mechanisms for destroying invading microbes in both the innate and adaptive immune response is their phagocytosis by macrophages and neutrophils, after which they are killed rapidly within the phagocytic vacuoles, or phagosomes. Uptake of microorganisms by phagocytes is enhanced by the opsonization of the particle—that is, coating it with complement, or, in the case of adaptive immune response, with antibody and complement (Fig. 2.1). The microbicidal actions of phagocytes occur through a variety of mechanisms (Fig. 2.2). One of the most important of these involves the production of hydrogen peroxide and superoxide radicals, which raises the pH of phagosomes and thereby enables the enzymes that are released into them to attack ingested microbes. A complex enzyme, NADPH oxidase, catalyzes the initial reaction, which results in the generation of superoxide radicals:

Topics bearing on this case:
Microbicidal action of phagocytes
Chronic inflammation
Leukocyte migration

Fig. 2.1 Activation of a neutrophil.
Microbes (red) are ingested by a phagocyte and enter the cytoplasm in a phagocytic vacuole, or phagosome. The phagosome fuses with a cytoplasmic granule containing microbicidal enzymes. The enzymes are released into the phagosome, where they kill and degrade the microbe.

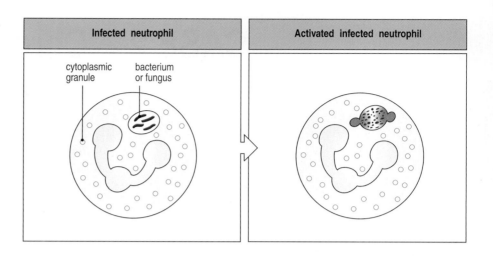

$$\text{NADPH} + 2O_2 \longrightarrow \text{NADP}^+ + 2O_2^- + \text{H}^+$$

The superoxide then dismutates to produce hydrogen peroxide:

$$2\text{H}^+ + 2O_2^- \longrightarrow \text{H}_2\text{O}_2 + \text{O}_2$$

NADPH oxidase is a large enzyme complex in the phagosome membrane that is assembled from component parts in the membrane and the cytosol in response to a phagocytic stimulus. Not all of the components of this multi-subunit complex have been defined clearly but at least two, p47^{phox} and p67^{phox}, reside in the cytoplasm in unstimulated phagocytes, whereas two others together comprise the membrane complex cytochrome b$_{558}$. The two chains of cytochrome b$_{558}$ are a heavy chain, gp91^{phox}, and a light chain, p21^{phox}, which contains the catalytic site (Fig. 2.3). The genes encoding p47^{phox}, p67^{phox}, and p21^{phox} map to autosomal chromosomes, whereas gp91^{phox} is encoded on the short arm of the X chromosome. Four different genetic defects affect various components of the NADPH oxidase enzyme but they all result in a disease with a common phenotype, called chronic granulomatous disease (CGD). The commonest form of the disease is X-linked CGD, which is caused by mutations in the gene encoding gp91^{phox}.

Fig. 2.2 Ingestion of bacteria triggers the production or release of many bactericidal agents in phagocytic cells.
Some of them are cytotoxic; others, such as lactoferrin, work by binding essential nutrients and preventing their uptake by bacteria. The same agents can be released by phagocytes interacting with large, antibody-coated surfaces such as parasitic worms or host tissues. As these mediators are also toxic to host cells, phagocyte activation can cause extensive tissue damage in infection.

Class of mechanism	Specific products
Toxic oxygen-derived products	Superoxide O_2^-, hydrogen peroxide H_2O_2, singlet oxygen $^1O_2^\bullet$, hydroxyl radical $OH^\bullet$, hypohalite OCl
Toxic nitrogen oxides	Nitric oxide NO
Antimicrobial peptides	Defensins and cationic proteins
Enzymes	Lysozyme—dissolves cell walls of some Gram-positive bacteria. Neutral proteases—further digest bacteria
Competitors	Lactoferrin (binds Fe) and vitamin B12 binding protein

Fig. 2.3 The NADPH oxidase complex. NADPH is a large, multisubunit enzyme complex whose assembly in the phagocyte membrane is triggered by a phagocytic stimulus. For simplicity, only four major components are shown here: gp91^{phox} and p21^{phox}, which together form the membrane complex known as cytochrome b$_{558}$, and p67^{phox} and p47^{phox}, which normally reside in the cytoplasm. On delivery of a phagocytic stimulus to the cell, p47^{phox} becomes hyperphosphorylated and binds to p67^{phox} and other cytosolic components, and these components migrate to the membrane to form the complete complex with cytochrome b$_{558}$.

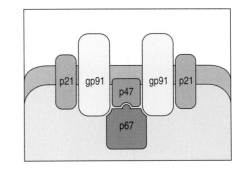

The diagnosis of CGD can be ascertained easily by taking advantage of the metabolic defect in the phagocytic cells. A dye called nitro blue tetrazolium (NBT) is pale yellow and transparent. When it is reduced, it becomes insoluble and turns a deep purple colour. To test for CGD, a drop of blood is suspended on a slide and is provided with a phagocytic stimulus (for example, phorbol myristate acetate (PMA), an activator of the oxidase); a drop of NBT is added to the blood. Neutrophils, which are the main phagocytes in blood, take up the NBT and the PMA at the same time. In normal blood the NBT is reduced to a dark purple, insoluble formazan, easily seen in the phagocytic cells; in CGD blood no dye reduction is seen (Fig. 2.4).

The case of Randy Johnson: a near death from exotic bacteria and fungi.

Randy Johnson, a 15-year-old high-school student, had a summer job working with a gardening crew. His job entailed spreading bark mulch in garden beds. At the end of August of that summer he rapidly developed severe shortness of breath, a persistent cough and chest pain. Because of impending respiratory failure he was admitted to the hospital. A radiological examination of Randy's chest was

Fig. 2.4 Neutrophils stimulated with phorbol myristate acetate in the presence of nitro blue tetrazolium (NBT) dye. The left upper panel shows two normal neutrophils stained purple. The left lower panel shows neutrophils from a patient with CGD that failed to reduce NBT. The neutrophils in the right panel are from a heterozygous carrier of chronic granulomatous disease. Half the neutrophils have reduced NBT (purple) and half do not stain. Photographs courtesy of P. Newberger.

Fifteen-year-old boy with impending respiratory failure.

Repeated respiratory infections.

performed and this revealed the presence of large 'cotton ball' densities in both lungs (Fig. 2.5). One of these lesions was aspirated with a fine needle and when the aspirate was stained numerous aspergillus hyphae were seen. A culture of the aspirate grew *Aspergillus fumigatus*.

Randy was started on intravenous amphotericin B and assisted mechanical ventilation through a tracheotomy. He slowly improved over the course of two months. During this time in the hospital, he contracted two further respiratory infections with *Pseudomonas aeruginosa* and *Streptococcus faecalis*. These infections were also treated with appropriate antibiotics.

At the time of his initial admission to the hospital, Randy's white blood cell count was 11,500 μl^{-1} (normal 5000–10,000 μl^{-1}). He had 65% neutrophils, 30% lymphocytes, and 5% monocytes: these proportions are normal.

Because of his infections, which were unusual for a seemingly healthy 15-year-old adolescent, his serum immunoglobulins were measured. The serum IgG level was 1650 mg dl^{-1} (upper limit of normal 1500 mg dl^{-1}). IgM and IgA were in the high normal range at 250 and 175 mg dl^{-1} respectively. Because no defect could be found in his humoral immunity, his white cell function was tested with an NBT slide test. His white blood cells failed to reduce NBT.

Randy had four older sisters and two brothers. They all had an NBT test performed on their blood. One, his brother Ralph, aged 9 years, also failed to reduce NBT. His mother reported that Ralph had had a perirectal abscess in infancy but had otherwise been well.

Randy's mother and one sister had a mixed population of neutrophils in the NBT test; about half the phagocytes reduced NBT and the other half did not (see Fig. 2.4). His other three sisters and other brother gave normal NBT reduction.

Further studies with Randy's and Ralph's granulocytes revealed that the rate of hydrogen peroxide production was 0.16 and 0.028 nmol min^{-1} 10^{-6} cells (normal 6.35 nmol min^{-1} 10^{-6} cells). The cytochrome b content in both brothers' granulocytes was <1.0 pmol mg^{-1} protein (normal control 101 pmol mg^{-1} protein). Both Ralph and Randy were started on treatment with injections of interferon-γ.

Fig. 2.5 A radiograph of aspergillus pneumonia. Arrows point to 'cotton ball' densities characteristic of fungi.

Chronic granulomatous disease (CGD).

Randy and Ralph Johnson present a phenotype of X-linked chronic granulomatous disease. Males affected with this defect commonly show undue susceptibility to infection in the first year of life. The most common infections are pneumonia, infection of the lymph nodes (lymphadenitis), and abscesses of the skin and of the viscera such as the liver. Granuloma formation occurs when microorganisms are opsonized and ingested by phagocytic cells but the infection cannot be cleared. The persistent presentation of microbial antigens by phagocytes induces a sustained cell-mediated response by CD4 T cells, which recruit other inflammatory cells and set up a chronic local inflammation called a granuloma. This can occur in normal individuals in response to intracellular bacteria that colonize macrophages (Fig. 2.6); in CGD, where the intracellular microbicidal mechanisms are defective, many infections can persist and set up chronic inflammatory reactions resulting in granulomas. The fungi and bacteria that are most frequently responsible for infections and granuloma formation in CGD are listed in Fig. 2.7.

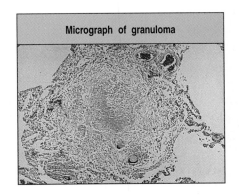

Fig. 2.6 Granulomas form when an intracellular pathogen or its constituents cannot be eliminated completely. The pathogen illustrated here is the intracellular bacterium *Mycobacterium tuberculosis*. In the left panel, mycobacteria (red) are shown resisting the effects of macrophage activation by a T$_H$1 CD4 T cell. The middle panel is a schematic diagram of a granuloma, which forms as the result of a localized inflammatory response that develops because of the sustained activation of chronically infected phagocytes. The granuloma consists of a central core of infected macrophages, and may include multinucleated giant cells, which are fused macrophages, surrounded by large macrophages often called epithelioid cells. These may, in turn, be surrounded by T cells, many of which are CD4 positive. Granulomas frequently form in CGD because macrophages are unable to destroy the bacteria they ingest and infected macrophages accumulate. The right panel shows a micrograph of a granuloma. Photograph courtesy of J. Orrell.

It was discovered quite by chance that interferon-γ improves the resistance of males with X-linked CGD to infection. The basis for this effect is still not understood. This cytokine does not increase superoxide or peroxide production in the deficient phagocytes; its positive effects remain to be determined.

Questions.

1 *50% of Randy's mother's neutrophils have reduced the NBT and 50% have not (see Fig. 2.4). How do you explain this?*

2 *Children with CGD do not have problems with pneumococcal infections. Could you hypothesize why this is so?*

3 *The immunoglobulin levels in Randy's blood were somewhat elevated. How do you explain this?*

4 *How does chronic granulomatous disease differ from leukocyte adhesion deficiency?*

Staphylococcus aureus
Aspergillus spp.
Chromobacterium violaceum
Pseudomonas cepacia
Nocardia spp.
Salmonella typhimurium
Serratia marcescens
Mycobacterium fortuitum
Klebsiella spp.
Escherichia coli
Actinomyces spp.
Legionella bosmanii
Clostridium difficile

Fig. 2.7 Table of fungal and bacterial infectious agents most commonly responsible for infections in CGD.

CASE 3 | Leukocyte Adhesion Deficiency

The traffic of white blood cells.

Newly differentiated blood cells continually enter the bloodstream from their sites of production: red blood cells, monocytes, granulocytes, and B lymphocytes from the bone marrow; and T lymphocytes from the thymus. Under ordinary circumstances red blood cells spend their entire life span of 120 days in the bloodstream. However, white blood cells are destined to emigrate from the blood to perform their effector functions. Lymphocytes recirculate through secondary lymphoid tissues, where they are detained if they encounter an antigen to which they can respond; macrophages migrate into the tissues as they mature from circulating monocytes; effector T lymphocytes and large numbers of granulocytes are recruited to extravascular sites in response to infection or injury. For example, it is estimated that, each day, three billion neutrophils enter the oral cavity, the most contaminated site in our body.

Topics bearing on this case:

Migration and homing of leukocytes

Phagocytic cell defects

Adhesion molecules in T-cell interactions

Bone marrow transplantation

Complement receptors

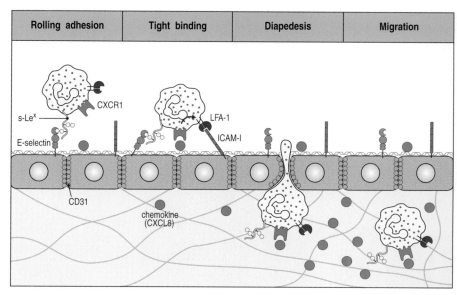

Fig. 3.1 Phagocytic leukocytes and effector T cells are directed to sites of infection through interactions between adhesion molecules induced by monokines. The first step (right panel) involves the reversible binding of leukocytes to vascular endothelium through interactions between selectins induced on the endothelium and their carbohydrate ligands on the leukocyte, shown here for E-selectin and its ligand the sialyl-Lewis^x moiety (s-Le^x). This interaction cannot anchor the cells against the shearing force of the flow of blood and instead they roll along the endothelium, continually making and breaking contact. The binding does, however, allow stronger interactions, which occur as a result of the induction of ICAM-1 on the endothelium and the activation of its receptors LFA-1 and Mac-1 (not shown) on the leukocyte. This is mediated by chemokines (e.g., CXCL8) retained on heparan sulfate proteoglycans on the endothelial cell surface. Tight binding between these molecules arrests the rolling and allows the leukocyte to squeeze between the endothelial cells forming the wall of the blood vessel (extravasate). The leukocyte integrins LFA-1 and Mac-1 are required for extravasation, and for migration toward chemoattractants. Adhesion between molecules of CD31, expressed on both the leukocyte and the junction of the endothelial cells, is also thought to contribute to diapedesis. Finally, the leukocyte migrates along a concentration gradient of chemokines (here shown as CXCL8) secreted by cells at the site of infection. The electron micrograph shows a neutrophil that has just started to migrate between two endothelial cells (bottom of photo). Note the pseudopod that the neutrophil has inserted between adjacent endothelial cells. The dark mass at the bottom right is an erythrocyte that has become trapped underneath the neutrophil. Photograph ($\times$ 5500) courtesy of I. Bird and J. Spragg.

The process by which white blood cells migrate from the bloodstream to sites of infection is fairly well understood (Fig. 3.1). First their flow is retarded by the interaction between selectins whose expression is induced on activated vascular endothelium and certain fucosylated glycoproteins on the white cell surface (for example sialyl-Lewis^x). Tight binding of leukocytes to the endothelial surface is then triggered by chemokines, such as CXCL8 (formerly known as IL-8), which activate an enhanced ability of the leukocyte integrins (for example LFA-1 and Mac-1) to adhere to their receptors. Crossing the endothelial cell wall also involves interactions between the leukocyte integrins and their receptors, while the subsequent direction of migration follows a concentration gradient of chemokines (for example CXCL8) produced by cells already at the site of infection or injury. The process by which lymphocytes home to secondary lymphoid tissue is very similar, except that it is initiated by mucin-like addressins on lymphoid venules binding to L-selectin on the surface of naive lymphocytes (Fig. 3.2). The various adhesion molecules involved in leukocyte trafficking and other important leukocyte cell–cell interactions are described in Fig. 3.3.

An excellent opportunity to study the role of integrins is provided by a genetic defect in CD18, the common β chain of the three β$_2$ integrins: LFA-1 (CD11a:CD18), Mac-1 (CD11b:CD18, also known as CR3), and p150,95 (CD11c:CD18, also known as CR4). Children with this genetic defect suffer from leukocyte adhesion deficiency. They have recurrent pyogenic infections,

Fig. 3.2 L-selectin and the mucin-like vascular addressins direct naive lymphocyte homing to lymphoid tissues. L-selectin is expressed on naive T cells, which bind to the vascular addressins CD34 and GlyCAM-1 on high endothelial venules to enter lymph nodes. The relative importance of CD34 and GlyCAM-1 in this interaction is not clear. GlyCAM-1 is expressed exclusively on high endothelial venules but has no transmembrane region and it is unclear how it is attached to the membrane; CD34 has a transmembrane anchor and is expressed on endothelial cells but not exclusively on high endothelial venules. The addressin MAdCAM-1 is expressed on mucosal endothelium and guides entry into mucosal lymphoid tissue. L-selectin recognizes carbohydrate moieties on the vascular addressins.

problems with wound healing, and if they survive long enough they develop severe inflammation of the gums (gingivitis). Surprisingly, children with leukocyte adhesion deficiency are not unduly susceptible to opportunistic infections. This implies normal T-cell function despite the absence of LFA-1, which was thought to be important for T-cell adhesion to antigen-presenting cells (Fig. 3.4). The capacity to form antibodies is also unimpaired, showing that adequate collaboration between T and B cells can also occur without LFA-1.

The case of Luisa Ortega: a problem of immobile white blood cells.

Luisa Ortega was born at full term and weighed 3.7 kg. She was the second child born to the Ortegas. At 4 weeks of age Luisa was taken by her parents to her pediatrician because she had swelling and redness around the umbilical cord stump (omphalitis), and a fever of 39°C. Her white blood cell count was 71,000 μl⁻¹ (normal 5000–10,000 μl⁻¹). She was treated in the hospital with intravenous antibiotics for 12 days and then discharged home with oral antibiotics. At the time of discharge her white blood count was 20,000 μl⁻¹. Cultures obtained from the inflamed skin about the umbilical stump before antibiotic treatment grew *Escherichia coli* and *Staphylococcus aureus*.

The Ortegas had a baby boy 3 years prior to Luisa's birth. At 2 weeks of age he developed a very severe infection of the large intestine (necrotizing enterocolitis). Separation of his umbilical cord was delayed. He subsequently suffered from multiple skin infections and he died of staphylococcal pneumonia at 1 year of age. Just before his death his white blood cell count was recorded at 75,000 μl⁻¹.

Because of the previous family history, Luisa was referred to the Children's Hospital. At the time of her admission to the Children's Hospital she seemed normal on physical examination, and radiographs of the chest and abdomen were normal.

Cultures of urine, blood, and cerebrospinal fluid were negative. Her white blood count was 68,000 μl⁻¹ (very elevated). Of her white cells, 73% were neutrophils, 22% lymphocytes, and 5% eosinophils (this distribution of cell types is in the normal range but the absolute count for each is abnormally high). Her serum IgG

Luisa, 4 weeks old, presents with acute omphalitis.

4-week-old infant with family history of overwhelming infection.

Fig. 3.3 Adhesion molecules in leukocyte interactions. Several structural families of adhesion molecules play a part in leukocyte migration, homing and cell–cell interactions: the selectins; mucin-like vascular addressins; the integrins; proteins of the immunoglobulin superfamily; and the protein CD44. The figure shows schematic representations of an example from each family, a list of other family members that participate in leukocyte interactions, their cellular distribution, and their partners in adhesive interactions. The nomenclature of the different molecules in these families is confusing because it often reflects the way in which the molecules were first identified rather than their related structural characteristics. Thus while all the ICAMs are immunoglobulin-related, and all the VLA molecules are β_1 integrins, the CD nomenclature reflects the characterization of leukocyte cell-surface molecules by raising monoclonal antibodies against them, and embraces adhesion molecules in all the structural families. The LFA molecules were defined through experiments in which cytotoxic T-cell killing could be blocked by monoclonal antibodies against cell-surface molecules on the interacting cells, and there are LFA molecules in both the integrin and the immunoglobulin families. Alternative names for each of the molecules shown are given in parentheses. Sialyl-Lewis^x, which is recognized by P- and E-selectin, is an oligosaccharide present on cell-surface glycoproteins of circulating leukocytes. The ligand DC-SIGN, which binds to ICAM-3, is expressed on dendritic cells. It is suspected to have a major role in interactions between dendritic cells and T cells and serves as a co-receptor for HIV.

		Name	Tissue distribution	Ligand
Selectins Bind carbohydrates. Initiate leukocyte–endothelial interaction	L-selectin	L-selectin (MEL-14, CD62L)	Naive and some memory lymphocytes, neutrophils, monocytes, macrophages, eosinophils	Sulfated sialyl Lewis^x, GlyCAM-1, CD34, MAdCAM-1
		P-selectin (PADGEM, CD62P)	Activated endothelium and platelets	Sialyl Lewis^x, PSGL-1
		E-selectin (ELAM-1, CD62E)	Activated endothelium	Sialyl Lewis^x
Mucin-like vascular addressins Bind to L-selectin. Initiate leukocyte–endothelial interaction	CD34	CD34	Endothelium	L-selectin
		GlyCAM-1	High endothelial venules	L-selectin
		MAdCAM-1	Mucosal lymphoid tissue venules	L-selectin, integrin $\alpha_4:\beta_7$
Integrins Bind to cell-adhesion molecules and extracellular matrix. Strong adhesion	LFA-1	$\alpha_L:\beta_2$ (LFA-1, CD11a/CD18)	Monocytes, T cells, macrophages, neutrophils, dendritic cells	ICAMs
		$\alpha_M:\beta_2$ (Mac-1, CR3, CD11b/CD18)	Neutrophils, monocytes, macrophages	ICAM-1, iC3b, fibrinogen
		$\alpha_X:\beta_2$ (CR4, p150.95, CD11c/CD18)	Dendritic cells, macrophages, neutrophils	iC3b
		$\alpha_4:\beta_1$ (VLA-4, LPAM-2, CD49d/CD29)	Lymphocytes, monocytes, macrophages	VCAM-1 Fibronectin
		$\alpha_5:\beta_1$ (VLA-5, CD49d/CD29)	Monocytes, macrophages	Fibronectin
		$\alpha_4:\beta_7$ (LPAM-1)	Lymphocytes	MAdCAM-1
		$\alpha_E:\beta_7$	Intraepithelial lymphocytes	E-cadherin
Immunoglobulin superfamily Various roles in cell adhesion. Ligand for integrins	CD2	CD2 (LFA-2)	T cells	LFA-3
		ICAM-1 (CD54)	Activated vessels, lymphocytes, dendritic cells	LFA-1, Mac-1
		ICAM-2 (CD102)	Resting vessels	LFA-1
		ICAM-3 (CD50)	Naive T cells	DC-SIGN
		LFA-3 (CD58)	Lymphocytes, antigen-presenting cells	CD2
		VCAM-1 (CD106)	Activated endothelium	VLA-4

concentration was 613 mg dl^{-1} (normal), her IgM was 89 mg dl^{-1} (normal), and her IgA 7 mg dl^{-1} (normal). The concentration of complement component C3 in her serum was 185 mg dl^{-1} and that of C4 was 28 mg dl^{-1} (both normal).

A Rebuck skin window was performed. In this procedure, the skin of the forearm is gently abraded with a scalpel blade and a cover slip is placed on the abrasion. After 2 hours the cover slip is removed and replaced by another every subsequent 2 hours for a total of 8 hours. In this way, the migration of immune cells into the damaged skin can be monitored. No white cells accumulated on the cover slips. All of Luisa's blood leukocytes, however, were present in abnormally high numbers. Of her blood lymphocytes 53% (7930 μl^{-1}) were T cells (as measured by CD3 expression); of these, 36% were CD4 and 16% CD8 (normal proportions); 25% (3740 μl^{-1}) were B cells (as measured by antibody to CD19), and 14% were NK cells (as measured by CD16 expression). These were both elevated.

T cells initially bind APC through low-affinity LFA-1:ICAM-1 interactions	Subsequent binding of T-cell receptors signals LFA-1	Conformational change in LFA-1 increases affinity and prolongs cell–cell contact

Fig. 3.4 Transient adhesive interactions between T cells and antigen-presenting cells are stabilized as specific antigen recognition induces high-affinity binding by LFA-1. When a T cell binds to its specific ligand on an antigen-presenting cell (APC), intracellular signaling through the T-cell receptor (TCR) induces a conformational change in LFA-1 that causes it to bind with higher affinity to ICAM-3 on the antigen-presenting cell. The cell shown here is a CD4 T cell.

Proliferation of Luisa's T cells in response to phytohemagglutinin (PHA) and concanavalin A (Con A) was slightly depressed. Further flow cytometric analysis revealed that, whereas 60% of Luisa's lymphocytes were stained by a monoclonal antibody to CD3, only 5% reacted with a monoclonal antibody to CD18, giving a CD18/CD3 ratio of 5%/60% compared with 62%/65% on testing cells from a control subject (Fig. 3.5). Her blood mononuclear cells were stimulated with PHA and examined after 3 days of incubation with a monoclonal antibody to CD11a (the α chain of LFA-1). No LFA-1 expression was found.

Luisa was treated with busulfan, cyclophosphamide and anti-thymocyte serum for 10 days. After this therapy, she was given bone marrow cells from her mother at a dose of 500×10^6 per kg body weight, and a short course of immunosuppressive therapy. Her mother's bone marrow donation had been depleted of mature T cells with a monoclonal antibody to mature T cells and complement. Twenty-eight days after the transplant, the lymphoid and myeloid cells in Luisa exhibited complete chimerism. She subsequently did well clinically and her white blood count remained at 7800 μl^{-1}.

Leukocyte adhesion deficiency.

Children like Luisa are subject to recurrent, severe bacterial infections that are eventually fatal. In these patients, encapsulated bacteria are coated with antibody and complement. However, the neutrophils and monocytes that would normally be recruited to the site of infection are entrapped in the bloodstream and cannot emigrate into the tissues because they lack LFA-1 (CD11a:CD18) and Mac-1/CR3 (CD11b:CD18). In a normal individual, the first cover slip in a Rebuck glass window would contain many neutrophils. Monocytes begin to appear at 4 hours and by 8 hours the cover slip contains predominantly monocytic cells and very few neutrophils. In Luisa's case the cover slips had no cells because her leukocytes were unable to emigrate from the bloodstream and onto the cover slip. For this reason, the levels of white cells in the bloodstream are very high: a very high white blood cell count is characteristic of leukocyte adhesion deficiency. The ability to deal with pyogenic bacteria is further compromised because of the vital role of CR3-mediated uptake of these opsonized bacteria by neutrophils. The role of CR4 or CD11c:CD18, the third member of the β2 integrin family, is less well understood but, like CR3, it binds complement fragments, and it is thought to have a role in uptake of bacteria by macrophages.

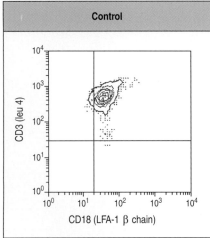

Fig. 3.5 Flow cytometric analysis of CD18 expression on Luisa's lymphocytes (top panel) compared with a control subject (bottom panel). The vertical axis measures fluorescence from labeled antibody to CD3; the horizontal axis measures fluorescence from labeled antibody to CD18. The shift of the plot to the left in the patient compared with the control represents a 10-fold decrease in the number of lymphocytes displaying CD18 on their surface.

The importance of the daily, massive neutrophil emigration into the oral cavity is well illustrated by individuals with leukocyte adhesion deficiency, who invariably develop severe gingivitis when they survive. Another, poorly understood, consequence of the lack of leukocyte emigration is the failure to heal wounds. Delayed separation of the umbilical cord is the earliest manifestation of this defect in wound healing. Subsequently, affected children may develop fistulas (abnormal connecting channels) in their intestine after bacterial infections of the gut.

Bone marrow transplantation has been very successful in rescuing severely affected infants from certain death.

Questions.

1   Can you surmise the inheritance pattern of the leukocyte adhesion deficiency from this case?

2   In some families with the leukocyte adhesion deficiency, a mild phenotype is observed and leukocytes of affected individuals express 10% of the normal amount of CD18. The mild phenotype is usually due to a splice site defect in the gene encoding CD18, whereas, in the severe form of the disease, there is usually no transcription of the CD18 gene. Do these observations help you to predict the likely effects of treatment with monoclonal antibodies to CD18? Can you suggest situations in which such treatment might be therapeutically useful?

3   Can you suggest why the homing of T cells to lymphoid tissue is normal in patients with leukocyte adhesion deficiency and why these individuals do not suffer from susceptibility to opportunistic infections?

4   What manifestation of defective wound healing occurred in this case?

5   Another rare form of leukocyte adhesion deficiency has been described in which affected infants cannot convert mannose to fucose. Can you explain why the failure to convert mannose to fucose should cause leukocyte adhesion deficiency?

6   Shortly after Luisa received a bone marrow transplant from her mother it was ascertained that she had complete chimerism of the myeloid and lymphoid cells. How was this done?

7   Why was Luisa treated with busulfan, anti-thymocyte serum, and cyclophosphamide before the transplantation?

8   Why were the responses of Luisa's T cells to PHA and Con A depressed?

CASE 4 | Hereditary Angioneurotic Edema

Regulation of complement activation.

Complement is a system of plasma proteins that participates in a cascade of reactions, generating active components that allow pathogens and immune complexes to be destroyed and eliminated from the body. Complement is part of the innate immune defenses of the body and is also activated via the antibodies produced in an adaptive immune response. Complement activation is generally confined to the surface of pathogens or circulating complexes of antibody bound to antigen.

Complement is normally activated by one of three routes—the classical pathway, which is triggered by antigen:antibody complexes or antibody bound to the surface of a pathogen; the lectin pathway, which is activated by mannose-binding lectin (MBL) and the ficolins; and the alternative pathway, in which complement is activated spontaneously on the surface of some bacteria. The early part of each pathway is a series of proteolytic cleavage events leading to the generation of a convertase, a serine protease that cleaves complement component C3 and thereby initiates the effector actions of complement. The C3 convertases generated by the three pathways are different, but evolutionarily homologous, enzymes. Complement components and activation pathways, and the main effector actions of complement, are summarized in Fig. 4.1.

The principal effector molecule, and a focal point of activation for the system, is C3b, the large cleavage fragment of C3. If active C3b, or the homologous but less potent C4b, accidentally becomes bound to a host cell surface instead of a pathogen, the cell can be destroyed. This is usually prevented by the rapid hydrolysis of active C3b and C4b if they do not bind immediately to the surface where they were generated. Protection against inappropriate activation of complement is also provided by regulatory proteins.

One of these, and the most potent inhibitor of the classical pathway, is the C1 inhibitor (C1INH). This belongs to a family of serine protease inhibitors (called serpins) that together constitute 20% of all plasma proteins. In addition to being the sole known inhibitor of C1, C1INH contributes to the regulation of serine proteases of the clotting system and of the kinin system, which is activated by injury to blood vessels and by some bacterial toxins.

Topics bearing on this case:
Classical pathway of complement activation
Inhibition of C1 activation
Alternative pathway of complement activation
Inflammatory effects of complement activation
Regulation of C4b

Fig. 4.1 Overview of the main components and effector actions of complement. The early events of all three pathways of complement activation involve a series of cleavage reactions that culminate in the formation of an enzymatic activity called a C3 convertase, which cleaves complement component C3 into C3b and C3a. The production of the C3 convertase is the point at which the three pathways converge and the main effector functions of complement are generated. C3b binds covalently to the bacterial cell membrane and opsonizes the bacteria, enabling phagocytes to internalize them. C3a is a peptide mediator of local inflammation. C5a and C5b are generated by cleavage of C5b by a C5 convertase formed by C3b bound to the C3 convertase (not shown in this simplified diagram). C5a is also a powerful peptide mediator of inflammation. C5b triggers the late events in which the terminal components of complement assemble into a membrane-attack complex that can damage the membrane of certain pathogens. Although the classical complement activation pathway was first discovered as an antibody-triggered pathway, it is now known that C1q can activate this pathway by binding directly to pathogen surfaces, as well as paralleling the lectin activation pathway by binding to antibody that is itself bound to the pathogen surface. In the lectin pathway, MASP stands for mannose-binding lectin-associated serine protease.

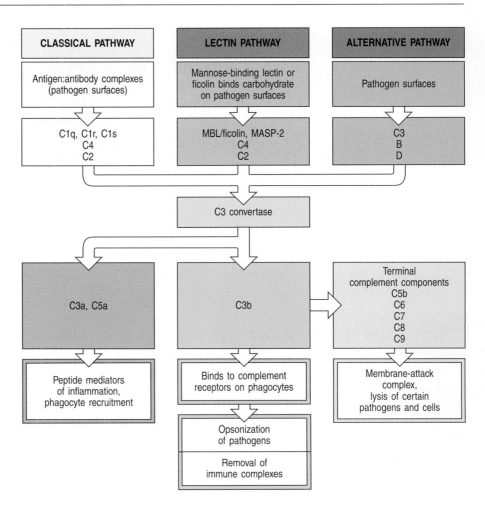

C1INH intervenes in the first step of the complement pathway, when C1 binds to immunoglobulin molecules on the surface of a pathogen or antigen:antibody complex (Fig. 4.2). Binding of two or more of the six tulip-like heads of the C1q component of C1 is required to trigger the sequential activation of the two associated serine proteases, C1r and C1s. C1INH inhibits both of these proteases, by presenting them with a so-called bait-site, in the form of an arginine–threonine bond that they cleave. When C1r and C1s attack the bait-site they covalently bind C1INH and dissociate from C1q. By this mechanism, the C1 inhibitor limits the time during which antibody-bound C1 can cleave C4 and C2 to generate C4b,2a, the classical pathway C3 convertase.

Activation of C1 also occurs spontaneously at low levels without binding to an antigen:antibody complex, and can be triggered further by plasmin, a protease of the clotting system, which is also normally inhibited by C1INH. In the absence of C1INH, further active complement components are produced. This is seen in hereditary angioneurotic edema (HANE), a disease caused by a genetic deficiency of C1INH.

Fig. 4.2 Activation of the classical pathway of complement and intervention by C1INH. In the left panel, one molecule of IgM, bent into the 'staple' conformation by binding several identical epitopes on a pathogen surface, allows binding by the globular heads of C1q to its Fc pieces on the surface of the pathogen. In the right panel, multiple molecules of IgG bound to the surface of the pathogen allow binding by C1q to two or more Fc pieces. In both cases, binding of C1q activates the associated C1r, which becomes an active enzyme that cleaves the proenzyme C1s, a serine protease that initiates the classical complement cascade. Active C1 is inactivated by C1INH, which binds covalently to C1r and C1s, causing them to dissociate from the complex. There are in fact two C1r and two C1s molecules bound to each C1q molecule, although for simplicity this is not shown here. It takes four molecules of C1INH to inactivate all the C1r and C1s.

The case of Richard Crafton: a failure of communication as well as of complement regulation.

Richard Crafton was a 17-year-old high-school senior when he had an attack of severe abdominal pain at the end of a school day. The pain came as frequent sharp spasms and he began to vomit. After 3 hours, the pain became unbearable and he went to the emergency room at the local hospital.

At the hospital, the intern who examined him found no abnormalities other than dry mucous membranes of the mouth, and a tender abdomen. There was no point tenderness to indicate appendicitis. Richard continued to vomit every 5 minutes and said the pain was getting worse.

A surgeon was summoned. He agreed with the intern that Richard had an acute abdominal condition but was uncertain of the diagnosis. Blood tests showed an elevated red blood cell count, indicating dehydration. The surgeon decided to proceed with exploratory abdominal surgery. A large midline incision revealed a moderately swollen and pale jejunum but no other abnormalities were noted. The surgeon removed Richard's appendix, which was normal, and Richard recovered and returned to school 5 days later.

Richard, age 17, presents as an acute abdominal emergency.

Appendectomy performed. Appendix appears normal.

Family history of colic.

Fig. 4.3 Hereditary angioneurotic edema. Transient localized swelling that occurs in this condition often affects the face.

What Richard had not mentioned to the intern or to the surgeon was that, although he had never had such severe pains as those he was experiencing when he went to the emergency room, he had had episodes of abdominal pain since he was 14 years old. No one in the emergency room asked him if he was taking any medication or took a family history or a history of prior illness. If they had, they would have learned that Richard's mother, a maternal uncle, and maternal grandmother also had recurrent episodes of severe abdominal pain, as did his only sibling, a 19-year-old sister.

As a newborn, Richard was prone to severe colic. When he was 4 years old, a bump on his head led to abnormal swelling. When he was 7, a blow with a baseball bat caused his entire left forearm to swell to twice its normal size. In both cases, the swelling was not painful, nor was it red or itchy, and it disappeared after 2 days. At age 14 years, he began to complain of abdominal pain every few months, sometimes accompanied by vomiting and, more rarely, by clear, watery diarrhea.

Richard's mother had taken him at age 4 years to an immunologist, who listened to the family history and immediately suspected hereditary angioneurotic edema. The diagnosis was confirmed on measuring key complement components. C1INH levels were 16% of the normal mean and C4 levels were markedly decreased, while C3 levels were normal.

When Richard turned up for a routine visit to his immunologist a few weeks after his surgical misadventure, the immunologist, noticing Richard's large abdominal scar, asked what had happened. When Richard explained, he prescribed daily doses of Winstrol (stanozolol). This caused a marked diminution in the frequency and severity of Richard's symptoms. When Richard was 20 years old, purified C1INH became available; he has since been infused intravenously on several occasions to alleviate severe abdominal pain, and once for swelling of his uvula, pharynx, and larynx. The infusion relieved his symptoms within 25 minutes.

Richard subsequently married and had two children. The C1INH level was found to be normal in both newborns.

Hereditary angioneurotic edema.

Individuals like Richard with a hereditary deficiency of C1INH are subject to recurrent episodes of circumscribed swelling of the skin (Fig. 4.3), intestine, and airway. When the swelling occurs in the intestine it causes severe abdominal pain, and obstructs the intestine so that the patient vomits. When the colon is affected, severe watery diarrhea may occur. Swelling in the larynx is the most dangerous symptom, as the patient may rapidly choke to death. Such episodes may be triggered by trauma, menstrual periods, excessive exercise, exposure to extremes of temperature, or mental stress. These types of event are associated with activation of four serine proteases, which are normally inhibited by C1INH. At the top of this cascade is Factor XII, which directly or indirectly activates the other three (Fig. 4.4). Factor XII is normally activated by injury to blood vessels, and initiates the kinin cascade, activating kallikrein, which generates the vasoactive peptide bradykinin. Factor XII also indirectly activates plasmin, which, as mentioned before, activates C1 itself. Plasmin also cleaves C2b to generate a vasoactive fragment called C2 kinin. In patients deficient in C1INH, the uninhibited activation of Factor XII leads to the activation of kallikrein and plasmin; kallikrein then catalyzes the formation of bradykinin, and plasmin activates C1, which cleaves C2, generating a small fragment, C2b, which is further cleaved by plasmin to produce C2 kinin. Bradykinin and C2 kinin increase the permeability of the

Activation of Factor XII

- Activation of kallikrein
- Activation of proactivator
- Cleavage of kininogen to generate bradykinin, vasoactive peptide
- Cleavage of C2b to generate C2 kinin, vasoactive peptide
- Cleavage of plasminogen to generate plasmin
- Cleavage of C2 to generate C2b
- Activation of C1

Fig. 4.4 Pathogenesis of hereditary angioneurotic edema. Activation of Factor XII leads to the activation of kallikrein, which cleaves kininogen to produce the vasoactive peptide bradykinin; it also leads to the activation of plasmin, which in turn activates C1. C1 cleaves C2, whose smaller fragment C2b is further cleaved by plasmin to generate the vasoactive peptide C2 kinin. The red bars represent inhibition by C1INH.

postcapillary venules by causing contraction of endothelial cells so as to create gaps in the blood vessel wall (Fig. 4.5). This is responsible for the edema; movement of fluid from the vascular space into another body compartment, such as the gut, causes the symptoms of dehydration as the vascular volume contracts.

Fig. 4.5 Contraction of endothelial cells creates gaps in the blood vessel wall.
A guinea pig was injected intravenously with India ink (a suspension of carbon particles). Immediately thereafter the guinea pig was injected intradermally with a small amount of activated C1s. An area of angiodema formed about the injected site, which was biopsied 10 minutes later. An electron micrograph reveals that the endothelial cells in post-capillary venules have contracted and formed gaps through which the india ink particles have leaked from the blood vessel. L is the lumen of the blood vessel. P is a polymorpho-nuclear leukocyte in the lumen. rbc is a red blood cell that has leaked out of the blood vessel. Micrograph courtesy of Kaethe Willms.

Questions.

1   *Activation of the complement system results in the release of histamine and chemokines, which normally produce pain, heat, and itching. Why is the edema fluid in HANE free of cellular components, and why does the swelling not itch?*

2   *Richard has a markedly decreased amount of C4 in his blood. This is because it is being rapidly cleaved by activated C1. What other complement component would you expect to find decreased? Would you expect the alternative pathway components to be low, normal or elevated? What about the terminal components?*

3   *Despite the complement deficiency in patients with HANE, they are not unduly susceptible to infection. Why not?*

4   *What is stanozolol, and why was it prescribed?*

5   *Emergency treatment for HANE cases is sometimes necessary because of airway obstruction. In most cases, however, a patient with obstruction of the upper airways is likely to be suffering from an anaphylactic reaction. The treatment in this case would be epinephrine. How might you decide whether to administer epinephrine or intravenous C1INH?*

6   *Fig. 4.6 shows Richard's family tree. What is the mode of inheritance (dominant or recessive, sex-linked or not) of HANE? Can Richard's two children pass the disease onto their offspring?*

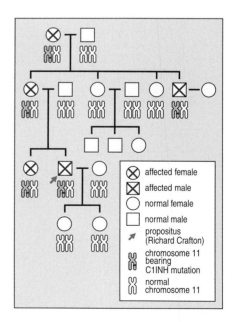

Fig. 4.6 The inheritance of HANE in Richard's extended family.

CASE 5 | Factor I Deficiency

The alternative pathway of complement activation is important in innate immunity.

The complement system plays a crucial part in the destruction and removal of microorganisms from the body. Pathogens coated with complement proteins are more efficiently phagocytosed by macrophages, and bacteria coated with complement can also be directly destroyed by complement-mediated lysis. The system of plasma proteins known collectively as complement can be activated in various ways (see Fig. 4.1), of which the so-called alternative pathway is important in innate or nonadaptive immunity. This pathway can be activated in the absence of antibody, although even low titers of IgM antibodies against an infecting microorganism will greatly amplify complement activation.

The complement protein C3 is the starting point of the alternative pathway. It is one of the more abundant globulins in blood and is continuously being cleaved at a fairly low rate into a smaller C3a and a larger C3b fragment by a variety of host or microbial proteinases—this is called the 'tickover' of C3. Cleavage exposes a highly reactive thioester bond in the C3b fragment, which enables C3b to bond covalently with the hydroxyl group of serine or threonine in a protein or the hydroxyl group of a sugar on a microbial surface. If C3b fails to attach to a microbial surface, the thioester bond is spontaneously hydrolyzed and the C3b is inactivated (Fig. 5.1).

Topics bearing on this case:
Alternative pathway of complement activation
Factor I cleavage of C3b
Opsonizing activity of C3b
C3a activation of mast cells

Fig. 5.1 Cleavage of C3 exposes a reactive thioester bond that enables the larger cleavage fragment C3b to bind covalently to the bacterial cell surface. Intact C3 has a shielded thioester bond that is exposed when C3 is cleaved by a proteinase to produce C3b. The highly reactive thioester bond in C3b can react with hydroxyl or amino groups to form a covalent linkage with molecules on the microbial surface. In the absence of such a reaction the thioester bond is rapidly hydrolyzed, inactivating C3b.

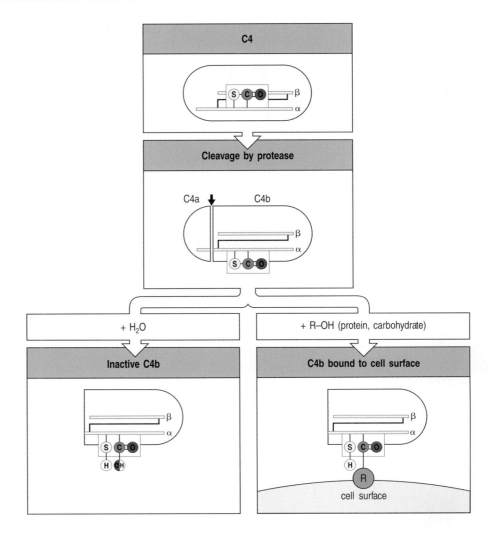

The binding of C3b to a microbial surface stimulates the cleavage of more C3 molecules. Another alternative pathway component, factor B, binds to C3b, and in this bound state is cleaved by a pre-existing blood proteinase, factor D, leaving the larger Bb fragment still bound to the C3b. The resulting C3b,Bb complex is an active serine protease, known as the alternative pathway C3 convertase, which specifically cleaves native C3 to make more C3b and C3a (Fig. 5.2).

Fig. 5.2 The alternative pathway of complement activation leads to amplification of C3 cleavage. C3b deposited on a microbial surface can bind factor B, making it susceptible to cleavage by factor D. The C3b,Bb complex is the C3 convertase of the alternative pathway of complement activation and its action results in the deposition of many more molecules of C3b on the pathogen surface.

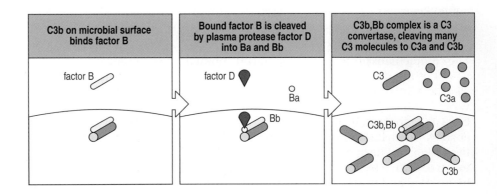

Fig. 5.3 The conversion of C3b to iC3b by factor I. On microbial surfaces, factor H displaces B from the C3b,Bb complex and cleaves C3b to produce iC3b. On host cell surfaces, the complement receptor CR1, which binds C3b, can substitute for factor H in this reaction.

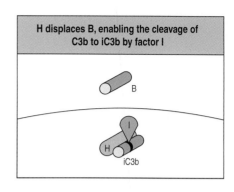

C3b bound to a microbial surface acts as an opsonin by binding to a specific receptor, the complement receptor 3 (CR3), on phagocytes, facilitating the ingestion of C3b-coated particles. But before C3b can act as a ligand for CR3, and thus as an effective opsonin, it has to undergo a further cleavage to a fragment called iC3b, which is effected by a blood serine protease called factor I, acting in conjunction with the blood protein factor H, components of the alternative pathway (Fig. 5.3). iC3b acting at the receptor CR3 can activate neutrophils and macrophages in the absence of antibody (Fig. 5.4). Cleavage of C3b by factor I also has another critical effect. It inhibits the C3 convertase activity of the C3b complex, thus ensuring that supplies of C3 do not become depleted. On host cell surfaces, complement receptor 1 (CR1) may bind to C3b and serve as the cofactor instead of factor H.

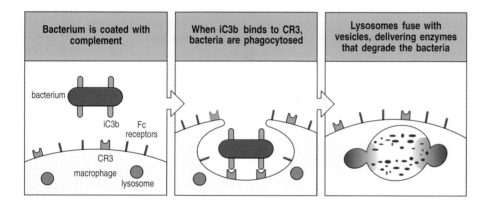

Fig. 5.4 iC3b binds to complement CR3 receptors on phagocytes and stimulates phagocytosis. iC3b on the surface of the pathogen binds to CR3 receptors on the phagocyte. Binding signals the phagocyte to engulf the particle and activates the internal destruction mechanisms.

The case of Morris Townsend: uncontrolled complement activation leads to susceptibility to infection and to hives.

Morris Townsend was admitted to the Brighton City Hospital at age 25 with pneumonia. This was his 28th admission to the hospital in his lifetime. From his first year onwards he had been repeatedly admitted for middle ear infections and mastoiditis. During these episodes, which were successfully treated with antibiotics, a variety of pyogenic (pus-forming) bacteria were cultured from his ears or mastoids, including *Staphylococcus aureus*, *Proteus vulgaris,* and *Pseudomonas aeruginosa.* At age 3 he had a tonsillectomy and adenoidectomy because of enlargement and chronic infection of his nasopharyngeal lymphoid tissue; at age 6 he had scarlet fever. He had also been admitted at other times with left lower lobe pneumonia (when *Haemophilus influenzae* had been cultured from his sputum), an abscess in the groin, acute sinusitis, a posterior ear abscess due to *Corynebacterium* species, skin abscesses with accompanying bloodstream infection (septicemia) due to β-hemolytic streptococci and, on one occasion, septicemia due to *Neisseria meningitidis* (meningococcemia).

25-year-old male with repeated bacterial infections. Complement or Ig deficiency?

Ig normal; check C3.

C3 synthesis normal; check factor I.

Genetic deficiency? Test family for factor I.

On physical examination at his latest admission, Morris was found to be slightly obese but otherwise normally developed. His hearing was poor in both ears and this was attributed to his recurrent ear infections and mastoiditis. He also told doctors that he developed hives all over his body after drinking alcohol or after taking a bath or shower.

A urine analysis yielded normal results. His hematocrit was 43% (normal) and his white cell count was 6000 μl^{-1}. His platelet count was 240,000 ml^{-1} (normal) and his blood clotted normally. His red blood cells gave a strong positive agglutination reaction with an antibody to C3 but no agglutination with an antibody to IgG or IgM. His serum IgG level was 915 mg dl^{-1}, IgA 475 mg dl^{-1} and IgM 135 mg dl^{-1} (all normal). Morris responded normally to an injection of tetanus toxoid; his antibody titer rose from 0.25 to 8.0 hemagglutinating units ml^{-1}. He gave a positive delayed-type skin reaction to mumps and monilia antigens.

Serum levels of C3 were 27 mg dl^{-1} (normal values 97–204 mg dl^{-1}); of this, 8 mg dl^{-1} was C3 and 19 mg dl^{-1} C3b. The serum levels of all other complement components were normal except for factor B, which was undetectable. His serum failed to kill a smooth strain of *Salmonella newport*, even after addition of C3 to the serum to render the C3 concentration normal. To investigate the turnover of C3, Morris was injected with a dose of C3 labeled with the radioactive tracer ^{125}I. The results of this investigation showed that the rate of synthesis of C3 was normal but that C3 was being broken down at four times the normal rate (Fig. 5.5). A test of his serum with an antibody to factor I showed that his serum lacked factor I.

Morris's family had no history of recurrent bacterial infections, but investigations showed reduced levels of factor I in both his parents and in several of his siblings (Fig. 5.6).

Factor I deficiency.

Patients such as Morris Townsend with a genetic deficiency in factor I were instrumental in deciphering the mechanism of activation of the alternative pathway of complement. Innate immunity is a first and highly effective means of defense against the common extracellular bacteria that cause pyogenic infections. The lack of factor I means that the alternative pathway C3 convertase is uninhibited and consumption of C3 is greatly accelerated, leading to C3 depletion. The lack of C3, and the nonproduction of iC3b, results in defective opsonization, which is the main means of removing and destroying these bacteria. Thus, factor I deficiency, like the genetic deficiency of C3 itself, results in a greatly increased susceptibility to infections with such bacteria. The clinical findings in factor I deficiency are not unlike those observed in X-linked agammaglobulinemia—a failure of opsonization results in frequent pyogenic infections.

The gene encoding factor I is on chromosome 4. The family of Morris Townsend provides a classic case of the inheritance of a recessive mutation in an autosomal gene (see Fig. 5.6). His parents, some of his siblings and two nephews are heterozygous for the defect; they produce roughly half the normal amounts of factor I, which is sufficient to prevent any clinical symptoms. Morris appears to be the only family member who is homozygous for the defect, and who thus exhibits symptoms.

Fig. 5.5 The rate of disappearance of ^{125}I-labeled C3 from the plasma. The rate of disappearance of radioactively labeled C3 from the patient's plasma (curve 1) is much faster than that in normal subjects (shaded area; 11 normal subjects). Curve 2 shows the rate of disappearance of C3 in the patient after the infusion of 500 ml of normal plasma.

Two interesting facts emerge from his clinical history. He sustained recurrent hives and had one bout of meningococcemia. It is easy to understand why he had hives. He was constantly cleaving C3 to C3a and C3b. C3a binds to mast cells and, among other things, causes the release of histamine and hence hives. The interesting question is why he did not have hives all the time and why they became problematic only after the ingestion of alcohol or exposure to hot and cold water. We must suppose that he had tachyphylaxis, or end-organ unresponsiveness, to histamine. Only when histamine release was increased, as by alcohol consumption or sudden changes in ambient body temperature, did the symptoms appear.

Morris Townsend's meningococcemia in particular is symptomatic of a deficiency in components of the alternative pathway of complement action. There are two common human pathogens in the bacterial genus *Neisseria*: *Neisseria gonorrhoeae* and *Neisseria meningitidis*. The former causes the sexually transmitted disease gonorrhea; the latter causes septicemia and meningitis and can be rapidly fatal. Patients have died from septic shock within 20 minutes of the onset of the symptoms of meningococcemia. Patients with genetic defects in the alternative pathway of complement activation or in the terminal components of complement sustain overwhelming and repeated infection with *Neisseria*. Deficiencies in the alternative pathway components factor D and properdin (factor B deficiency has never been observed in humans) were discovered because these patients developed recurrent meningococcemia. Similar observations have been made in patients with deficiencies of the later-acting C5, C6, C7, C8, and C9 components. These clinical observations highlight the importance of the bactericidal action of complement in controlling septicemia due to *Neisseria*.

Fig. 5.6 Autosomal recessive inheritance of the factor I defect. The numbers indicate percentages of normal level of factor I in serum. No number indicates normal levels. The blue arrow indicates the patient, Morris Townsend.

Questions.

1 Morris Townsend's clinical course has improved with age and he now has far fewer infections than he had as a child and adolescent. How do you explain this?

2 From the radiolabeled C3 experiment, we found that Morris Townsend catabolized the C3 very quickly but that his rate of synthesis of C3 was normal. What do you anticipate would happen if we repeated the experiment with radiolabeled factor B?

3 Morris Townsend was given a large dose of pure factor I intravenously. What changes would you predict to occur in his serum proteins?

4 What other genetic defect in the alternative pathway might lead to the same clinical and laboratory results as factor I deficiency?

5 Why did Morris' red blood cells agglutinate with antibody to C3?

CASE 6 | Deficiency of the C8 Complement Component

A loss of the lytic function of complement.

The complement system of plasma proteins is an effector mechanism of both innate and adaptive immunity that tags pathogens for destruction (see Cases 4 and 5). The assembly of the so-called terminal components of the complement system (C5 through C9; Fig. 6.1) on the surface of a bacterial cell or a human cell results in the formation of protein complexes that make pores in the cell membrane, leading to cell lysis and death.

Assembly of the terminal complex is initiated when the fifth component of complement, C5, binds to the cell surface and is cleaved by C5 convertase. This reaction releases C5a, a peptide with potent chemotactic activity, from the α, or heavy chain, of C5. The rest of the molecule, C5b, binds C6 to initiate the formation of the so-called membrane-attack complex of terminal complement components. The C5b6 complex binds one C7 molecule, resulting in the exposure of a hydrophobic site on C7 that enables the complex to sink partway into the lipid bilayer of a cell membrane. Complement component C8, unlike the single-chain C6 and C7, is composed of three chains, C8α, C8β, and C8γ. The C8γ chain binds to the C5a67 complex and enables the hydrophobic portion of C8 to embed itself in the cell membrane (Fig. 6.2). This last event induces the polymerization of 10–16 molecules of C9 to make a cylindrical structure, the membrane-attack complex, which forms a pore in the cell membrane. The diameter of the pore is approximately 100 Å, and through this channel sodium and water enter the cell, which swells until it bursts (lysis). The cell-surface protein CD59, which is found in most mammalian cell membranes, inhibits the action of C8 on C9, thereby preventing formation of the membrane-attack complex on the cells of the body but not on those of bacteria.

Topics bearing on this case:
Classical pathway of complement activation
Terminal complement components

Fig. 6.1 The terminal complement components that assemble to form the membrane-attack complex.

The terminal complement components that form the membrane-attack complex		
Native protein	Active component	Function
C5	C5a	Small peptide mediator of inflammation (high activity)
	C5b	Initiates assembly of the membrane-attack system
C6	C6	Binds C5b; forms acceptor for C7
C7	C7	Binds C5b6; amphiphilic complex inserts in lipid bilayer
C8	C8	Binds C5b67; initiates C9 polymerization
C9	C9$_n$	Polymerizes to C5b678 to form a membrane-spanning channel, lysing cell

Curiously, humans with a genetic deficiency in C5, C6, C7, C8, or C9 have an increased susceptibility to systemically invasive infection with bacteria of the genus *Neisseria*. Two common pathogens, *N. meningitidis* and *N. gonorrhoeae*, belong to this genus; the former causes endemic and epidemic meningitis and the latter the sexually transmitted disease gonorrhea.

The case of Dolly Oblonsky: recurrent meningococcal infection leads to the discovery of a complement deficiency.

Female student, temperature, stiff neck, reddish-purple rash. Call physician urgently.

Meningitis? Start antibiotics immediately.

Dolly Oblonsky was doing well in her first year at university when she developed a cough and diarrhea. She felt very tired and achy and went to bed early in her dormitory room. The next morning she woke early with a severe headache. She felt sick and her neck seemed to be stiff. Her roommate took her to the university infirmary emergency room. She seemed ill and somewhat confused to the nurse on duty, who found that Dolly had a blood pressure of 70/40 (low), a pulse of 124 (fast), a respiratory rate of 24 (increased) and a temperature of 39.2°C (elevated). The nurse noticed a petechial rash (small areas of reddish-purple discoloration) on Dolly's chest and urgently summoned the physician on call, Dr Tolstoy. He found that Dolly had a red throat with moderate enlargement of the tonsils. No other physical symptoms were apparent except for the neck stiffness and the petechial rash on her palate, trunk, and extremities. Dr Tolstoy immediately placed an intravenous needle into a vein on the back of Dolly's left hand, obtained blood for blood counts and bacterial cultures, and started intravenous administration of the antibiotic Cephtriaxone because he suspected meningococcal meningitis. He then performed a lumbar puncture to obtain cerebrospinal fluid (CSF).

The blood cultures grew *Neisseria meningitidis* (meningococcus) serogroup C. The CSF proved sterile but contained 20 white blood cells μl^{-1} (an abnormally high number; CSF usually contains five or fewer white blood cells μl^{-1}). Dolly's hematocrit was 36.5% (slightly low) and her blood white-cell count was 8700 cells μl^{-1}, of which 90% were neutrophils and 10% were lymphocytes.

Fig. 6.2 Assembly of the membrane-attack complex generates a pore in the lipid bilayer membrane. The sequence of steps and their approximate appearance are shown here in schematic form. C5b triggers the assembly of a complex of one molecule each of C6, C7, and C8, in that order. C7 and C8 undergo conformational changes that expose hydrophobic domains that insert into the membrane. This complex causes moderate membrane damage in its own right, and also serves to induce the polymerization of C9, again with the exposure of a hydrophobic site. Up to 16 molecules of C9 are then added to the assembly to generate a channel 100 Å in diameter in the membrane. This channel disrupts the bacterial cell membrane, killing the bacterium. The electron micrographs show erythrocyte membranes with membrane-attack complexes in two orientations, end on and side on. Photographs courtesy of S. Bhakdi and J. Tranum-Jensen.

Dolly quickly improved on continued intravenous antibiotic therapy. The fever disappeared, she became alert, and her neck stiffness resolved over the next 72 hours. Blood taken after 24 hours of antibiotic treatment proved sterile on culture and she was discharged from the infirmary.

Dolly told Dr Tolstoy that she had had meningococcal meningitis in her third year of high school. She was told at the time that she had positive CSF and blood cultures for *N. meningitidis* serogroup Y. Dr Tolstoy suspected that Dolly might have a complement deficiency and sent a blood sample for a CH_{50} assay, which tests the ability of the blood to perform complement-mediated hemolysis. A week later he received the report: Dolly's CH_{50} was zero, which means that her serum contained no hemolytic complement activity.

Dolly was one of five sisters. Blood samples were obtained from all of them to assay their CH_{50}, which turned out also to be zero in her three younger sisters, aged 9, 14, and 17. The two youngest sisters were well; the 17-year-old sister had had meningococcal meningitis the year before. An older sister, aged 20, had a normal CH_{50}. The father was unavailable for testing, but the mother proved to have a half-normal CH_{50}. The sera of all the sisters and the mother were tested in a double-diffusion assay in agar to ascertain whether C5, C6, C7, C8, and C9 were present in their blood. Dolly's serum lacked C8, as did that of her three sisters with CH_{50} values of zero. Her mother had a half-normal level of C8. Her older sister had no evident deficiency of these complement components. All the family were given tetravalent meningococcal vaccine.

Complement component deficiencies.

Inherited deficiencies of all the components of the classical and alternative pathways have been found in humans, with the singular exception of factor B of the alternative pathway. Deficiencies of C1q, C1r, C1s, C4, or C2 of the classical pathway result in immune-complex disease, which frequently proves fatal. Inherited defects in factor D, or properdin, of the alternative pathway, as well as defects in components of the membrane-attack complex, C5, C6, C7, C8, or C9, on which both pathways converge, result in increased susceptibility to neisserial infections, as illustrated by this case. C3 deficiency causes increased susceptibility to all pyogenic infections.

There is a high frequency of particular complement deficiencies in certain populations. For example, 1 in 40 Japanese is heterozygous for C9 deficiency. One in 100 Caucasians is heterozygous for C2 deficiency, and, as one would expect, 1 in 10,000 is completely deficient in C2. This prediction was verified by testing blood donors in Manchester, UK. The gene encoding C2 is in the major histocompatibility complex (MHC). An MHC haplotype of HLA-B18, HLA-DR2 is tightly linked to C2 deficiency in approximately 90% of affected Caucasians, all of whom carry the same mutation in the C2 gene. The genes encoding C4 are also in the MHC complex. There are two genes for C4—C4A and C4B—which are present in variable numbers of copies in the human population. Thirty percent of humans fail to express the products of one, two, or three of these genes, and thus may have a low C4 titer in their blood.

Deficiency of C8 in Caucasian populations is caused by a few well-known mutations in the gene encoding the β chain. C8 deficiency in blacks in the USA, in contrast, is commonly due to mutations in the gene encoding the α chain. More than 90% of the mutations in C8β are due to C→T transitions in the C8β gene. Such a mutation was carried by Dolly and her sisters and was presumably inherited from her Caucasian father, who was not available for testing. The mother was of Filipino origin and was heterozygous for another type of mutation in the C8β gene not found in Caucasians—hence her half-normal levels of C8. Dolly and her three affected sisters thus have a compound heterozygous deficiency for the C8β gene.

Questions.

1 Hemolytic complement levels were determined by measuring the CH_{50}. What is that and how is the test performed?

2 Bacteria of the genus Neisseria are encapsulated in a thick carbohydrate capsule. Clinical evidence shows that the membrane-attack complex is vital in host defense against these bacteria. It is obvious that the hydrophobic components of the membrane-attack complex cannot penetrate the polysaccharides of the bacterial capsule. How, then, is complement involved in bacterial killing?

3 Patients who are completely deficient in C1q, C1r, C1s, or C4 sustain persistent and severe immune-complex disease such as glomerulonephritis which can ultimately be fatal. In contrast, patients who are C2-deficient usually sustain only minor forms of immune-complex disease. How do you explain this difference?

CASE 7 | Hereditary Periodic Fever Syndromes

Defects in innate immunity leading to an uncontrolled inflammatory response.

One of the first responses to infection is the generation of an inflammatory response by the innate immune system, which causes pain, redness, heat, and swelling (edema) in the infected tissues. Inflammation at the site of infection augments the killing of invading microorganisms by the front-line macrophages and causes changes in the walls of local blood vessels that enable additional effector molecules and cells, such as neutrophils, to enter the infected tissue from the blood (Fig. 7.1). Inflammation also induces local blood clotting, making a physical barrier to the spread of infection through the bloodstream. These changes are induced by a variety of inflammatory mediators that are released by macrophages as a consequence of their recognition of pathogens.

Topics bearing on this case:
Inflammatory responses
Acute-phase responses
The autoinflammatory syndromes

| Healthy mucosal surface is not inflamed | Bacteria crossing the mucosa activate resident macrophages to secrete cytokines | Vasodilation and increased vascular permeability allow fluid, protein, and inflammatory cells to leave blood and enter tissue | The infected tissue becomes inflamed, causing redness, heat, swelling, and pain |

Fig. 7.1 Innate immune mechanisms establish a state of inflammation at sites of infection. Pathogens penetrating the skin or mucosal epithelium invade the underlying connective tissue. Here they activate resident effector cells such as macrophages to secrete chemokines, cytokines, and other pro-inflammatory mediators, which trigger the events leading to inflammation. The cells primarily attracted into the inflamed tissues are monocytes/macrophages and neutrophils, which contribute to enhancing and sustaining the state of inflammation.

Among the inflammatory mediators are pro-inflammatory cytokines that help to induce and maintain an inflammatory response. The main pro-inflammatory cytokines are interleukin-1β (IL-1β), tumor necrosis factor-α (TNF-α), and IL-6. In addition to inducing tissue inflammation, these cytokines also induce fever. Fever is generally beneficial to host defense; most pathogens grow better at lower temperatures, and adaptive immune responses are more intense at elevated temperatures. IL-1 and IL-6 also enhance the production of innate immunity proteins called the acute-phase reactants, which are synthesized by the liver and released into the blood.

A group of clinical syndromes called the autoinflammatory syndromes are manifested as inflammation in the absence of any evidence of microbial infection. They are characterized by episodes of fever accompanied by systemic and localized inflammation, affecting abdominal structures, joints, and the skin in particular. They do not involve self-reactive T cells or self-reactive antibodies (autoantibodies) and so are not due to autoimmune reactions, another cause of unexplained inflammation. As the name 'autoinflammatory' suggests, these syndromes are driven by unchecked inflammatory responses. In normal individuals, the inflammatory cascade of reactions is tightly controlled, in great part by regulating the production of pro-inflammatory cytokines such as IL-1β (Fig. 7.2). In contrast, individuals affected by autoinflammatory syndromes develop adaptive immune responses to stimuli such as infection or vaccination, but they are unable to control inflammation once it is under way, and inflammation may also flare up in response to minor stimuli that would not elicit an inflammatory response in unaffected individuals.

The autoinflammatory syndromes include relatively common conditions such as Crohn's disease and systemic-onset juvenile rheumatoid arthritis, but one category comprises the rarer diseases known as the hereditary periodic fever syndromes, which are due to mutations in single genes. The vari-

Fig. 7.2 Systemic effects of IL-1β. Active IL-1β is secreted by many cell types including monocytes and macrophages. It enters the circulation and triggers IL-1 receptors on the hypothalamic vascular network, resulting in the synthesis of cyclooxygenase-2, which causes levels of prostaglandin E2 in the brain to rise, thus activating the thermoregulatory center for fever production. In the periphery, IL-1β activates IL-1 receptors on endothelium, resulting in rashes and the production of IL-6. Circulating IL-6 stimulates liver cells to synthesize several acute-phase proteins, which cause an increased erythrocyte sedimentation rate. IL-1β also acts on the bone marrow to increase the mobilization of granulocyte progenitors and mature neutrophils, resulting in peripheral neutrophilia. It also indirectly increases IL-6-induced platelet production, resulting in thrombocytosis, and causes a decreased response to erythropoietin, which results in anemia.

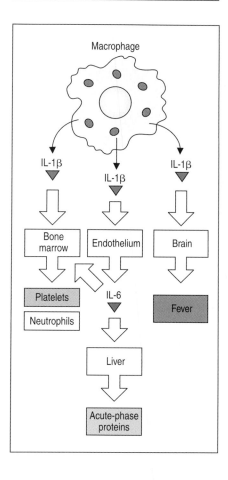

ous hereditary syndromes differ clinically and in the genes that are mutated, but are unified by the involvement of both systemic and focal (localized) inflammation. Although classified as 'periodic' fever syndromes, the recurrent bouts of inflammation often do not occur at regular intervals. The genetic defects underlying these syndromes have been identified and have proved to be defects in pathways that normally control the inflammatory response by regulating the production of the pro-inflammatory cytokines.

Currently, five distinct disorders are classified as hereditary periodic fever syndromes (Fig. 7.3). Familial Mediterranean fever (FMF) is the most common and is associated with mutations in the gene *MEFV*, which codes for a protein called pyrin, named after the fever characteristic of the disease. TNF receptor-associated periodic syndrome (TRAPS, originally known as familial Hibernian fever) presents with longer episodes of fever, abdominal pain, and

Disease	Clinical features	Functional defects	Inheritance	Defective gene (protein)
Familial Mediterranean fever (FMF)	Recurrent fever, serositis (inflammation of the pleural and/or peritoneal cavity) responsive to colchicine, amyloidosis. Predisposes to vasculitis and inflammatory bowel disease	Decreased production of pyrin leading to increased IL-1 processing and secretion; decreased macrophage apoptosis	Autosomal recessive	*MEFV* (pyrin/marenostrin)
TNF-receptor associated periodic syndrome (TRAPS)	Recurrent fever, serositis, rash, and ocular or joint inflammation	Reduced levels of soluble TNF receptor leading to excess TNF-α in the circulation	Autosomal dominant	*TNFRSF1A* (TNF 55kDa receptor/TNFR-I)
Hyper IgD syndrome (HIDS)	Periodic fever and leukocytosis with high levels of IgD in serum	Mevalonate kinase deficiency affecting cholesterol synthesis; pathogenesis of disease unclear	Autosomal recessive	*MVK* (mevalonate kinase)
Pyogenic arthritis, pyoderma gangrenosum and acne (PAPA)	Periodic fever, myalgia (muscle pain), rash, acute-phase response	Sequestration of pyrin by CD2BP1 leading to increased IL-1β processing and secretion	Autosomal dominant	*PSTPIP* (CD2-binding protein 1/CD2BP1)
Muckle–Wells syndrome (MWS)*	Urticaria, hearing loss	Constitutive activation of cryopyrin leading to increased caspase-1 activation and IL-1β processing and secretion	Autosomal dominant	*CIAS1* (cryopyrin/NALP3/PYPAF1)
Familial cold autoinflammatory syndrome (FCAS)*	Non-pruritic urticaria, arthritis, chills, fever, and leukocytosis after exposure to cold			
Neonatal-onset multisystem inflammatory disease/chronic infantile neurologic cutaneous and articular syndrome (NOMID/CINCA)*	Neonatal-onset rash, chronic meningitis, hearing loss, arthropathy with fever and inflammation responsive to the IL-1R antagonist anakinra			

Fig. 7.3 The spectrum of hereditary autoinflammatory syndromes. The cells primarily affected in all these disease are mainly granulocytes, especially neutrophils, and monocytes and macrophages. Chondrocytes are also affected in NOMID/CINCA.

*The cryopyrinopathies. All three syndromes are associated with similar mutations in *CIAS1*, the gene for cryopyrin; the disease phenotype in any individual seems to depend on the modifying effects of other genes and environmental factors.

a rash. This disorder is due to mutations in the gene encoding the TNF receptor 55 kDa subunit (*TNFRSF1A*) that lead to excess TNF-α in the circulation. Hyperimmunoglobulinemia D with periodic fever syndrome (HIDS) was originally differentiated from FMF by the detection of elevated levels of IgD. Attacks of HIDS tend to last 7 days and recur every 4–8 weeks. Mutations in the gene encoding CD2-binding protein 1 (CD2BP1), a pyrin-interacting protein, are associated with the syndrome of pyogenic arthritis, pyoderma gangrenosum, and acne (PAPA). These mutations accentuate the interaction between pyrin and CD2BP1, and it has been suggested that the sequestration of pyrin results in excess secretion of the pro-inflammatory cytokine IL-1β.

The fifth class of periodic fevers are the cryopyrinopathies (see Fig. 7.3), which are characterized by excessive production of IL-1β. They are three related but clinically distinct syndromes that all result from mutations in the gene *CIAS1*, which encodes cryopyrin, a protein closely related to pyrin. We shall look here at a case of neonatal-onset multisystem inflammatory disease/chronic infantile neurologic cutaneous and articular syndrome (NOMID/CINCA). Infants with NOMID/CINCA have intermittent fevers associated with persistent urticaria, arthritis-like symptoms and central nervous system (CNS) involvement that ranges through papilledema (swelling of the optic disk, where the sensory fibers from the retina exit the eye), increased numbers of inflammatory cells in the cerebrospinal fluid (CSF), causing a chronic meningitis, to sensorineural hearing loss (deafness due to a defect in nerve conduction or perception by the brain of the auditory impulse).

The case of Chancellor Krook: fever and inflammation in the absence of infection.

Infant with recurrent fever and irritability. No infection detectable.

Chancellor was born at full term with no complications, but on routine screening at birth he was found to have evidence of sensorineural hearing loss in both ears. At 1 week old, he developed intermittent fever accompanied by irritability, and a severe bacterial infection was suspected. His white blood cell count was elevated, at 21,000 µl⁻¹ (normal <10,000 µl⁻¹). CSF was sampled and also had an abnormally high white blood cell count, of 500 ml⁻¹ (normal for age is <20 ml⁻¹). Although no bacteria could be cultured from samples of blood, urine, and CSF, Chancellor was kept in hospital and treated with antimicrobial drugs because bacterial or viral meningitis was suspected. Despite this treatment the fevers continued and no cause could be found. His mother noticed a raised, red, itchy rash all over Chancellor's body, which his physicians diagnosed as urticaria (hives), attributing it to a reaction to the antibiotics he had been given. The rash persisted, however, and Chancellor continued to be intermittently feverish. He was eventually sent home without a diagnosis.

The episodes of fever and irritability with no apparent source continued throughout Chancellor's first year, and his growth and development appeared unusually slow. Tests revealed progressive hearing loss, and he was given a hearing aid. In view of his continuing symptoms, Chancellor was referred to the rheumatology department at the Children's Hospital when he was 1 year old for testing for inflammatory disease.

Hearing loss, arthritis, recurrent fever. Autoinflammatory disease?

The examining physician noted Chancellor's prominent forehead and saddle-shaped nose (Fig. 7.4). His liver and spleen were enlarged, and examination of his joints revealed evidence of arthritis in the knees. A diffuse, red, urticaria-like rash covered his trunk, back, legs, and arms. Laboratory tests revealed an erythrocyte

Fig. 7.4 Clinical features of NOMID/CINCA. Panel a, Dysmorphic facial features: frontal bossing, saddle nose, and long philtrum of upper lip; panel b, urticarial rash; panel c, enlarged liver and spleen with distended abdomen (markings show the lower margins of the spleen (right) and the liver (left)); panel d, fundoscopic image of the optic disk showing papilledema.

sedimentation rate (ESR) of 35 mm h^{-1} (normal 0–20 mm h^{-1}). Sedimentation of red blood cells is hastened when the concentration of certain blood proteins is increased, among them the acute-phase reactants. One of these reactants, C-reactive protein, was measured and was elevated, at 8 mg dl^{-1} (normal <0.50 mg dl^{-1}).

On the basis of the constellation of symptoms shown by Chancellor—fever, rash, arthritis, hearing loss, and irritability—a diagnosis of NOMID/CINCA was made. Genetic testing revealed a mutation in the *CIAS1* gene, which confirmed the diagnosis. Chancellor was treated with anakinra (Kineret), a recombinant human IL-1 receptor antagonist, and his symptoms disappeared. Now 2 years old, Chancellor continues to require treatment with anakinra. His growth and development are now normal for his age, and his blood markers of inflammation are also at normal levels.

Mutation in CIAS1 gene.

Hereditary periodic fever syndromes.

Mutations in several genes responsible for the regulation of inflammatory pathways have been implicated in the pathogenesis of hereditary periodic fever syndromes (see Fig. 7.3). In the cryopyrinopathies, the mutations are in the *CIAS1* gene, which encodes cryopyrin (also known as NALP3). Cryopyrin has been shown to be involved in regulating the activation of the protease caspase-1, which processes IL-1β to the mature form and thus enables its secretion. In some way not yet completely worked out, the mutations in cryopyrin abolish the regulation of the caspase, leading to unchecked processing and secretion of IL-1β. The effects of excess IL-1β include fever, rash, excess neutrophils in the blood (neutrophilia), high levels of blood platelets (thrombocytosis), and the production of acute-phase reactants, all of which are seen in Chancellor's case.

Because dysregulated inflammation is harmful to the host, the production of pro-inflammatory cytokines is normally tightly controlled. Cryopyrin and caspase-1 are part of larger cytoplasmic protein complexes called inflammasomes (Fig. 7.5), which are responsible for the processing and secretion of IL-1β and of IL-18, another pro-inflammatory cytokine. Inflammasosomes are found in monocytes/macrophages and neutrophils. Two types of inflammasomes have been identified, which contain different types of cryopyrin and different caspases. One is composed of the cryopyrin NALP1, an adaptor protein called ASC, and caspases-1 and -5. NALP3 is part of the so-called NALP2/3 inflammasome, which includes NALP3, the related protein NALP2, ASC, caspase-1, and a protein called Cardinal.

Inflammasomes containing normal cryopyrins are thought to become activated only in the presence of an infection. Both NALP1 and NALP3 contain

Fig. 7.5 Schematic structure of the NALP1 and NALP2/3 inflammasomes. NALPs are the chief platforms of the inflammasomes. Their amino-terminal sequence is characterized by the presence of pyrin domain (PYD), followed by NACHT domain, a NACHT-associated domain (NAD), and several leucine-rich repeats (LRRs). The LLR domain is thought to bind to pathogen-associated molecular patterns (PAMPs). The NACHT domain is responsible for the oligomerization of the inflammasomes. NALP1 contains an FIIND domain, and a caspase-recruitment domain (CARD).

a leucine-repeat rich (LRR) domain that recognizes pathogen-associated molecular patterns (PAMPs), and binding of PAMPs is proposed to trigger mutual proteolytic activation of the associated caspases. The activation of caspase-1 leads to cleavage of the precursor IL-1β and IL-18 and their further processing and secretion as active cytokines, leading to a powerful inflammatory response. Bacterial RNA, bacterial cell-wall components (such as muramyl dipeptide), the bacteria *Staphylococcus aureus* and *Listeria monocytogenes*, and gout-associated crystals have all been found to activate this NALP pathway.

The exact mechanism by which the mutations found in patients with cryopyrinopathies promote inflammation is still under investigation. The cryopyrin (NALP3) mutants that have been found are constitutively active and are hyperresponsive to microbial antigens or to physical stimuli such as cold. In either case this results in exaggerated secretion of IL-1β and IL-18.

One of the most serious long-term complications of the periodic fever syndromes is the development of systemic amyloidosis, which occurs when misfolded fragments of serum amyloid, an acute-phase reactant, get deposited in tissues. The kidneys, gastrointestinal tract, adrenal glands, spleen, testes, and lung are most often affected, but the liver, heart, and thyroid gland can also be involved. Amyloidosis occurs most commonly in patients with familial Mediterranean fever, but it can also affect patients with TRAPS, Muckle–Wells syndrome, or NOMID/CINCA. Treatment with the drug colchicine, and to a lesser extent with TNF-α inhibitors and the IL-1 receptor antagonist anakinra, has been shown to prevent the amyloidosis developing.

Questions.

1   Why has the drug anakinra proved a successful treatment for some of the autoinflammatory syndromes?

2   The symptoms of systemic-onset juvenile rheumatoid arthritis are fever, rash, and arthritis. Is it considered to be an autoinflammatory syndrome?

3   What is the proposed mechanism of action for colchicine in the treatment of the autoinflammatory syndromes?

4   What cytokine, other than IL-1 and IL-18, would you expect to be elevated in the setting of caspase-1 activation?

CASE 8 | Interleukin 1 Receptor-Associated Kinase 4 Deficiency

An inability of the innate immune system to detect the presence of pathogens.

The presence of an invading microorganism is first recognized by tissue macrophages at the site of infection. Pathogens carry characteristic chemical structures on their surface and in their nucleic acids that are not present in human cells. These pathogen-associated molecular patterns (PAMPs) are recognized by invariant pattern recognition receptors (PRRs) on macrophages, and also on neutrophils, which take up the pathogen and destroy it. PPRs are also present on dendritic cells, whose function is to present antigen to and activate naive T cells; dendritic cells form the link between the innate and the adaptive immune response. Among the PRRs are the Toll-like receptors (TLRs), which are present on the cell surface and in the membranes of endosomes, and which enable the cells of the innate immune system to detect and respond to a wide variety of pathogens.

Topics bearing on this case:
Innate immunity
Toll-like receptors
Toll-like receptor/IL-1R signaling pathways
Dendritic cells and the initiation of adaptive immune responses

Fig. 8.1 Toll-like receptors and cytokine production. Toll-like receptors (TLRs) and their representative ligands are shown. Simplified signaling pathways are shown below each receptor, leading to the production of cytokines characteristic of each TLR. TLR-7, TLR-8, and TLR-9 are present in the membranes of endosomes, where they recognize ligands generated in the endosomal pathway from microbes engulfed by phagocytosis or macropinocytosis. The other TLRs are expressed on the cell surface. GPI, glycosylphosphatidylinositol (*Trypanosoma cruzi*); CpG, unmethylated DNA; LPS, bacterial lipopolysaccharide.

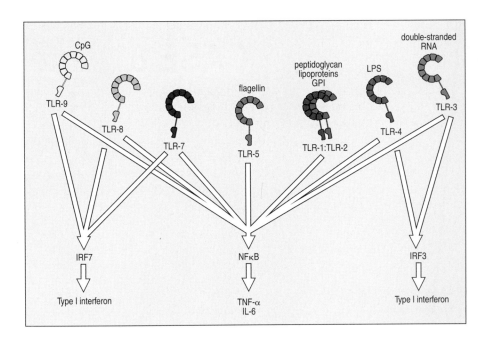

There are 10 different TLRs in humans, which between them recognize microbial nucleic acids, such as unmethylated bacterial DNA and long double-stranded RNA, as well as molecules specific to particular classes of microorganisms, such as the bacterial lipopolysaccharide (LPS or endotoxin) characteristic of Gram-positive bacteria and the protein flagellin of bacterial flagella. All these molecules are essential for the microorganism's survival and are therefore relatively invariant (Fig. 8.1). The limited number of TLRs can thus detect the presence of many different microorganisms by recognizing features that are typical of the different groups of pathogens—bacteria, viruses, fungi, and parasites.

The TLRs are signaling receptors, and their activation leads to enhancement of the antimicrobial activity of macrophages and neutrophils, and the secretion of cytokines by macrophages, which help attract more macrophages and neutrophils out of the blood and into the site of infection. Upon engagement of their TLRs, dendritic cells undergo a brief period of macropinocytosis that is important for taking up and processing microbial antigens for subsequent presentation to naive T cells. Signaling via TLRs also induces the maturation of tissue dendritic cells and their migration to peripheral lymphoid tissues such as lymph nodes, where they encounter and activate antigen-specific T cells. During maturation, conventional dendritic cells upregulate the production of co-stimulatory molecules such as CD40, B7.1 (CD80), and B7.2 (CD86), which are essential for T-cell activation, and produce chemokines and cytokines that help induce adaptive immune responses, such as the pro-inflammatory cytokines tumor necrosis factor-α (TNF-α), interleukin-1 (IL-1), IL-6, and IL-12 (see Case 7). Another type of dendritic cell, the plasmacytoid dendritic cell, produces large amounts of the antiviral type I interferons (IFN-α and IFN-β) in response to viral infections.

The stimulation of dendritic cells by PAMPs such as LPS and unmethylated CpG in bacterial DNA is particularly critical for their ability to induce the development of T$_H$1 effector cells. The exposure of dendritic cells to pro-inflammatory cytokines is sufficient to enable them to induce the proliferation (clonal expansion) of antigen-specific naive T cells, but the upregulation of co-stimulatory molecules and of IL-12 production as a result of TLR stimulation is essential for dendritic cells to be able to induce the differentiation of T_H1 effector functions, such as the secretion of IFN-γ, in CD4 T cells. Thus, TLRs have a vital role in the development of both innate and adaptive immune responses.

The TLRs and the IL-1 receptor signal through a common intracellular signaling pathway, and this case shows how a defect in this pathway has serious effects on a person's ability to respond to some classes of pathogens.

The case of Douglas Mooster: a 6-year-old boy with recurrent pneumococcal meningitis.

Douglas is an only child and was referred to the immunology clinic at the Boston Children's Hospital at the age of 6 years after two episodes of pneumococcal meningitis, the first a year previously, caused by serotypes 3 and 14, respectively. As a result of the meningitis, Douglas had suffered a stroke and seizures, and had been left with mild hearing loss and mild learning impairment. When he was first seen at the immunology clinic, his overall neurological function was good.

Frequent middle-ear infections (otitis media) were a significant part of Douglas's medical history, and myringotomy tubes had been inserted in both ears when he was 2 years old to aerate the middle ear. At 11 months old he had developed intestinal intussusceptions (telescoping of a segment of the intestine into an adjacent segment), complicated by perforation of the intestine and formation of an abscess in the peritoneum. He had also suffered from boils (furuncles) of the scalp and several skin infections, all of which had responded well to treatment with antibiotics. His medical history also noted an unusually weak febrile response, as Douglas usually developed only low-grade fevers late in the course of these illnesses. He seemed to have no particular susceptibility to viral infections.

Recurrent pyogenic infections. No fever.

On physical examination, he appeared thin but otherwise well. His tonsils were slightly enlarged, as were the lymph nodes of the anterior cervical (neck) chains (lymphadenopathy). His abdomen was soft and not tender, although the liver could be felt roughly 3 cm below the lower rib on the right (the liver is normally non-palpable). The spleen was not enlarged and his skin showed no rash or other unusual features.

Laboratory tests showed a normal complete blood count, complement function, and immunoglobulin titer. Specific antibody titers showed that Douglas had made protective immune responses to protein antigens, such as tetanus toxoid, after his routine immunizations. Douglas had been given the vaccine Pneumovax as part of the evaluation at the immunology clinic; this vaccine contains polysaccharide antigens from 23 different serotypes of *Streptococcus pneumoniae* (pneumococcus), and he had not made responses to any of the 14 serotypes that were tested for at the clinic. He had also failed to mount a protective antibody titer after immunization with polysaccharide antigens from *Streptococcus meningitidis* (meningococcus), a common cause of meningitis. He had, however, normal absolute numbers of B cells and T cells.

Adaptive immunity mostly normal. Test for TLR function.

Given Douglas's susceptibility to pneumococcal infections, with no obvious underlying immunological cause, and his poor febrile response to infection, his Toll-like receptor function was evaluated. A sample of peripheral blood cells (PBCs) produced little or no TNF-α in response to known TLR ligands when tested in the

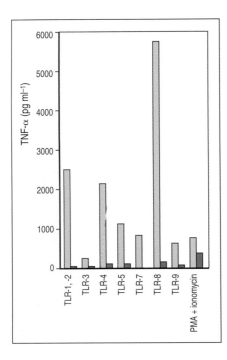

Fig. 8.2 Impaired TLR-induced TNF-α production in IRAK4 deficiency.

Peripheral blood mononuclear cells from Douglas (dark blue) and a normal control (light blue) were stimulated by ligands for the TLR receptors indicated or by phorbol myristate acetate (PMA) plus ionomycin as a positive control. The latter compounds bypass receptor signaling by activating protein kinase C and causing an influx of calcium, respectively. TNF-α production was measured from cell supernatants 12 hours after stimulation. The histogram shows that Douglas's cells produced no or very little TNF-α in comparison with the response of normal cells to TLR ligands.

laboratory (Fig. 8.2). The integrity of the signal transduction pathways leading from the TLRs was evaluated *in vitro* by testing the cells' responses to IL-1, as the IL-1 receptor (IL-1R) stimulates the same signaling pathway. Stimulation of the cells with IL-1 produced no activation of MAP kinases and no phosphorylation of IκBα, important intermediate steps in these pathways (Fig. 8.3). These results were consistent with a defect very early in the pathway, and a defect in the IL-1R-associated kinase 4 (IRAK4) was suspected. Sequencing of Douglas's DNA for the *IRAK4* gene revealed a homozygous nonsense mutation within the kinase domain, resulting in the production of a truncated protein with no kinase activity.

Interleukin 1 receptor-associated kinase 4 deficiency.

The IL-1R and the TLRs share a common signal transduction pathway that involves the adaptor protein MyD88, the signaling intermediate TRAF-6, and the receptor-associated protein kinases IRAK1 and IRAK4. Activation of these receptors leads to recruitment of MyD88 to the receptor, followed by the recruitment and activation of IRAK4, the initial protein kinase in the TLR signal transduction pathway. This is followed by recruitment of IRAK1 and TRAF-6 to the pathway. This pathway can then diverge, one branch activating the MAP kinases, the other leading to activation of the NFκB transcription factor (Fig. 8.4). The steps leading to the activation of NFκB can be seen in more detail in Fig. 9.1. Activation of the classical NFκB pathway results in the production of pro-inflammatory cytokines such as IL-6, IL-12, and TNF-α (see Case 7). A functional IRAK4 is critical to the ability to make responses to TLR ligands, as signaling via the NFκB pathway is blocked in its absence.

There are exceptions, however, as TLR-3 and TLR-4 can also signal via a pathway that does not use IRAK4. They can use a second, MyD88-independent, pathway involving the adaptor proteins TRIF (TLR-3) or TRIF/TRAM (TLR-4). This pathway results in the activation of the interferon response factor 3 (IRF3) and the production of type I interferons (see Fig. 8.4). Interferon production is an essential function of the innate response to certain viruses, including the herpes virus family and the human immunodeficiency virus (HIV).

Type I interferon production is also stimulated in plasmacytoid dendritic cells and monocytes by the activation of TLR-7, TLR-8, and TLR-9. Like the other TLRs, these receptors can signal via the IRAK/TRAF-6 pathway, leading to the production of pro-inflammatory cytokines. IRAK1 can, however, also phosphorylate and activate the transcription factor IRF7, which induces the expression of the gene for IFN-α. Stimulation of TLRs 7, 8, and 9 in these cells can thus also induce interferon production. Indeed, herpes simplex virus infection in mice in which IRF7 is completely absent is lethal.

Fig. 8.3 Impaired IL-1-induced phosphorylation of p38 MAP kinase and IκBα in IRAK4 deficiency.

Peripheral blood mononuclear cells from Douglas and a normal control were stimulated with IL-1 or PMA plus ionomycin (P/I) as a positive control for the times indicated. Cell lysates were subjected to Western blotting with antibodies against phosphorylated p38 MAP kinase or phosphorylated IκBα. No phospho-IκBα or phospho-p38 were present in Douglas's blood cells. Antibody against TRAF-6 is shown as a protein-loading control. Courtesy of D. McDonald.

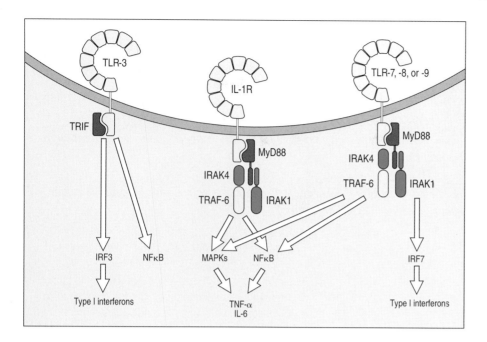

Fig. 8.4 TLR signaling pathways.
Signaling through IL-1R and all TLRs can activate IRAK4 and IRAK1 (as shown here for IL-1R and the TLRs 7, 8, or 9) through the adaptor protein MyD88, leading to activation of the transcription factor NFκB and cytokine production. Stimulation of TLRs 7, 8, or 9 can also lead to type I interferon production via activation of the interferon response factor 7 (IRF7). When TLR-3 uses the adaptor protein TRIF instead of MyD88, interferon response factor 3 (IRF3) is activated, which leads to the synthesis of type I interferons.

As well as its role in innate immunity, IRAK4 may also be directly involved in the development of adaptive immune responses, because it has been shown to interact with the tyrosine kinase ZAP-70 in T cells, which is an intermediate in the signaling pathway leading from the T-cell receptor. IRAK4 knockout mice also have a deficiency of IL-2 and thus clonal expansion of activated T cells is impaired. An important point to note is that IRAK4-deficient mice are susceptible to both bacterial and viral infections, whereas the IRAK4-deficient patients identified so far seem to be primarily susceptible to pyogenic bacteria. Up to now, only 18 patients with IRAK4 deficiency have been identified, as a result of their severe, recurrent infections with pyogenic bacteria. By routine immunological criteria they appear normal, with normal numbers of B and T cells, normal serum immunoglobulin levels, and protective antibody titers to protein antigens, but impaired antibody titers to polysaccharide antigens. One IRAK4-deficient patient has impaired antibody responses to both protein and polysaccharide antigens, however. To completely understand the role of IRAK4 in human innate and adaptive immunity will require the identification and evaluation of more patients with this defect.

Questions.

1 Douglas's immune deficiency is characterized by susceptibility to severe, invasive pneumococcal infection. Are there other immunodeficiencies associated with similar susceptibility to pneumococci?

2 What clue in the history of a patient with recurrent pneumococcal or staphylococcal infections would favor a possible diagnosis of IRAK4 deficiency?

3   What is the clinical course of patients with known defects in IRAK4?

4   Why might IRAK-deficient patients fail to respond to immunization with polysaccharide antigens?

5   IRAK4 seems to be involved in signaling from all TLRs. Why has increased susceptibility to viral infections not yet been documented in human patients?

6   How should IRAK4-deficient patients be managed to prevent the recurrence of invasive infection?

CASE 9 | X-linked Hypohydrotic Ectodermal Dysplasia and Immunodeficiency

Immunodeficiency due to a defective component in an intracellular signaling pathway required for both innate and adaptive immunity.

Intracellular signaling molecules are vital to the normal maturation of B and T lymphocytes. In Case 10 we will see that X-linked agammaglobulinemia results from a failure of normal B-cell maturation because of mutations in a tyrosine kinase (Bruton's tyrosine kinase, Btk) that is vital to normal calcium movements in pre-B cells. In its absence, pre-B cells fail to progress to mature B cells and thus never become immunoglobulin-secreting plasma cells. Other signaling molecules are involved in the transmission of co-stimulatory signals to lymphocytes. In Case 11 we will see that X-linked hyper IgM syndrome results from the failure of B cells to undergo class-switch recombination and somatic hypermutation because they do not receive a signal from helper T cells, owing to a failure of these cells to express CD40 ligand.

Although some signaling molecules, such as CD40 ligand, are peculiar to lymphocytes, others are found in many cell types. A crucial step in the activation of many different cells is the induction of the transcription of specific genes by transcription factors of the NFκB family. When not required, NFκB and other family members reside in the cytoplasm, where they are maintained in an inactive state in a complex with their inhibitor protein, IκB. When the cell is activated by the appropriate external signal, a cascade of protein kinase activity is induced in the cytoplasm. This causes the assembly and activation of the IκB kinase (IKK) complex (Fig. 9.1) which consists of three components—the kinases IKKα and IKKβ along with

Topics bearing on this case:

Macrophage activation

Isotype switching

Activation of NFκB

T-cell help in antibody response

CD40 ligand and isotype switching

Fig. 9.1 The NFκB activation pathway.
Various receptors can activate a series of intracellular adaptor molecules and kinases that ultimately induce the formation of the IKK complex (consisting of IKKα, IKKβ, and NEMO/IKKγ). The IKK complex then phosphorylates the inhibitory protein IκB, resulting in its dissociation from cytoplasmic NFκB and its ultimate ubiquitination and degradation. Once free from the inhibitory influence of IκB, NFκB translocates to the nucleus, where it induces transcription of specific genes.

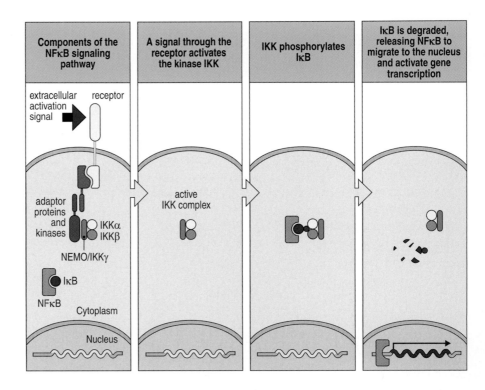

another protein, the NFκB essential modifier (NEMO), also known as IKKγ. The active IKK complex phosphorylates IκB, which targets IκB for degradation by the ubiquitin–proteasome system, thus freeing NFκB from its inhibitor. A nuclear localization signal is uncovered on NFκB that enables the transcription factor to move into the nucleus, where it binds to particular nucleotide sequences in the promoters of selected genes and initiates their transcription.

NFκB is activated by the ligation of a variety of receptors, including many with fundamental roles in inflammation and in the generation of immune responses. For example, the binding of the pro-inflammatory cytokines tumor necrosis factor-α (TNF-α) or interleukin 1 (IL-1) to their receptors results in NFκB activity. The Toll-like receptors (TLRs) on macrophages, which are key components of the innate immune system and recognize microbial components and other danger signals, also work through NFκB activation. And NFκB is activated in B cells after stimulation via CD40.

Functional NFκB is measured in the laboratory by an electrophoretic mobility-shift assay (EMSA) that detects its presence in the nucleus. Cell nuclei are isolated by ultracentrifugation, and their NFκB content is measured by assessing the capacity of the nuclear extract to bind to a synthetic radiolabeled oligonucleotide probe and thus to retard its migration in gel electrophoresis (Fig. 9.2).

Fig. 9.2 Electrophoretic mobility-shift assay (EMSA). Nuclei were isolated from normal B cells by ultracentrifugation and nuclear extracts were mixed with a ^{32}P-radiolabeled DNA probe that binds to NFκB. Free probe, not associated with NFκB, can be seen at the bottom of the gel, whereas probe bound to NFκB is retarded in the gel as a result of the higher molecular weight of the protein–probe complex. Although unstimulated B cells contain a baseline level of NFκB in their nuclei (as seen in the left lane), ligation of CD40 by CD40 ligand (CD40L) results in a higher level of NFκB in the nucleus (as evidenced by the darker band in the right lane).

The case of Robert Teixiera: severe bacterial infections and pointed teeth.

Robert was born after an uneventful pregnancy. He received his normal immunizations at 2, 4 and 6 months, including a pneumococcal vaccine that contained seven different serotypes of *Streptococcus pneumoniae*, without any ill effects. At 9 months old he developed recurrent fever associated with nasal discharge, which was diagnosed as an upper respiratory tract viral infection. He began to show less interest in his bottle, cried frequently, and became increasingly difficult to console.

Three weeks after the fevers began he had a seizure and was taken to the local emergency department. The seizure was controlled with an intravenous anti-convulsive medicine and the physician noted that his skin felt warm and had a reticular (lacy) pattern of blood vessels. Robert's neck was rigid and he screamed when his head was moved. A complete blood count revealed an abnormally high white blood cell count of 48,200 μl^{-1}, 90% of which were neutrophils (normal 5000 μl^{-1}). Spinal fluid was sampled by lumbar puncture and was free-flowing and cloudy. Microscopic and chemical analysis revealed 12 red blood cells per high-power field (normal 0); 11,382 white blood cells per high-power field (normal less than 6 per high-power field), of which 85% were neutrophils; a protein concentration of 410 mg dl^{-1} (elevated compared with normal); and a glucose concentration of 39 mg dl^{-1} (low compared with normal).

A Gram stain of the spinal fluid showed paired Gram-positive cocci. Robert was treated with ceftriaxone (an extended-spectrum cephalosporin antibiotic) and admitted to intensive care. He responded rapidly to the antibiotic and improved daily. A culture of his spinal fluid grew *S. pneumoniae*, which expressed the p14 cell-wall polysaccharide. An immunology consultation was requested because children receiving the seven-valent pneumococcal vaccine should be protected against strains of *S. pneumoniae* carrying this polysaccharide.

Laboratory analysis of Robert's serum immunoglobulin showed 170 mg dl^{-1} IgG (normal 400–1300 mg dl^{-1}), 34 mg dl^{-1} IgM (normal 30–120 mg dl^{-1}) and 184 mg dl^{-1} IgA (normal 20–230 mg dl^{-1}). He had a normal absolute lymphocyte count and normal percentages of T cells and B cells. Specific antibodies to 15 different pneumococcal polysaccharides or to tetanus toxoid (with which he had also been immunized) were not present in his serum. Because of the severity of his infection, his hypogammaglobulinemia and lack of specific antibody, intravenous immunoglobulin replacement therapy was begun at a dose of 400 mg kg^{-1} body weight every 3 weeks.

As Robert grew, it became clear that he had dysplastic ectoderm (that is, an abnormality in the growth of the structures produced from the ectoderm). His hair never grew properly; his first teeth erupted at 21 months and were pointed and conical (Fig. 9.3). A skin biopsy revealed a lack of eccrine sweat glands (hypohydrosis). On the basis of these findings, his hypogammaglobulinemia and his clinical history, the diagnosis of ectodermal dysplasia with immunodeficiency was made. A mutation in the *NEMO* gene was suspected. DNA sequence analysis revealed a point mutation that had caused an amino acid change at position 417 at the carboxy-terminal end of the NEMO protein.

Robert remained well until he was 2 years old, when he developed a hyperpigmented lacy rash on his back. A biopsy revealed diffuse granulomatous inflammation, and an acid-fast stain was positive for bacilli. A culture of the skin sample grew *Mycobacterium avium*. The cutaneous atypical mycobacterial

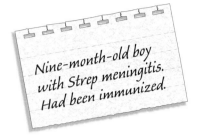

Nine-month-old boy with Strep meningitis. Had been immunized.

Hypogamma-globulinemia and no anti-strep antibodies.

Ectodermal dysplasia apparent. Query: NEMO deficiency.

Fig. 9.3 Infant with X-linked ectodermal dysplasia and immunodeficiency due to NEMO deficiency. The features of ectodermal dysplasia include frontal bossing, deep-set eyes, fine or sparse hair and conical or missing teeth.

infection was treated with a combination of antibiotics chosen on the basis of the *in vitro* sensitivity of the isolated mycobacteria. After several months of antibiotic treatment the rash had disappeared, but repeated attempts to discontinue the antibiotics failed, because the rash always returned. Robert is currently well on immunoglobulin replacement therapy and continuous treatment with anti-mycobacterial antibiotics.

NEMO deficiency and X-linked hypohydrotic ectodermal dysplasia with immunodeficiency.

X-linked ectodermal dysplasia with immunodeficiency is caused by mutations in the *NEMO* gene. This results in immunodeficiency because the end point of the intracellular signaling pathways leading from a variety of receptors important in activating immune responses is activation of NFκB and the induction of gene transcription. In the absence of any component of IKK (such as NEMO) IκB is not properly phosphorylated and it remains bound to NFκB, thus preventing the translocation of NFκB to the nucleus. The receptors leading to NFκB activation include the Toll-like receptors on macrophages, which recognize bacterial lipopolysaccharide (LPS), peptido-glycans, and bacterial DNA (CpG), and whose engagement stimulates innate immunity. Other receptors include CD40, which is important in adaptive immunity. In patients with a *NEMO* mutation the interaction between CD40 on B cells and CD40 ligand on T cells occurs but fails to transduce an effective intracellular signal; there is therefore no activation of mature naive B cells to make antibody and to switch antibody class (as assessed by a lack of upregulation of the surface markers CD23 and CD54 (ICAM-1); Fig. 9.4).

Patients with mutations in *NEMO* have ectodermal dysplasia because a receptor that is required for ectodermal development also depends on IKK function and NFκB activation to transmit a signal to the nucleus. The receptor is the ectodysplasin A receptor (a TNF receptor superfamily member), which binds ectodysplasin A (a member of the TNF superfamily). Although defects in the genes for these two proteins result in ectodermal dysplasia, there is no concomitant immunodeficiency because the function of the IKK complex is intact.

The *NEMO* gene is located on the X chromosome and thus the disease is transmitted by mothers to their sons. The gene contains 10 exons and is 419 amino acids long. Most of the mutations in the *NEMO* gene that cause ectodermal dysplasia are located in the tenth exon, which encodes a zinc finger domain that is a site of protein–protein interaction.

Women who carry a single copy of a *NEMO* gene with a point mutation generally do not have overt symptoms, although they express both normal NEMO and mutant NEMO. Some may have missing teeth, mildly abnormal patterns of hair or unusual birthmarks. However, a major frameshift or deletion mutation in *NEMO* results in a condition known as incontinentia pigmenti, which is characterized by dermal scarring and hyperpigmentation in heterozygotes. This disorder does not affect the immune system, because the normal *NEMO* gene allows appropriate immune-system development and function. In fact, the immune cells of these women display nonrandom

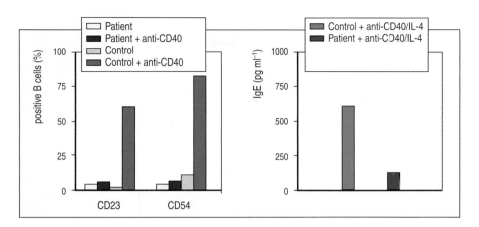

Fig. 9.4 Impairment of NFκB activation and B-cell activation in a patient with ectodermal dysplasia and immunodeficiency due to NEMO mutation. Left panel: B-cell activation can be assessed by increases in surface expression of CD23 (the low-affinity IgE receptor) or CD54 (ICAM-1). The presence of these molecules on the surface of B cells is measured by flow cytometry. In a healthy individual, stimulation through CD40 results in the majority of B cells expressing CD23 and CD54 on the cell surface. The gray bars show the unstimulated B cells, and the blue bars show the stimulated B cells. No increased expression of CD23 and CD54 after CD40 ligation is seen in the patient's B cells: compare the unstimulated and stimulated patient's B cells (yellow and red, respectively). Right: CD40 ligation causes IgE isotype switching and IgE synthesis in normal B cells (green) but not in B cells from a NEMO patient (purple), further demonstrating their inability to transmit a signal from CD40.

X-chromosome inactivation and all their immune-system cells contain the normal *NEMO* gene. Any male offspring of women with incontinentia pigmenti who receive the mutant *NEMO* gene die *in utero*.

Questions.

1 | *Why are boys with NEMO mutations susceptible to infection with opportunistic mycobacteria such as M. avium?*

2 | *Patients with NEMO mutations also suffer from chronic viral infections such as cytomegalovirus (CMV). Why is this?*

3 | *Although B cells from some patients with NEMO mutations fail to undergo immunoglobulin class switching in response to CD40 ligation, these patients have residual serum levels of IgG and IgA. Why is this?*

4 | *Recently a patient has been discovered to have a gain-of-function mutation in IκBα. This patient, who also has ectodermal dysplasia, has a more severe immunological defect than patients with the NEMO defect. The gain-of-function mutation results from the change of a serine residue to an isoleucine residue, so that the IκBα chain cannot be phosphorylated. Why do you think the immunodeficiency is more severe in this defect?*

CASE 10 | X-linked Agammaglobulinemia

An absence of B lymphocytes.

One of the most important functions of the adaptive immune system is the production of antibodies. It is estimated that a human being can make over one million different specific antibodies. This remarkable feat is accomplished through a complex genetic program carried out by B lymphocytes and their precursors in the bone marrow (Fig. 10.1). Every day about 2.5 billion (2.5×10^9) early B-cell precursors (pro-B cells) take the first step in this genetic program and enter the body pool of pre-B cells. From this pool of rapidly dividing pre-B cells 30 billion daily mature into B cells, which leave the bone marrow as circulating B lymphocytes, while 55 billion fail to mature successfully and undergo programmed cell death. This process continues throughout life, although the numbers gradually decline with age.

Topics bearing on this case:

B-cell maturation

Effector functions of antibodies

Humoral versus cell-mediated immunity

Effector mechanisms of humoral immunity

Methods for measuring T-cell function

Actions of complement and complement receptors

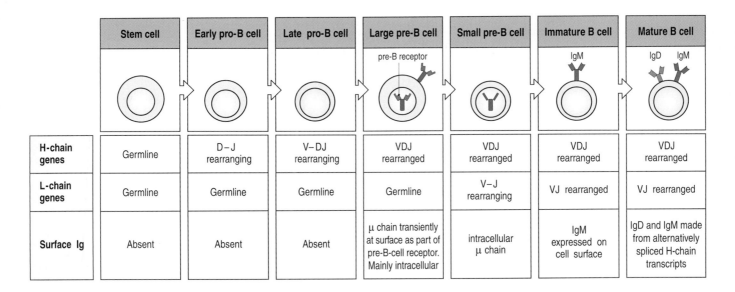

	Stem cell	Early pro-B cell	Late pro-B cell	Large pre-B cell	Small pre-B cell	Immature B cell	Mature B cell
H-chain genes	Germline	D–J rearranging	V–DJ rearranging	VDJ rearranged	VDJ rearranged	VDJ rearranged	VDJ rearranged
L-chain genes	Germline	Germline	Germline	Germline	V–J rearranging	VJ rearranged	VJ rearranged
Surface Ig	Absent	Absent	Absent	μ chain transiently at surface as part of pre-B-cell receptor. Mainly intracellular	intracellular μ chain	IgM expressed on cell surface	IgD and IgM made from alternatively spliced H-chain transcripts

Fig. 10.1 The development of B cells proceeds through several stages marked by the rearrangement of the immunoglobulin genes. The bone marrow stem cell that gives rise to the B-lymphocyte lineage has not yet begun to rearrange its immunoglobulin genes; they are in germline configuration. The first rearrangements of D gene segments to J_H gene segments occur in the early pro-B cells, generating late pro-B cells. In the late pro-B cells, a V_H gene segment becomes joined to the rearranged DJ_H, producing a pre-B cell that is expressing both low levels of surface and high levels of cytoplasmic μ heavy chain. Finally, the light-chain genes are rearranged and the cell, now an immature B cell, expresses both light chains (L chains) and μ heavy chains (H chains) as surface IgM molecules. Cells that fail to generate a functional surface immunoglobulin, or those with a rearranged receptor that binds a self-antigen, die by programmed cell death. The rest leave the bone marrow and enter the bloodstream.

Two-year-old boy, two maternal uncles died in infancy from infection.

Mature circulating B cells proliferate on encounter with antigen and differentiate into plasma cells, which secrete antibody. Antibodies, which are made by the plasma cell progeny of B cells, protect by binding to and neutralizing toxins and viruses, by preventing the adhesion of microbes to cell surfaces and, after binding to microbial surfaces, by fixing complement and thereby enhancing phagocytosis and lysis of pathogens (Fig. 10.2).

This case concerns a young man who has an inherited inability to make antibodies. His family history reveals that he has inherited this defect in antibody synthesis as an X-linked recessive abnormality. This poses an interesting puzzle because the genes encoding the structure of the immunoglobulin polypeptide chains are encoded on autosomal chromosomes and not on the X chromosome. Further inquiry reveals that he has no B cells, so that some gene on the X chromosome is critical for the normal maturation of B lymphocytes.

The case of Bill Grignard: a medical student with scarcely any antibodies.

Bill Grignard was well for the first 10 months of his life. In the next year he had pneumonia once, several episodes of otitis media (inflammation of the middle ear) and on one occasion developed erysipelas (streptococcal infection of the skin) on his right cheek. These infections were all treated successfully with antibiotics but it seemed to his mother, a nurse, that he was constantly on antibiotics.

His mother had two brothers who had died 30 years prior to Bill's birth from pneumonia in their second year of life, before antibiotics were available. She also has two sisters who are well; one has a healthy son and daughter and the other a healthy daughter.

Bill was a bright and active child who gained weight, grew, and developed normally but he continued to have repeated infections of the ears and sinuses and twice again had pneumonia. At 2 years and 3 months his local pediatrician tested his serum immunoglobulins. He found 80 mg dl⁻¹ IgG (normal 600–1500 mg dl⁻¹), no IgA (normal 50–125 mg dl⁻¹), and only 10 mg dl⁻¹ IgM (normal 75–150 mg dl⁻¹).

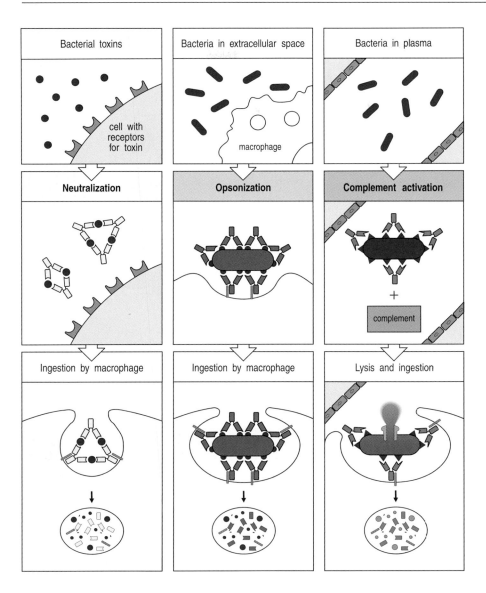

Fig. 10.2 Antibodies can participate in host defense in three main ways. The left column shows antibodies binding to and neutralizing a bacterial toxin, preventing it from interacting with host cells and causing pathology. Unbound toxin can react with receptors on the host cell, whereas the toxin:antibody complex cannot. Antibodies also neutralize complete virus particles and bacterial cells by binding to them and inactivating them. The antigen:antibody complex is eventually scavenged and degraded by macrophages. Antibodies coating an antigen render it recognizable as foreign by phagocytes (macrophages and polymorphonuclear leukocytes), which then ingest and destroy it; this is called opsonization. The central column shows the opsonization and phagocytosis of a bacterial cell. The right column shows the activation of the complement system by antibodies coating a bacterial cell. Bound antibodies form a receptor for the first protein of the complement system, which eventually forms a protein complex on the surface of the bacterium that favors its uptake and destruction by phagocytes and can, in some cases, directly kill the bacterium. Thus, antibodies target pathogens and their products for disposal by phagocytes.

Bill was started on monthly intramuscular injections of gamma globulin; his serum IgG level was maintained at 200 mg dl^{-1}. He started school at age 5 years and performed very well (he was reading at second grade level at age 5 years) despite prolonged absences because of recurrent pneumonia and other infections.

At 9 years of age he was referred to the Children's Hospital because of atelectasis (collapse of part of a lung) and a chronic cough. On physical examination he was found to be a well-developed, alert boy. He weighed 33.5 kg and was 146 cm in height (this height and weight is normal for his age). The doctor noted that he had no visible tonsils (he had never had a tonsillectomy). With a stethoscope the doctor also heard rales (moist crackles) at both lung bases.

Further family history revealed that Bill had one younger sibling, John, a 7-year-old brother, who also had contracted pneumonia on three occasions. John had a serum IgG level of 150 mg dl^{-1}.

Laboratory studies at the time of Bill's visit to the Children's Hospital gave a white blood cell count of 5100 μl^{-1} (normal) of which 45% were neutrophils (normal), 43% were lymphocytes (normal), 10% were monocytes (elevated) and 2% were eosinophils (normal).

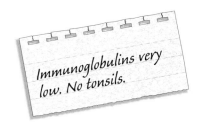

Immunoglobulins very low. No tonsils.

Complete absence of B cells.

Fig. 10.3 The FACS™ allows individual cells to be identified by their cell-surface antigens and to be sorted. Cells to be analyzed by flow cytometry are first labeled with fluorescent dyes (top panel). Direct labeling uses dye-coupled antibodies specific for cell-surface antigens (as shown here), while indirect labeling uses a dye-coupled immunoglobulin to detect unlabeled cell-bound antibody. The cells are forced through a nozzle in a single-cell stream that passes through a laser beam (second panel). Photo-multiplier tubes (PMTs) detect the scattering of light, which is a sign of cell size and granularity, and emissions from the different fluorescent dyes. This information is analyzed by computer (CPU). By examining many cells in this way, the number of cells with a specific set of characteristics can be counted and levels of expression of various molecules on these cells can be measured. The lower part of the figure shows how these data can be represented, using the expression of two surface immunoglobulins, IgM and IgD, on a sample of B cells from a mouse spleen. The two immunoglobulins have been labeled with different-colored dyes. When the expression of just one type of molecule is to be analyzed (IgM or IgD), the data are usually displayed as a histogram, as in the left-hand panels. Histograms display the distribution of cells expressing a single measured parameter (e.g., size, granularity, fluorescence color). When two or more parameters are measured for each cell (IgM and IgD), various types of two-color plot can be used to display the data, as shown in the right-hand panel. All four plots represent the same data. The horizontal axis represents intensity of IgM fluorescence and the vertical axis the intensity of IgD fluorescence. Two-color plots provide more information than histograms; they allow recognition, for example, of cells that are 'bright' for both colors, 'dull' for one and bright for the other, dull for both, negative for both, and so on. For example, the cluster of dots in the extreme lower left portions of the plots represents cells that do not express either immunoglobulin, and are mostly T cells. The standard dot plot (upper left) places a single dot for each cell whose fluorescence is measured. It is good for picking up cells that lie outside the main groups but tends to saturate in areas containing a large number of cells of the same type. A second means of presenting these data is the color dot plot (lower left), which uses color density to indicate high-density areas. A contour plot (upper right) draws 5% 'probability' contours, with 5% of the cells lying between each contour providing the best monochrome visualization of regions of high and low density. The lower right plot is a 5% probability contour map that also shows outlying cells as dots.

Mixture of cells is labeled with fluorescent antibody

Green photomultiplier tube (PMT)

Stream of fluid containing antibody-labeled cells

Red PMT

CPU

Side scatter

Laser

Forward scatter

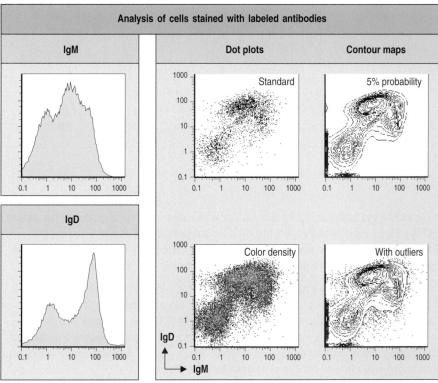

Analysis of cells stained with labeled antibodies

IgM

IgD

Dot plots

Contour maps

Standard

5% probability

Color density

With outliers

IgD

IgM

Flow cytometry (Fig. 10.3) showed that 85% of the lymphocytes bound an antibody to CD3, a T-cell marker (normal); 55% were helper T cells reacting with an anti-CD4 antibody; and 29% were cytotoxic T cells reacting with an anti-CD8 antibody (normal). However, none of Bill's peripheral blood lymphocytes bound an antibody to the B-cell marker CD19 (normal 12%) (Fig. 10.4).

Fig. 10.4 Clinical FACS analysis of a normal individual (top panel) and a patient with X-linked agammaglobulinemia (bottom panel). Blood lymphocytes from a normal individual bind labeled antibody to both the B-cell marker CD19 and the T-cell marker CD3 (see top panel). However, blood lymphocytes from an individual with X-linked agammaglobulinemia such as Bill show only binding to antibodies against the T-cell marker CD3. This indicates an absence of B cells in these patients.

T-cell proliferation indices in response to PHA, concanavalin A, tetanus toxoid, and diphtheria toxoid were 162, 104, 10, and 8, respectively (all normal). Serum IgG remained low at 155 mg dl^{-1}, while serum IgA and IgM were undetectable.

Bill was started on a preparation of gamma globulin rendered suitable for intravenous administration. He was given a dose of gamma globulin intravenously so as to maintain his IgG level at 600 mg dl^{-1}. He improved remarkably. The rales at his lung bases disappeared. He continued to perform well in school and ultimately entered medical school. Except for occasional bouts of conjunctivitis or sinusitis, which respond well to oral antibiotic treatment, he remains in good health and leads an active life. He became skilled at inserting a needle into a vein on the back of his hand and he infuses himself with 10 g gamma globulin every weekend.

X-linked agammaglobulinemia.

Males such as Bill with a hereditary inability to make antibodies are subject to recurrent infections. However, the infections are due almost exclusively to common extracellular bacterial pathogens—*Haemophilus influenzae, Streptococcus pneumoniae, Streptococcus pyogenes,* and *Staphylococcus aureus.* An examination of scores of histories of boys with this defect has established that they have no problems with intracellular infections, such as those caused by the common viral diseases of childhood. T-cell number and function in males with X-linked agammaglobulinemia are normal, and they therefore have normal cell-mediated responses, which are able to terminate viral infections and infections with intracellular bacteria such as those causing tuberculosis.

The bacteria that are the major cause of infection in X-linked agamma-globulinemia are all so-called pyogenic bacteria. Pyogenic means pus-forming, and pus consists largely of neutrophils. The normal host response to pyogenic infections is the production of antibodies that coat the bacteria and fix complement, thereby enhancing rapid uptake of the bacteria into phagocytic cells such as neutrophils and macrophages, which destroy them. Since antibiotics came into use, it has been possible to treat pyogenic infections successfully. However, when they recur frequently, the excessive release of proteolytic enzymes (for example elastase) from the bacteria and from the host phagocytes causes anatomical damage, particularly to the airways of the lung. The bronchi lose their elasticity and become the site of chronic inflammation (this is called bronchiectasis). If affected males do not receive replacement therapy—gamma globulin—to prevent pyogenic infections, they eventually die of chronic lung disease.

Gamma globulin is prepared from human plasma. Plasma is pooled from approximately one thousand or more blood donors and is fractionated at very cold temperatures (–5°C) by adding progressively increasing amounts of ethanol. This method was developed by Professor Edwin J. Cohn at the

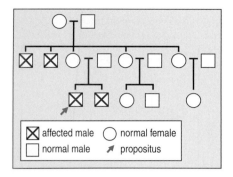

Fig. 10.5 Bill's family tree.

Harvard Medical School during the Second World War. The five plasma fractions obtained are still called Cohn Fractions I, II, III, IV, and V. Cohn Fraction I is mainly composed of fibrinogen. Cohn Fraction II is almost pure IgG and is called gamma globulin. Cohn Fraction III contains the beta globulins, including IgA and IgM, Fraction IV, the alpha globulins, and Fraction V, albumin. Cohn Fraction II, or gamma globulin, is commercially available as a 16% solution of IgG. During the processing of the plasma some of the gamma globulin aggregates and, for this reason, the 16% solution cannot be given intravenously. Aggregated gamma globulin acts like immune complexes and causes a reaction of shaking chills, fever, and low blood pressure when given intravenously. Gamma globulin can be disaggregated with low pH or insoluble proteolytic enzymes. It can then be safely administered intravenously as a 5% solution.

The gene defect in X-linked agammaglobulinemia was identified when the gene was mapped to the long arm of the X chromosome at Xq22 and subsequently cloned. The gene encodes a cytoplasmic protein tyrosine kinase called Bruton's tyrosine kinase (Btk), which is found in pre-B cells, B cells, and neutrophils. Btk is activated at different stages of B-cell development by the engagement of both the pre-B-cell receptor and the B-cell receptor. Btk is required to mediate the survival and further differentiation of the progenitor B cells in which successful rearrangement of their heavy-chain genes has occurred. It is also required for the survival of mature B cells.

Questions.

1   *Fig. 10.5 shows Bill's family tree. It can be seen that only males are affected and that the females who carry the defect (Bill's mother and maternal grandmother) are normal. This inheritance pattern is characteristic of an X-linked recessive trait. We do not know whether Bill's aunts are carriers of the defect because neither of them has had an affected male child. Now that the btk gene has been mapped, it is possible in principle to detect carriers by testing for the presence of a mutant btk gene. But there is a much simpler test that was already available at the time of Bill's diagnosis, which is still used routinely. Can you suggest how we could have determined whether Bill's aunts were carriers?*

2   *Bill was well for the first 10 months of his life. How do you explain this?*

3   *Patients with immunodeficiency diseases should never be given live viral vaccines! Several male infants with X-linked agammaglobulinemia have been given live oral polio vaccine and have developed paralytic poliomyelitis. What sequence of events led to the development of polio in these boys?*

4 Bill has a normal number of lymphocytes in his blood (43% of a normal concentration of 5100 white blood cells µl⁻¹). Only by phenotyping these lymphocytes do we realize that they are all T cells (CD3⁺) and that he has no B cells (CD19⁺). What tests were performed to establish that his T cells function normally?

5 Bill's recurrent infections were due almost exclusively to Streptococcus and Haemophilus species. These bacteria have a slimy capsule composed primarily of polysaccharide polymers, which protect them from direct attack by phagocytes. Humans make IgG2 antibodies to these polysaccharide polymers. The IgG2 antibodies 'opsonize' the bacteria by fixing complement on their surface, thereby facilitating the rapid uptake of these bacteria by phagocytic cells (Fig. 10.6). What other genetic defect in the immune system might clinically mimic X-linked agammaglobulinemia?

6 The doctor noted that Bill had no tonsils though he had never had his tonsils removed surgically. How do you explain this absence of tonsils, an important diagnostic clue in suspecting X-linked agammaglobulinemia?

7 It was found by trial and error that Bill would stay healthy and have no significant infections if his IgG level were maintained at 600 mg dl⁻¹ of plasma. He was told to take 10 g of gamma globulin every week to maintain that level. How was the dose calculated?

8 Females with a disease exactly mimicking X-linked agammaglobulinemia have been found. Explain how this might happen.

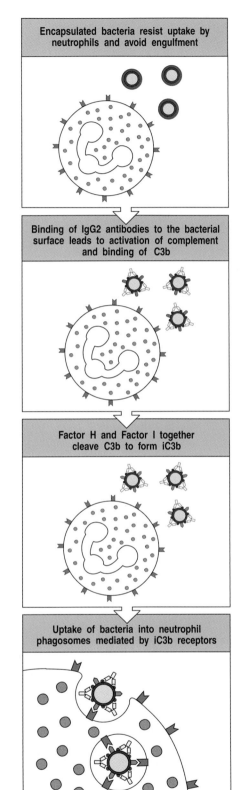

Fig. 10.6 Encapsulated bacteria are efficiently engulfed by phagocytes only when they are coated with complement. Encapsulated bacteria resist ingestion by phagocytes unless they are recognized by antibodies that fix complement. IgG2 antibodies are produced against these bacteria in humans, and lead to the deposition of complement component C3b on the bacterial surface, where it is cleaved by Factor H and Factor I to produce iC3b, still bound to the bacterial surface. iC3b binds a specific receptor on phagocytes and induces the engulfment and destruction of the iC3b bacterium. Phagocytes also have receptors for C3b, but these are most effective when acting in concert with Fc receptors for IgG1 antibodies, whereas the iC3b receptor is potent enough to act alone, and is the most important receptor for the phagocytosis of pyogenic bacteria.

CASE 11 | X-linked Hyper IgM Syndrome

Failure of immunoglobulin isotype switching.

After exposure to an antigen, the first antibodies to appear are IgM. Later, antibodies of other classes appear: IgG predominates in the serum and extravascular space, while IgA is produced in the gut, and IgE may be secreted at other epithelial surfaces. The different effector functions of these different classes, or isotypes, are summarized in Fig. 11.1. The changes in the isotype of antibody produced in the course of an immune response reflect the occurrence of isotype switching in the B cells that synthesize antibody, so that the

Functional activity	IgM	IgD	IgG1	IgG2	IgG3	IgG4	IgA	IgE
Neutralization	+	–	++	++	++	++	++	–
Opsonization	–	–	+++	*	++	+	+	–
Sensitization for killing by NK cells	–	–	++	–	++	–	–	–
Sensitization of mast cells	–	–	+	–	+	–	–	+++
Activates complement system	+++	–	++	+	+++	–	+	–

Distribution	IgM	IgD	IgG1	IgG2	IgG3	IgG4	IgA	IgE
Transport across epithelium	+	–	–	–	–	–	+++ (dimer)	–
Transport across placenta	–	–	+++	+	++	+/–	–	–
Diffusion into extravascular sites	+/–	–	+++	+++	+++	+++	++ (monomer)	+
Mean serum level (mg ml^{-1})	1.5	0.04	9	3	1	0.5	2.1	3×10^{-5}

Fig. 11.1 Each human immunoglobulin isotype has specialized functions and a unique distribution. The major effector functions of each isotype (+++) are shaded in dark red, while lesser functions (++) are shown in dark pink, and very minor functions (+) in pale pink. The distributions are similarly marked with actual average levels in serum shown in the bottom row. *IgG2 can act as an opsonin in the presence of Fc receptors of a particular allotype, found in about 50% of Caucasians.

Topics bearing on this case:

Antibody isotypes

Isotype switching

CD40 ligand and isotype switching

Antibody-mediated bacterial killing

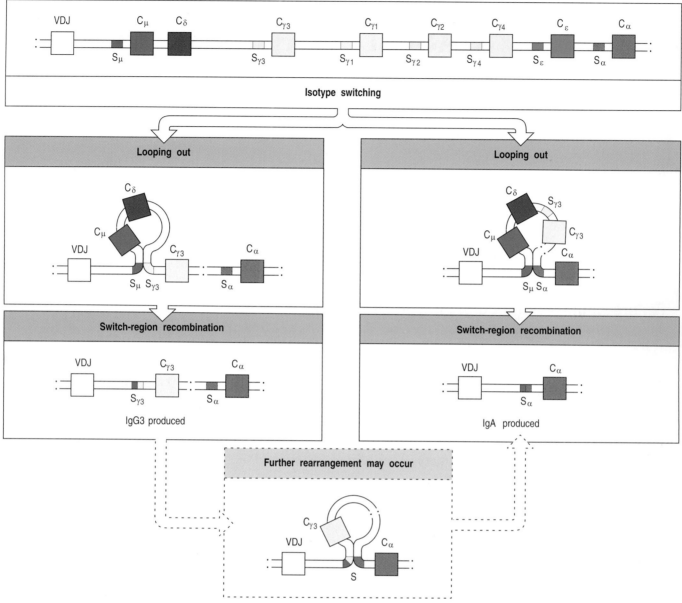

Fig. 11.2 Isotype switching involves recombination between specific signals. Repetitive DNA sequences that guide isotype switching are found upstream of each of the immunoglobulin C-region genes, with the exception of the δ gene. Switching occurs by recombination between these repetitive sequences or switch signals, with deletion of the intervening DNA. The initial switching event takes place from the μ switch region: switching to other isotypes can take place subsequently from the recombinant switch region formed after μ switching. S, switch region.

variable region, which determines the specificity of an antibody, becomes associated with the constant regions of different isotypes as the immune response progresses (Fig. 11.2).

Isotype switching is induced by T cells. T cells are also required to initiate B-cell responses to many antigens: the only exceptions are responses triggered by some microbial antigens, or by certain antigens with repeating epitopes. T-cell help is delivered in the context of an antigen-specific interaction with the B cell (Fig. 11.3). This interaction activates the T cell to express the CD40 ligand (CD40L), which then delivers an activating signal to the B cell by binding CD40. Activated T cells also secrete cytokines, which are required at the

| B cell binds virus through viral coat protein. Virus particle is internalized, and degraded | Peptides from internal proteins of the virus are presented to the T cell, which activates the B cell | Activated B cell produces antibody to viral coat protein |

Fig. 11.3 B cells are activated by helper T cells that recognize antigenic peptide bound to class II molecules on their surface. An epitope on a viral coat (spike) protein is recognized by the surface immunoglobulin on a B cell and the virus is internalized and degraded. Peptides derived from viral proteins are returned to the B-cell surface bound to MHC class II molecules, where they are recognized by previously activated helper T cells that activate the B cells to produce antibody against the virus.

initiation of the humoral immune response to drive the proliferation and differentiation of naive B cells, and are later required to induce isotype switching (Fig. 11.4). In humans, isotype switching to IgE synthesis is best understood and is known to require interleukin-4 or interleukin-13, as well as contact with CD40L expressed by an activated T cell.

The gene for the CD40 ligand is on the X chromosome at position Xq26. In males with a defect in the CD40 ligand gene, isotype switching fails to occur; such individuals make only IgM and IgD and cannot switch to IgG, IgA, or IgE synthesis. This defect can be mimicked in mice in which the genes for CD40 or CD40 ligand have been disrupted by gene targeting: B cells in these animals also fail to undergo isotype switching. The underlying defect in patients with X-linked hyper IgM syndrome can be demonstrated easily by using

Fig. 11.4 Isotype switching to IgE production by human B cells. Purified human B cells in culture transcribe the μ and δ loci at a low rate, giving rise to surface IgM and IgD. On co-culture with T cells activated with ionomycin and phorbol myristate acetate (PMA), IgM is secreted. The presence of IL-4 or IL-13 stimulates an isotype switch to IgE. Purified B cells cultured alone with these cytokines transcribe the Cε gene at a low rate, but the transcripts originate in the switch region preceding the gene and do not code for protein. On co-culture with activated T cells in addition to IL-4 or IL-13, an isotype switch occurs, mature ε RNA is expressed and IgE synthesis and secretion proceed.

Fig. 11.5 Flow cytometric analysis showing that activated T cells from hyper IgM patients do not express the CD40 ligand. T cells from two patients and one healthy donor were activated *in vitro* with a T-cell mitogen, incubated with soluble CD40 protein and analyzed by flow cytometry (see Fig. 10.3). The results are shown in the top three panels. In the normal individual, there are two populations of cells: one that does not bind CD40 (the peak to the left, with low-intensity fluorescence) and one that does (the peak to the right, with high-intensity fluorescence). The dotted line is the negative control, showing nonspecific binding of a fluorescently labeled protein to the same cells. In the X-linked hyper IgM patients (middle and right-hand panels), CD40 fluorescence exactly coincides with the nonspecific control, showing that there is no specific binding to CD40 by these cells. The bottom panels show that the T cells have indeed been activated by the mitogen for the T cells of both the normal and the two patients have increased expression of the interleukin-2 receptor, CD25, as anticipated after T-cell activation. The dotted line is a negative control of fluorescent goat anti-mouse immunoglobulin.

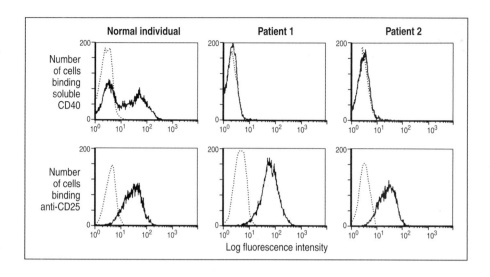

soluble CD40. Soluble CD40 can be made by engineering the extracellular domain of CD40 onto the constant region of IgG. The soluble protein can then be labeled with a dye. Activated T cells from X-linked hyper IgM patients fail to bind fluorescently labeled, soluble CD40 (Fig. 11.5).

CD40 is expressed not only on B cells but also on the surfaces of macrophages, dendritic cells, follicular dendritic cells, and mast cells. Macrophages and dendritic cells are antigen-presenting cells that can trigger the initial activation and expansion of antigen-specific T cells at the start of an immune response. Recent experiments in CD40L-deficient mice indicate a role for the CD40–CD40L interaction in this early priming event, because in the absence of CD40L the initial activation and expansion of T cells in response to protein antigens is greatly reduced.

The case of Dennis Fawcett: a failure of T-cell help.

Five-year-old boy fails to make antibody to strep infection.

Dennis Fawcett was 5 years old when he was referred to the Children's Hospital with a severe acute infection of the ethmoid sinuses (ethmoiditis). His mother reported that he had had recurrent sinus infections since he was 1 year old. Dennis had pneumonia from an infection with *Pneumocystis carinii* when he was 3 years old. These infections were treated successfully with antibiotics. While he was in the hospital with ethmoiditis, group A β-hemolytic streptococci were cultured from his nose and throat. The physicians caring for Dennis expected that he would have a brisk rise in his white blood cell count as a result of his severe bacterial infection, yet his white blood cell count was 4200 μl^{-1} (normal count 5000–9000 μl^{-1}). 26% of his white blood cells were neutrophils, 56% lymphocytes, and 28% monocytes. Thus his neutrophil number was very low, whereas his lymphocyte number was normal and the number of monocytes was elevated.

Seven days after admission to the hospital, during which time he was successfully treated with intravenous antibiotics, his serum was tested for antibodies to streptolysin O, an antigen secreted by streptococci. When no antibodies to the streptococcal antigen were found, his serum immunoglobulins were measured. The IgG level was 25 mg dl^{-1} (normal 600–1500 mg dl^{-1}), IgA was undetectable (normal 150–225 mg dl^{-1}) and his IgM level was elevated at 210 mg dl^{-1} (normal 75–150 mg dl⁻¹). A lymph-node biopsy showed poorly organized structures with an absence of secondary follicles and germinal centers (Fig. 11.6).

Dennis was given a booster injection of diphtheria toxoid, pertussis, and tetanus toxoid (DPT) as well as typhoid vaccine. Fourteen days later, no antibody was detected to tetanus toxoid nor to typhoid O and H antigens. Dennis had red blood cells of group O. People with type O red blood cells make antibodies to the A substance of type A red cells and antibodies to the B substance of type B red cells. This is because bacteria in the intestine have antigens that are closely related to A and B antigens. Dennis' anti-A titer was 1:3200 and his anti-B titer 1:800, both very elevated. His anti-A and anti-B antibodies were of the IgM class only.

His peripheral blood lymphocytes were examined by FACS analysis and normal results were obtained: 11% reacted with an antibody to CD19, a B-cell marker, 87% with anti-CD3, a T-cell marker, and 2% with anti-CD56, a marker for natural killer (NK) cells. However, all of his B cells (CD19⁺) had surface IgM and IgD and none were found with surface IgG or IgA. His activated T cells did not bind soluble CD40.

Dennis had an older brother and sister. They were both well. There was no family history of unusual susceptibility to infection.

Dennis was treated with intravenous gamma globulin, 600 mg kg⁻¹ body weight each month, and subsequently remained free of infection.

X-linked Hyper IgM syndrome.

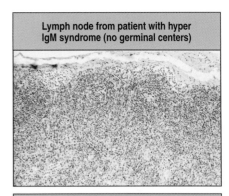

Lymph node from patient with hyper IgM syndrome (no germinal centers)

Lymph node with germinal centers

Fig. 11.6 Photographs of lymph nodes. Bottom panel courtesy of A. Perez-Atayde.

Males with a hereditary deficiency of the CD40 ligand exhibit consequences of a defect in both humoral and cell-mediated immunity. As we saw in Case 10, defects in antibody synthesis result in susceptibility to the so-called pyogenic infections. These infections are caused by pus-forming (pyogenic) bacteria such as *Haemophilus influenzae, Streptococcus pneumoniae, Streptococcus pyogenes,* and *Staphylococcus aureus,* which are resistant to destruction by phagocytic cells unless they are coated (opsonized) with antibody and complement. On the other hand, defects in cellular immunity result in susceptibility to opportunistic infections. Bacteria, viruses, and fungi that normally reside in our bodies and only cause disease when cell-mediated immunity in the host is defective are said to cause opportunistic infections.

Dennis revealed susceptibility to both kinds of infection. His recurrent sinusitis, as we have seen, was caused by *Streptococcus pyogenes,* a pyogenic infection. He also had pneumonia caused by *Pneumocystis carinii,* a fungus that is ubiquitous and causes opportunistic infections in individuals with a defect in cell-mediated immunity.

Males with a CD40 ligand deficiency can make IgM in response to T-cell independent antigens but they are unable to make antibodies of any other isotype, and they cannot make antibodies to T-cell dependent antigens, leaving them largely unprotected from many bacteria. They also have a defect in cell-mediated immunity that strongly suggests a role for CD40L in the T cell-mediated activation of macrophages.

The failure of this interaction also explains why affected males are unable to develop leukocytosis (a significant rise in the white count) in the face of severe infections. At times these males may become profoundly deficient in neutrophils, and this is a very prominent feature of their disease. As a consequence of this neutropenia they develop severe sores and blisters in their mouth and throat, a site normally infested with many bacteria. This defect can be overcome by giving them granulocyte–macrophage cell-stimulating factor (GM-CSF). GM-CSF is secreted by macrophages and to a lesser extent by T cells. The interaction of the CD40 ligand with CD40 on macrophages is required for the secretion of GM-CSF by macrophages, thus explaining why these patients cannot develop a leukocytosis.

Questions.

1   Dennis' B cells expressed IgD as well as IgM on their surface. Why did he not have any difficulty in isotype switching from IgM to IgD?

2   Normal mice are resistant to Pneumocystis carinii. SCID mice, which have no T or B cells but normal macrophages and monocytes, are susceptible to this infection. In normal mice, Pneumocystis carinii are taken up and destroyed by macrophages. Macrophages express CD40. When SCID mice are reconstituted with normal T cells they acquire resistance to Pneumocystis infection. This can be abrogated by antibodies to the CD40 ligand. What do these experiments tell us about this infection in Dennis?

3   Why did Dennis make antibodies to blood group A and B antigens but not to tetanus toxoid, typhoid O and H, and streptolysin antigens? Would he have made any antibodies in response to his Streptococcus pyogenes infection?

4   Most IgM is in the blood and less than 30% of IgM molecules get into the extravascular fluid. On the other hand, over 50% of IgG molecules are in the extravascular space. Furthermore we have 30 to 50 times more IgG molecules than IgM molecules in our body. Why are IgG antibodies more important in protection against pyogenic bacteria?

5   Newborns have difficulty in transcription of the CD40 ligand gene. Does this help to explain the susceptibility of newborns to pyogenic infections? Cyclosporin A, a drug widely used for immunosuppression in graft recipients, also inhibits the transcription of the CD40 ligand gene. What does this imply for patients taking this drug?

6   Dennis has an X-linked immunodeficiency and yet there is no informative family history. Is he a new mutation or is his mother a carrier of the defect in the CD40 ligand gene? Can we test his sister to determine whether she is a carrier? In the case of X-linked agammaglobulinemia we learned of the usefulness of examining random or nonrandom X-chromosome inactivation to answer these questions about the carrier state. Would it be useful to examine the T cells of Dennis' mother and sister for nonrandom X inactivation?

CASE 12 | Activation-induced Cytidine Deaminase (AID) Deficiency

An intrinsic B-cell defect prevents immunoglobulin isotype switching.

Immunoglobulin isotype switching, or class switching, is a complex process. After antigen has been encountered and the mature B cell is activated, the rearranged immunoglobulin variable (V) region can become progressively associated with different constant (C)-region genes by a form of somatic recombination (see Fig. 11.2). This class switch recombination requires signals from T lymphocytes as well as engagement of the appropriate biochemical machinery in B lymphocytes. In the preceding case of X-linked hyper IgM syndrome (see Case 11) a failure of isotype switching resulted from a defect in the gene that encodes the cell-surface molecule CD40 ligand (CD40L, also known as CD154), which is expressed by T cells. The engagement of CD40 ligand on activated CD4 T cells with its receptor, CD40, on B cells is required to initiate immunoglobulin class switching. CD40 ligand is encoded on the X chromosome and so this type of hyper IgM syndrome is seen only in males.

An inability to undergo isotype switching can also be caused by a defect in the B cell's own biochemical pathways. The biochemical events underlying isotype switching have only recently been clarified. While studying a cultured B-cell line that was being induced to undergo class switching from IgM to IgA synthesis, immunologists in Japan observed the marked upregulation of a new enzyme, *a*ctivation-*i*nduced cytidine *d*eaminase (AID). This enzyme

Topics bearing on this case:
Antibody isotypes
Isotype switching
Somatic hypermutation

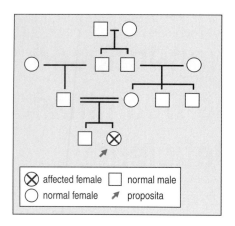

Fig. 12.1 A pedigree of a family with AID deficiency. As the parents of the affected child (the proposita) show no signs of disease themselves, the defective AID gene must be recessive. They both carry a single copy of this gene and, as they are first cousins, the most likely source of the defective gene is their shared grandfather or grandmother. If one of these were heterozygous for AID deficiency, they could have transmitted the defective gene to both their sons. The fact that the affected child is a girl, and that neither her father nor any other males in her extended family show signs of disease, indicates that the gene is carried on an autosome and not the X chromosome.

converts cytidine to uridine, and it is now known to trigger a DNA repair mechanism that causes isotype switching. The contribution of AID to class switching was confirmed by 'knocking out' the gene encoding AID by homologous recombination in mice; the mutant animals developed hyper IgM syndrome and were unable to make IgG, IgA, or IgE.

In humans, hyper IgM syndrome is also encountered in females, in whom it has an inheritance pattern in many families that suggests autosomal recessive inheritance (Fig. 12.1). The defective gene of the autosomal recessive form was mapped in several informative families to the short arm of chromosome 12, in a region that corresponds to the region containing the *AID* gene in mice. This prompted the search for a link between hyper IgM syndrome and AID deficiency in humans, and several cases of the autosomal recessive form were found to have mutations in the *AID* gene.

The case of Daisy Miller: a failure of a critical B-cell enzyme.

At 3 years old Daisy Miller was admitted to the Boston Children's Hospital with pneumonia. Her mother had taken her to Dr James, a pediatrician, because she had a fever and was breathing fast. Her temperature was high, at 40.1°C, her respiratory rate was 40 per minute (normal 20), and her blood oxygen saturation was 88% (normal >98%). Dr James also noticed that lymph nodes in Daisy's neck and armpits (axillae) were enlarged. A chest X-ray was ordered. It revealed diffuse consolidation (whitened areas of lung due to inflammation, indicating pneumonia) of the lower lobe of her left lung and she was admitted to the hospital.

Daisy had had pneumonia once before, at 25 months of age, as well as 10 episodes of middle-ear infection (otitis media) that had required antibiotic therapy. Tubes (grommets) had been placed in her ears to provide adequate drainage and ventilation of the ear infections.

In the hospital a blood sample was taken and was found to contain 13,500 white blood cells μl^{-1}, of which 81% were neutrophils and 14% lymphocytes. A blood culture grew the bacterium *Streptococcus pneumoniae*.

Because of Daisy's repeated infections Dr James consulted an immunologist. He tested Daisy's immunoglobulin levels and found that her serum contained 470 mg dl^{-1} of IgM (normal 40–240 mg dl^{-1}), undetectable IgA (normal 70–312 mg dl^{-1}), and 40 mg dl^{-1} of IgG (normal 639–1344 mg dl^{-1}). Although Daisy had been vaccinated against tetanus and *Haemophilus influenzae*, she had no specific IgG antibodies against tetanus toxoid or to the polyribosyl phosphate (PRP) polysaccharide antigen of *H. influenzae*. Because her blood type was A, she was tested for anti-B antibodies (isohemagglutinins). Her IgM titer of anti-B antibodies was positive at 1:320 (upper limit of normal), whereas her IgG titer was undetectable.

Daisy was started on intravenous antibiotics. She improved rapidly and was sent home on a course of oral antibiotics. Intravenous immunoglobulin (IVIG) therapy was started, which resulted in a dramatic decrease in the frequency of infections.

Analysis of Daisy's peripheral blood lymphocytes revealed normal expression of CD40 ligand on T cells activated by anti-CD3 antibodies, and normal expression of CD40 on B cells (Fig. 12.2). Nevertheless, her blood cells completely failed to secrete IgG and IgE after stimulation with anti-CD40 antibody (to mimic the effects of engagement of CD40 ligand) and interleukin-4 (IL-4), a cytokine that also helps to stimulate isotype switching, although the blood cells proliferated normally in

Normal individual   Patient 1

Anti-CD40 log fluorescence intensity

Anti-CD40L log fluorescence intensity

Fig. 12.2 Flow cytometric analysis showing normal expression of CD40 and CD40 ligand in a patient with AID deficiency. Top row: CD40 measured by the binding of flurorescently tagged anti-CD40 antibodies to CD40 on B cells from (left panel) a normal individual and (right panel) a patient with AID deficiency. Bottom row: measurement of CD40 ligand (CD40L) on T cells from (left panel) a normal individual and (right panel) the same patient activated *in vitro* with the mitogen phorbol ester (PMA) and ionomycin.

response to these stimuli (Fig. 12.3). cDNAs for CD40 and for the activation-induced cytidine deaminase (AID) were made and amplified by the reverse transcription–polymerase chain reaction (RT–PCR) on mRNA isolated from blood lymphocytes activated by anti-CD40 and IL-4. Sequencing of the cDNAs revealed a point mutation in the AID gene that introduced a stop codon into exon 5, leading to the formation of truncated and defective protein. The CD40 sequence was normal.

Activation-induced cytidine deaminase (AID) deficiency.

It is now apparent that there are two distinct phenotypes of hyper IgM syndrome, resulting from four different genetic defects. One phenotype, which results from defects in the genes encoding CD40 or CD40 ligand (see Case 11), manifests itself as susceptibility to both pyogenic and opportunistic infections. The other phenotype, which results from defects in the AID gene or the gene encoding uracil-DNA glycosylase (UNG), resembles X-linked agammaglobulinemia in that these patients have increased susceptibility to pyogenic infections only.

When CD40 and the IL-4 receptor on B cells are ligated by CD40 ligand and IL-4, the AID gene is transcribed and translated to produce AID protein. At the same time, transcription of cytidine-rich regions at isotype-switch sites is induced, which involves separation of the two DNA strands at these sites. AID, which can deaminate cytidine in single-stranded DNA only, then proceeds to convert the cytidine at the switch sites to uridine. As uridine is

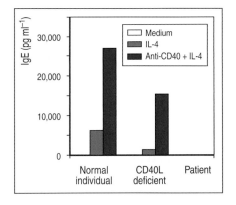

Fig. 12.3 Comparison of isotype switching, as judged by IgE secretion, in a normal individual, a patient with CD40L deficiency, and a patient with AID deficiency. IgE production from peripheral blood mononuclear cells *in vitro* was measured after no stimulation ('medium'; negative control), stimulation with IL-4 alone, or stimulation with anti-CD40 and IL-4. The anti-CD40 and IL-4 together can compensate to some degree for the lack of stimulation of B cells by CD40 ligand on T cells, but cannot compensate at all for the defect in AID.

Fig. 12.4 Histology of a lymph node from a patient with AID deficiency stained with hematoxylin and eosin. Note the large follicles.

not normally present in DNA, it is recognized by the enzyme uracil-DNA glycosylase, which removes the uracil base from the rest of the nucleotide. The damaged DNA strand is now recognized by DNA repair endonucleases, which excise the damaged region. The repair of these breaks in the switch regions results in class switching.

Interestingly, AID is not only required for class switching but also for somatic hypermutation in B cells. This is the process that underlies the production of antibodies of increasingly higher affinity for the antigen as an immune response proceeds. When a B cell is activated by the combination of binding of its surface IgM and signals from other cells, especially T cells (for example via CD40–CD40-ligand interaction), not only cell division and class switching are initiated. In addition, point mutations are introduced into the DNA that codes for the immunoglobulin V region. A process of selection then occurs. Cells expressing mutated surface immunoglobulin with a stronger affinity for antigen compete best for binding the available antigen and receive stronger signals via this cell-surface antigen receptor; they consequently proliferate. Cells with lower-affinity immunoglobulins are less likely to bind and be stimulated by antigen, and if they do not receive these stimulatory signals, they die. This leads to selection of cells with mutations that result in high-affinity antibodies, a process referred to as 'affinity maturation' of the antibody response. B cells in mice and humans that lack functional AID are unable to generate these mutations after activation and therefore cannot undergo affinity maturation. The only response these B cells can make to the activating signals is to proliferate. This results in the accumulation of IgM⁺ B cells in the lymphoid organs, giving rise to an enlarged spleen (splenomegaly) and enlarged lymph nodes (lymphadenopathy) (Fig. 12.4).

Questions.

1   *Daisy and other patients with hyper IgM syndrome caused by a mutation in AID do not seem to suffer from opportunistic infections, such as Pneumocystis carinii and Cryptosporidium, which are characteristic of the X-linked hyper IgM syndrome. Why?*

2   *A 3-year-old girl presents with hyper IgM syndrome, recurrent bacterial infection and a history of Pneumocystis carinii pneumonia. Her CD40L and AID genes are normal. What is the potential gene defect in this patient?*

3   *A 6-year-old girl presents with hyper IgM syndrome, recurrent bacterial infection and no history of opportunistic infections. Her CD40 and AID gene sequences are normal. What is the potential gene defect in this patient?*

4   *Can you think of a simple test to distinguish hyper IgM syndrome caused by an intrinsic B-cell defect from that caused by CD40 ligand deficiency without recourse to DNA sequencing?*

5   *Dr James noted that Daisy had enlarged lymph nodes. Patients with X-linked hyper IgM syndrome do not have enlarged lymph nodes. Can you explain the difference?*

CASE 13 | Common Variable Immunodeficiency

A failure to produce antibodies against particular antigens.

The ability to produce antibody of all different classes after exposure to antigen is an important aspect of successful and comprehensive humoral immunity. Although antibody production and class switching in response to most protein antigens require helper T cells (see Case 11), responses to some other antigens do not. This is because the special properties of certain bacterial polysaccharides, polymeric proteins, and lipopolysaccharides enable them to stimulate naive B cells in the absence of T-cell help. These antigens are known as thymus-independent antigens (TI antigens) because they stimulate strong antibody responses in athymic individuals. They fall into two classes: TI-1 antigens, such as bacterial lipopolysaccharide, can directly induce B-cell division; TI-2 antigens, such as bacterial capsular polysaccharides, do not have this property but have highly repetitive structures and probably stimulate the B cell by cross-linking a critical number of B-cell receptors. In particular, B-cell responses to thymus-independent antigens

Topics bearing on this case:
Thymus-independent antigens
Antibody classes
Class switching

Fig. 13.1 BAFF and APRIL, and their receptors on B cells. BAFF-R is expressed on resting B cells and mediates the effects of BAFF on B-cell development and survival but has only a minor role in isotype switching. Studies in mice suggest that interaction of BAFF with BAFF-R has an important role in B-cell survival and is required for the development of mature B cells *in vivo*. TACI seems to be the receptor that mainly mediates isotype switching by BAFF and APRIL. TACI may also have other functions such as promoting plasma cell differentiation and survival. BCMA is poorly expressed on resting B cells but is upregulated on germinal-center B cells and plasma cells. It has the ability to upregulate the expression of co-stimulatory molecules on the B-cell surface that enhance antigen presentation and T-cell activation, and has a role in maintaining the survival of long-lived plasma cells. BCMA has no detectable role in isotype switching by BAFF and APRIL.

provide a prompt and specific IgM response to an important class of pathogen—encapsulated bacteria. Pyogenic bacteria such as streptococci and staphylococci are surrounded by a polysaccharide capsule that enables them to resist phagocytosis. Antibody that is produced rapidly in response to this polysaccharide capsule can coat these bacteria, promoting their ingestion and destruction by phagocytes.

Whereas TI-1 antigens are inefficient inducers of affinity maturation and memory B cells, TI-2 antigens can induce both IgM and some class-switched responses. As we saw in Case 11, the initiation of B-cell class switching usually depends on the interaction of B cells and helper T cells via CD40 and CD40 ligand. Class switching in response to TI antigens is thought to involve other members of the TNF/TNFR family—the recently discovered TNF-like proteins BAFF (B-cell activating factor belonging to the TNF family) and APRIL (a proliferation-inducing TNF ligand), and their receptors on B cells. In human B cells, BAFF and APRIL induce class switching to IgA and IgG in the presence of TGF-β or IL-10 and to IgE in the presence of IL-4. In mice, in contrast, BAFF and APRIL on their own can switch B cells to IgG and IgA production.

BAFF is secreted by cells in the follicles of peripheral lymphoid tissue and can bind to several receptor proteins on B cells, which include BAFF receptor (BAFF-R), transmembrane activator and calcium-modulating cyclophilin ligand interactor (TACI), and B-cell maturation antigen (BCMA) (Fig. 13.1). In addition to its role in class switching, BAFF is involved in B-cell development and in promoting the survival of mature B cells. Mice carrying a BAFF transgene, leading to BAFF overexpression, develop high titers of autoantibodies and a systemic lupus erythematosus-like condition, whereas BAFF-deficient mice have a severe defect in B-cell development and an almost complete loss of mature B cells and of follicular and marginal zone (MZ) B cells in lymph nodes and spleen. A similar phenotype is observed in mice lacking BAFF-R, or in the A/WySnJ mouse strain, which carries a naturally occurring mutation in BAFF-R.

APRIL is the closest relative of BAFF in the TNF family, sharing 33% sequence identity, and was initially reported as a protein that stimulated the proliferation of tumor cells. Unlike BAFF, it does not seem to have an effect on B-cell development and survival, because APRIL-deficient mice have normal numbers of B cells in all developmental stages. This is because APRIL binds to TACI and BCMA, but not to BAFF-R. The only detectable immune defects in APRIL homozygous null mice are diminished serum IgA levels and an impaired antibody response to immunization by the oral route.

The receptors BAFF-R, TACI, and BCMA are all members of the TNF receptor (TNFR) superfamily and have different roles in immune responses. TACI seems to be the receptor that mainly mediates isotype switching by BAFF and APRIL. BCMA, and possibly TACI, may promote plasma-cell differentiation and survival. TACI-deficient mice have low levels of serum IgA and have a deficient antibody response to immunization with the TI-2 antigens Pneumovax, which contains polysaccharide antigens from a number of *Streptococcus pneumoniae* serotypes, and NP-Ficoll, a polysaccharide. They also have enlarged spleens and lymph nodes, with more cells than usual and an increased number of B cells. They develop autoimmunity, suggesting that TACI may play a regulatory role in B cells. Some cases of common variable immunodeficiency (CVID) in humans have recently been shown to be associated with mutations in TACI, as we shall see in this case study.

The case of Mary Johnson: impaired ability to generate all classes of antibodies leads to frequent and unusual infections.

Mrs Johnson was 40 years old when she was referred to the immunology clinic for evaluation of her immune system after suffering throughout her life from recurrent respiratory and gastrointestinal infections. As a child she was frequently diagnosed with otitis, sinusitis, and tonsillitis and had intermittent diarrhea, and from an early age she was hospitalized several times for pneumonia and gastrointestinal infections. The year before she came to the immunology clinic she had been in hospital with severe diarrhea caused by an infection with the protozoan parasite *Giardia lamblia*. When she was 25 years old, Mrs Johnson had been diagnosed with thyroid insufficiency and placed on thyroid hormone replacement therapy.

40-year-old woman. Recurrent infections.

Physical examination revealed an enlarged spleen, the edge of which extended 8 cm below the left mid-costal margin. Blood tests showed that Mrs Johnson had lower than normal levels of all immunoglobulin isotypes: IgM 18 mg dl^{-1} (normal 100–200 mg dl^{-1}), IgG 260 mg dl^{-1} (normal 600–1000 mg dl^{-1}), and IgA 24 mg dl^{-1} (normal 60–200 mg dl^{-1}). Although she had been immunized several times with pneumococcal vaccines, she had been unable to respond, as shown by her lack of antibodies against all pneumococcal serotypes present in the vaccine. Numbers of B cells and T cells were normal. No antinuclear autoantibodies or rheumatoid factor (an autoantibody against IgG) were detected, but Mrs Johnson had high levels of antithyroid antibodies.

Mrs Johnson's twin sister and her mother were both dead. They had also had hypogammaglobulinemia and an inability to make specific antibodies against polysaccharide vaccines. From her early teens onward, the sister had had recurrent viral and bacterial infections. She developed hemolytic anemia and granulomatous vasculitis, and died at the age of 31 years from gastrointestinal cancer. The mother had died of non-Hodgkin's lymphoma. Mrs Johnson's only brother had been diagnosed with common variable immune deficiency (CVID) and suffered from chronic sinusitis and recurrent chest infections. Her father was still alive.

Family history of CVID.

On the basis of the blood tests and her family history, Mrs Johnson was diagnosed with CVID and was placed on intravenous immunoglobulin 35 g every 2 weeks. This caused a remarkable improvement in her condition with a dramatic decrease in the frequency of infections. Sequencing of the *TACI* gene revealed that Mrs Johnson had a mutation in one of the alleles. The same mutation was also found in her brother's *TACI* gene but was absent from their father's DNA, implying that she and her brother had inherited the mutation from their mother.

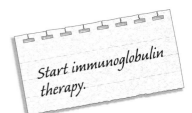

Start immunoglobulin therapy.

Common variable immunodeficiency (CVID).

CVID is an immunodeficiency disorder characterized by low serum levels of all switched immunoglobulin isotypes (IgG, IgA, and IgE), an impaired ability to produce specific antibodies, even of the IgM class after exposure to certain antigens, and increased susceptibility to infections of the respiratory and gastrointestinal tract, the latter due to the combined decrease in IgG and IgA.

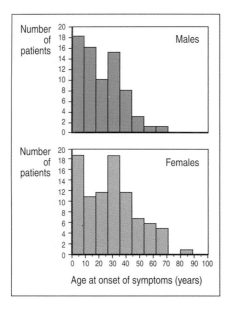

Fig. 13.2 The age at onset of clinical symptoms of immunodeficiency in CVID. The age at onset of clinical symptoms of immunodeficiency is shown in centiles. The median age at onset of symptoms for males is 23 years and for females 28 years.

It is relatively common (hence the name), being the most common primary immunodeficiency that comes to medical attention. The clinical course, as well as the degree of deficiency of serum immunoglobulins, varies from patient to patient (hence the name 'variable').

Like other immunodeficiency diseases, the symptoms of CVID are frequent and unusual infections. Symptoms may appear during early childhood, adolescence, or adult life, but the age of onset of symptoms is usually in the 20s or 30s (Fig. 13.2). Patients with CVID suffer from recurrent infections of the sinuses and lungs, the ears, and the gastrointestinal tract. If therapy is delayed, the air passages in the lung may become irreversibly damaged and chronic infections may develop (bronchiectasis). Patients with CVID also are at greater risk of developing autoimmune diseases such as autoimmune thyroiditis, hemolytic anemia, autoimmune thrombocytopenia (which is due to antiplatelet antibodies), and pernicious anemia (which is caused by antibodies against the intrinsic factor that is required for the absorption of vitamin B_{12}). CVID patients also have a 300-fold increased risk of lymphoma and a 50-fold increased risk of gastric carcinoma. Some patients with CVID develop granulomatous lesions (see Case 2) in the lungs and the skin that are characterized by the presence of T cells and macrophages. Human herpes virus 6 (HHV6) has been isolated from these lesions, which usually respond to immunosuppressive treatment with corticosteroids or cyclosporin.

Most cases of CVID are sporadic; that is, they are not due to an inherited defect. Studies in European populations show that around 20% of CVID cases are familial, commonly associated with autosomal dominant inheritance. Genetic analysis has shown a high degree of familial clustering of single IgA deficiency (a deficiency of IgA antibodies only, which usually does not produce any detectable symptoms) and CVID, suggesting that defects in the same genes might underlie both diseases (Fig. 13.3). Moreover, patients may first present with IgA deficiency and, after several years, may develop a full-blown picture of CVID.

The genetic and immunologic causes of CVID have been largely unknown. The heterogeneous nature of the disease is demonstrated by documentation of defects in T cells, B cells, and antigen-presenting cells, suggesting that many genes are involved. It has, however, recently been demonstrated that some familial CVID cases might be monogenic disorders, in that mutations in *TACI* are associated with up to 10% of CVID cases. Both homozygosity and heterozygosity for mutations in *TACI* have been shown to be associated with familial and sporadic cases of CVID (see Fig. 13.3). Mutations in *TACI* have been detected in a small number of individuals who do not suffer recurrent infections, including family members of index cases. This reflects the variable penetrance of the gene defect, possibly as a result of the influence of different alleles at other genes. Given the variability in the age of onset of CVID, it remains to be seen whether any of these individuals will develop CVID later in life.

Naive B cells from CVID patients with *TACI* mutations are severely impaired in their ability to secrete IgG and IgA in response to APRIL *in vitro*. Like Mrs Johnson, patients with *TACI* mutations exhibit hypogammaglobulinemia of all immunoglobulin isotypes. Consequently, they are susceptible to chronic and recurrent infections by encapsulated bacteria such as *S. pneumoniae* (pneumococcus) or *Haemophilus influenzae*. In addition, patients with *TACI* mutations demonstrate an impaired antibody response to the pneumococcal vaccine Pneumovax.

Fig. 13.3 Illustrative pedigree of a family with IgA deficiency (IgAD) and CVID. The index CVID case is marked with an arrow. Note the presence of both CVID and IgAD in family members with the same *TACI* genotype. Two different mutations in TACI have been found in this pedigree: C104R, which destroys ligand binding, and an insertion, A204bp, which results in a frameshift and premature termination of translation. Squares indicate males, circles females. Fully shaded symbols indicate homozygosity, half-shaded symbols heterozygosity.

Questions.

1 Like Mrs Johnson, most CVID patients with TACI mutations are heterozygous for the mutation. These patients express both mutant and normal TACI on the surface of their B cells. With one normal copy of TACI, why would a heterozygous patient have the disease?

2 Mrs Johnson and a maternal cousin had the same TACI mutation. However, Mrs Johnson had full-blown CVID whereas her cousin only had selective IgA deficiency. How can two individuals with the same TACI mutation present with different disorders (CVID versus IgA deficiency)?

3 What other monogenic defects have been described in CVID patients?

CASE 14 | X-linked Severe Combined Immunodeficiency

The maturation of T lymphocytes.

Without T cells, life cannot be sustained. In Case 10 we learned that an absence of B cells was compatible with a normal life as long as infusions of immunoglobulin G were maintained. When children are born without T cells, they appear normal for the first few weeks or months. Then they begin to acquire opportunistic infections and die while still in infancy. An absence of functional T cells causes severe combined immunodeficiency (SCID). It is called severe because it is fatal, and called combined because, in humans, B cells cannot function without help from T cells, so that even if the B cells are not directly affected by the defect, both humoral and cell-mediated immunity are lost. Unlike X-linked agammaglobulinemia, which results from a monogenic defect, SCID is a phenotype that can result from any one of several different genetic defects. The incidence of SCID is three times greater in males than in females and this male:female ratio of 3:1 is due to the fact that the most common form of SCID is X-linked. Approximately 55% of individuals with SCID have the X-linked form of the disease.

T-cell precursors migrate to the thymus to mature, at first from the yolk sac of the embryo, and subsequently from the fetal liver and bone marrow (Fig. 14.1). The rudimentary thymus is an epithelial anlage derived from the third and fourth pharyngeal pouches. By 6 weeks of human gestation, the invasion of precursor T cells, and of dendritic cells and macrophages, has transformed the gland into a central lymphoid organ (Fig. 14.2). T-cell precursors undergo rapid maturation in the thymus gland (Fig. 14.3), which becomes the site of the greatest mitotic activity in the developing fetus. By 20 weeks of gestation, mature T cells start to emigrate from the thymus to the peripheral lymphoid organs. In all common forms of SCID the thymus fails to become a central lymphoid organ. A small and dysplastic thymus, as revealed by biopsy, used to be the confirming diagnostic indicator of SCID. Now that the various gene defects that underlie SCID are better understood, confirmation usually involves a screen for the mutations in the genes likely to be involved.

Topics bearing on this case:

Defects in T-cell function result in severe combined immunodeficiency

The development of T cells in the thymus

Testing T-cell responses with polyclonal mitogens

Signal transduction through the T-cell receptor

Interleukin-2 receptor

Fig. 14.1 T-cell precursors migrate to the thymus to mature. T cells derive from bone marrow stem cells, whose progeny migrate from the bone marrow to the thymus (left panel), where the development of the T cell takes place. Mature T cells leave the thymus and recirculate from the bloodstream through peripheral lymphoid tissues (right panel) such as lymph nodes, spleen or Peyer's patches, where they may encounter antigen.

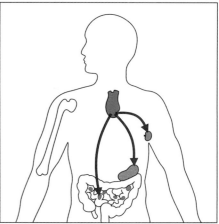

Defects that result in SCID are classified into three general categories. One comprises those defects that impair lymphocyte survival; the prototype of this class is adenosine deaminase (ADA) deficiency (see Case 15), in which adenosine metabolites that are toxic to T cells and B cells accumulate.

Normal thymus

Schematic diagram of a normal thymus

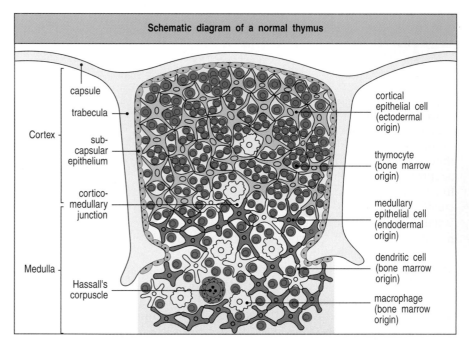

SCID thymus

Fig. 14.2 The cellular organization of the thymus. The thymus, which lies in the midline of the body above the heart, is made up from several lobules, each of which contains discrete cortical (outer) and medullary (central) regions. The cortex consists of immature thymocytes (dark blue), branched cortical epithelial cells (pale blue), with which the immature cortical thymocytes are closely associated, and scattered macrophages (yellow) involved in clearing apoptotic thymocytes. The medulla consists of mature thymocytes (dark blue), and medullary epithelial cells (orange), along with macrophages (yellow) and dendritic cells (yellow) of bone marrow origin. Hassall's corpuscles found in the human thymus are probably also sites of cell destruction. The thymocytes in the outer cortical cell layer are proliferating immature cells, while the deeper cortical thymocytes are mainly cells undergoing thymic selection. The upper photograph shows the equivalent section of a human thymus, stained with hematoxylin and eosin. The lower photograph shows a SCID thymus. There is no corticomedullary differentiation, very low thymocytis and no Hassall's corpuscles. Photographs courtesy of C.J. Howe (upper) and A. Perez-Atayde (lower).

Fig. 14.3 Changes in cell-surface molecules allow thymocyte populations at different stages of maturation to be distinguished. The most important cell-surface molecules for identifying thymocyte subpopulations have been CD4, CD8, and T-cell receptor complex molecules (CD3, and α and β chains). The earliest cell population in the thymus does not express any of these. As these cells do not express CD4 or CD8, they are called 'double-negative' thymocytes. (The γ:δ T cells found in the thymus also lack CD4 or CD8 but these are a minor population.) Maturation of α:β T cells occurs through stages where both CD4 and CD8 are expressed by the same cell, along with the pre-T-cell receptor (pTα:β) and, later, low levels of the T-cell receptor (α:β) itself. These α:β cells are known as 'double-positive' thymocytes. Most thymocytes (about 97%) die within the thymus after becoming small double-positive cells. Those whose receptors bind self-MHC molecules lose the expression of either CD4 or CD8 and increase the level of expression of the T-cell receptor. The outcome of this process is the 'single-positive' thymocytes, which, after maturation, are exported from the thymus as mature single-positive T cells.

A second category consists of defects in the machinery for somatic gene rearrangement that assembles the immunoglobulin and T-cell receptor genes during lymphocyte development. This category can be divided into lymphocyte-specific defects, namely defects in the *RAG* genes that encode the lymphocyte-specific recombinase (see Case 16), and defects in ubiquitously expressed DNA repair genes that are also involved in this recombination.

The third category consists of defects in cytokine-mediated signals for lymphocyte maturation and proliferation, and includes the X-linked form of SCID that is the subject of this case. The genetic defect responsible for the X-linked form of SCID has been mapped to the long arm of the X chromosome at Xq11. From this region the gene encoding the γ chain (CD132) of the interleukin-2 receptor (IL-2R) (Fig. 14.4) was cloned and was found to be mutated in X-linked SCID.

The case of Martin Causubon: without T cells life cannot be sustained.

Mr and Mrs Causubon had a normal daughter 3 years after they were married. Two years later they had a son and named him Martin. He weighed 3.5 kg at birth and seemed to be perfectly normal. At 3 months of age, Martin developed a runny nose and a persistent dry cough. One month later he had a middle ear infection (otitis media) and his pediatrician treated him with amoxicillin. At 5 months of age Martin had a recurrence of otitis media. His cough persisted and a radiological examination of his chest revealed the presence of pneumonia in both lungs. He was treated with another antibiotic, clarithromycin. Mrs Causubon noticed that Martin had thrush (*Candida* spp.) in his mouth (Fig. 14.5) and an angry red rash in the diaper area. He was not gaining weight; he had been in the 50th percentile for

Fig. 14.4 IL-2 receptors are three-chain structures composed of α, β, and γ chains. Resting mature T cells express only the β and γ chains, which bind IL-2 with moderate affinity, allowing resting T cells to respond to very high concentrations of IL-2. Activation of T cells induces the synthesis of the α chain and the formation of the heterotrimeric receptor, which has a high affinity for IL-2 and allows the T cell to respond to very low concentrations of IL-2. The β and γ chains show amino-acid similarities to cell-surface receptors for growth hormone and prolactin, all of which regulate cell growth and differentiation.

Fig. 14.5 An infant with SCID suffering from *Candida albicans* in the mouth.

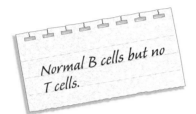

Normal B cells but no T cells.

Lack of IL-2Rγ on B cells indicates X-linked SCID.

No HLA-identical sibling. Mother agrees to donate bone marrow.

weight at age 4 months but by 6 months he had fallen to the 15th percentile. His pediatrician had given him oral polio vaccine at ages 4 and 5 months and, at the same time, diphtheria–pertussis–tetanus (DPT) immunizations.

Martin's pediatrician referred him to the Children's Hospital for further studies. On admission to the hospital, he was fretful and had tachypnea (fast breathing). A red rash was noted in the diaper area as well as white flecks of thrush on his tongue and buccal mucosa. His tonsils were very small. He had a clear discharge from his nose, and cultures of his nasal fluid grew *Pseudomonas aeruginosa*. Coarse, harsh breath sounds were heard in both lungs. His liver was slightly enlarged.

Martin's white blood count was 4800 cells μl^{-1} (normal 5000–10,000 cells μl^{-1}) and his absolute lymphocyte count was 760 cells μl^{-1} (normal 3000 lymphocytes μl^{-1}). None of his lymphocytes reacted with anti-CD3 and it was concluded that he had no T cells. Ninety-nine percent of his lymphocytes bound antibody against the B-cell molecule CD20 and 1% were natural killer (NK) cells reacting with anti-CD16. His serum contained IgG at a concentration of 30 mg dl^{-1}, IgA at 27 mg dl^{-1}, and IgM at 42 mg dl^{-1} (IgG levels are normally 400 mg dl^{-1}; the IgA and IgM levels were at the low end of the normal range for Martin's age). His blood mononuclear cells were completely unresponsive to phytohemagglutinin (PHA), concanavalin A (ConA), and pokeweed mitogen (PWM) (Fig. 14.6), as well as to specific antigens to which he had been previously exposed by immunization or infection—tetanus and diphtheria toxoids, and *Candida* antigen. His red cells contained normal amounts of adenosine deaminase and purine nucleoside phosphorylase. His B lymphocytes did not react with an antibody against the γ chain of the interleukin-2 receptor (IL-2Rγ). Cultures of sputum for bacteria and viruses revealed the abundant presence of respiratory syncytial virus (RSV).

At this point a blood sample was obtained from Martin's mother to examine her T cells for random inactivation of the X chromosome (a diagnostic test explained in Case 10). It was found that her T cells exhibited complete nonrandom X-chromosome inactivation. HLA typing showed that Martin's sister had no matching HLA alleles. His parents, as expected, each shared one HLA haplotype with Martin.

Martin was treated with intravenous gamma globulin at a dose of 2 g kg^{-1} body weight and his serum IgG level was maintained at 600 mg dl^{-1} by subsequent IgG infusions. He was given aerosolized ribavirin to control his RSV infection and trimethoprim-sulfamethoxazole intravenously for prophylaxis against *Pneumocystis carinii*. Without any further preparation, Martin was given 500×10^6 bone marrow cells from his mother. The bone marrow donation from his mother was depleted of mature T cells by treatment with a monoclonal antibody against T cells together with complement. After the transplant of the maternal bone marrow cells, Martin was given cyclosporin and prednisone to suppress any graft-versus-host disease. Sixty days after receiving the maternal bone marrow, Martin's blood

Fig. 14.6 Polyclonal mitogens, many of plant origin, stimulate lymphocyte proliferation in tissue culture. Many of these mitogens are used to test the ability of lymphocytes in human peripheral blood to proliferate.

Mitogen	Abbreviation	Source	Responding cells
Phytohemagglutinin	PHA	*Phaseolus vulgaris* (Red kidney beans)	T cells (human)
Concanavalin A	ConA	*Canavalia ensiformis* (Jack bean)	T cells
Pokeweed mitogen	PWM	*Phytolacca americana* (pokeweed)	T and B cells
Lipopolysaccharide	LPS	*Escherichia coli*	B cells (mouse)

contained 1000 maternal CD3+ T cells μl⁻¹, which responded to PHA. His immune system was slowly reconstituted over the ensuing 3 months.

Severe combined immunodeficiency.

Engraftment is successful.

Severe combined immunodeficiency, or SCID, presents the physician with a medical emergency. Unless there is a known family history, which provides the opportunity to take corrective therapeutic measures before the onset of infections, children with SCID come to medical attention only after they have been infected with a serious opportunistic infection. Because these infants die rapidly from such infections, even when treated adequately with anti-biotics or antiviral agents, measures must be taken quickly to reconstitute their immune system. In most cases of SCID, the first symptoms are those of thrush in the mouth and diaper area. A persistent cough usually betrays infection with *Pneumocystis carinii*. The third most common symptom of SCID is intractable diarrhea, usually due to enteropathic coliform bacilli.

As previously noted, SCID has many known genetic causes. The autosomal recessive form of SCID is most commonly caused by mutations in the purine degradation enzyme adenosine deaminase (ADA) (see Case 15), and, more rarely, by mutations in another such enzyme, purine nucleoside phosphory-lase (PNP). Defects in these enzymes lead to an accumulation of their nucleotide substrates that are highly toxic to developing T cells, and to a less-er extent to developing B cells. The X-linked form of SCID differs from these autosomal recessive forms in that males with this form of the disease have normal numbers of B cells, but they fail to function in the absence of T cells.

Other cases of autosomal recessive SCID resemble the phenotype of X-linked SCID and have been ascribed to defects in the protein Jak-3, which trans-duces the signal from many different interleukin receptors, and to defects in the α chain of the IL-7 receptor. SCID can also be caused by defects in the γ and ε components of the T-cell receptor or in other proteins that transduce signals from the T-cell receptor, such as the tyrosine kinase ZAP-70 and the calcium channel protein ORAI1, a CRAC channel (calcium-release activated calcium channel) protein (Fig. 14.7). In these cases B cells are normal and at least some T cells are present, but they fail to activate in an adaptive immune response, so that a combined immunodeficiency is seen.

Mice lacking IL-2 do not develop SCID, which argues against an important function for IL-2 signaling in thymocyte development. The discovery that mutations in the IL-2 receptor γ chain (IL-2Rγ) caused X-linked SCID in humans seemed inconsistent with this finding in mice. This led to a search for the γ chain in other interleukin receptors that might be more important for lymphocyte development. It was found that this chain also forms part of the receptors for IL-4, IL-7, IL-9, IL-15, and IL-21, and it was renamed the com-mon gamma (γc) chain. Jak-3 transduces the signal from all these receptors by binding the γc chain. IL-7 is known to be involved in T-cell development because mutations in the IL-7Rα chain result in SCID.

Most infants with SCID can be rescued by a successful bone marrow trans-plant. Continued gamma-globulin therapy is usually necessary, but with this, and with successfully engrafted T cells, SCID infants survive to lead a rela-tively normal life. Gene therapy has also been tried successfully in some patients with X-linked SCID.

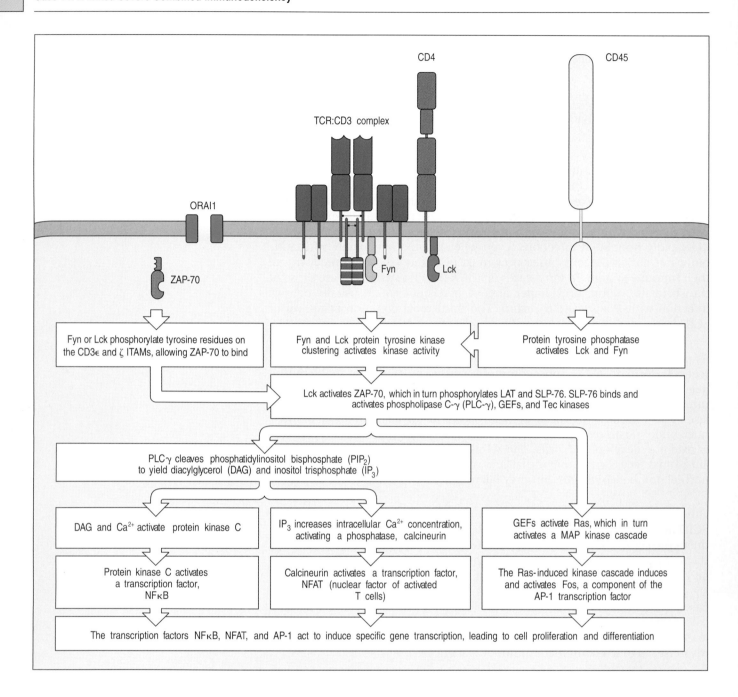

Fig. 14.7 Binding of antigen to the T-cell receptor initiates a series of biochemical changes within the T cell. It is thought that ligand binding to the T-cell receptor and the co-receptor CD4 (in this example) brings together CD4, the T-cell receptor:CD3 complex, and CD45, thus allowing the CD45 tyrosine phosphatase to remove inhibitory phosphate groups and thereby activate the Lck and Fyn tyrosine kinases associated with CD4 and the T-cell receptor:CD3 complex. One effect of tyrosine kinase activation is the activation of the cytosolic tyrosine kinase ZAP-70. Three important signaling pathways result. Two are initiated by the activation of phospholipase C-γ, which cleaves membrane phosphatidylinositol bisphosphate (PIP$_2$) to yield the second messengers diacylglycerol (DAG) and inositol trisphosphate (IP$_3$). DAG activates the serine–threonine protein kinase C, which results in the activation of the transcription factor NFκB. Meanwhile, IP$_3$ acts to release calcium ions (Ca^{2+}) from intracellular stores. In addition, a calcium-specific ion channel (the CRAC channel ORAI1) is opened in the T-cell membrane to allow the influx of Ca^{2+} from extracellular sources. The elevated concentration of Ca^{2+} activates a cytoplasmic phosphatase, calcineurin, which activates one part of the transcription factor NFAT. However, full NFAT activity also requires a member of the AP-1 family of transcription factors; these are dimers of members of the Fos and Jun families of transcription regulators. A third signaling pathway involves activation of the G protein Ras, and the subsequent activation of a cascade of protein kinases, culminating in the activation of Fos and hence of the AP-1 transcription factors. The combination of the transcriptional regulators NFκB, NFAT, and AP-1 act on the T-cell chromosomes, initiating new gene transcription that results in the differentiation, proliferation, and effector actions of T cells.

Questions.

1 Martin was suspected to have X-linked SCID because he had normal numbers of B cells and a total absence of T cells in his blood. The diagnosis was further confirmed on finding no IL-2Rγ on the B cells in his blood. What final confirmation proved that he had X-linked SCID?

2 What is known about the normal functions of the receptors of which the γ_c chain forms a part, and how might these account for the phenotype of X-linked SCID?

3 Martin received a haploidentical transplant from his mother. What was the advantage and the disadvantage of using his mother as a donor as opposed to his father?

4 Why was it necessary to treat the maternal marrow donation with a monoclonal antibody against mature T cells and complement before the transplant?

5 Frequently infants with SCID get very ill with Pneumocystis carinii pneumonia after a successful transplant. For this reason they are treated prophylactically with antibiotics to get rid of any P. carinii organisms that may be present in their lungs. Why is this a wise therapeutic maneuver and how do you explain the worsening of pneumonia after a transplant?

6 We have already discussed the risks associated with giving polio virus vaccine to immunodeficient infants in the case of X-linked agammaglobulinemia (see Case 10). Martin would not have been able to clear the poliovirus he received until he was started on gamma-globulin therapy, but luckily a pathogenic variant did not arise during that time. Fortunately, Martin escaped being given any other live vaccines before he was diagnosed. In many countries (but not the United States) infants are universally given bacille Calmette–Guérin (BCG), an attenuated form of the tuberculosis bacillus, which provides partial protection against tuberculosis infection. BCG incites a cell-mediated immune response and after receiving it infants become tuberculin-positive, which means they show a delayed-type hypersensitivity response to a skin-prick with minute quantities of tuberculin. In the United States, the tuberculin test is considered so diagnostically valuable for the detection of new tuberculosis infections that BCG is not given. What do you think happens to infants with SCID after they are given BCG?

7 Before smallpox was eradicated in the world, vaccinia virus was routinely administered to all children. What happened to infants with SCID who were vaccinated with vaccinia virus?

8 What would be another strategy for reconstituting immune function in Martin?

CASE 15 Adenosine Deaminase Deficiency

The purine degradation pathway and lymphocytes.

In Case 14 we learned about severe combined immunodeficiency (SCID) and examined the most common form—that due to an X-linked mutation in the gamma chain (γ_c) common to several interleukin receptors. In this type of SCID, T cells are virtually absent while B cells are present in normal numbers although they are not functional. Further examination of this phenotype also reveals an absence of natural killer (NK) cells. Thus, X-linked SCID is classified as $T^-B^+NK^-$ SCID.

SCID patients with an almost complete absence of B cells as well as T cells are also encountered; their phenotype is $T^-B^-NK^+$. This phenotype is exclusively associated with SCID transmitted as an autosomal recessive condition and has a quite different biochemical cause from X-linked SCID. The most common genetic defect encountered in these patients is mutation in the gene encoding

<table>
<tr><td>Topics bearing on this case:</td></tr>
<tr><td>Defects in lymphocyte function result in severe combined immunodeficiency</td></tr>
<tr><td>Bone marrow transplantation</td></tr>
<tr><td>Mixed lymphocyte reaction</td></tr>
<tr><td>Minor histocompatibility antigens</td></tr>
<tr><td>Graft-versus-host reaction</td></tr>
</table>

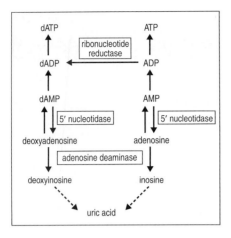

Fig. 15.1 Purine metabolic pathways. Adenosine deaminase catalyzes the conversion of deoxyadenosine and adenosine to deoxyinosine and inosine, which are eventually converted to waste products and excreted. In the absence of ADA in lymphocytes, adenosine and deoxyadenosine metabolites accumulate. In cells other than lymphocytes, the enzyme 5′ nucleotidase can convert AMP and dAMP to adenosine and deoxyadenosine, thus preventing the build-up of potentially toxic metabolites.

the purine degradation enzyme adenosine deaminase (ADA), which leads to no enzymes being produced. ADA is a ubiquitous housekeeping enzyme found in all mammalian cells and in blood serum; it converts the purine nucleosides adenosine and deoxyadenosine to inosine and deoxyinosine respectively, and hence to waste products that are excreted (Fig. 15.1). In its absence, cells can accumulate excessive amounts of adenosine and deoxyadenosine metabolites. As these tend to be toxic in excess one might expect an absence of ADA activity to seriously affect all the cells of the body. Quite surprisingly, however, it turns out that lymphocytes are particularly susceptible to the toxic effects of these metabolites, and thus ADA deficiency leads to SCID.

The case of Roberta Alden: lymphocytes poisoned by toxic metabolites.

3-week-old female infant with thrush. Family history of SCID.

The Aldens are an African-American family from a remote rural area of the state of Georgia. Mr and Mrs Alden are probably distantly related to each other. At the time they moved to Boston they had seven healthy children—four boys and three girls. Their eighth child was a boy, who developed severe pneumonia at 3 months old and died at the City Hospital. An autopsy revealed that he had SCID. His thymus had a fetal appearance with only rare thymocytes and no Hassall's corpuscles (see Fig. 14.2). Two years later the Aldens had a daughter, named Roberta. She appeared healthy at birth but at 3 weeks old she developed deeply pigmented spots on her trunk and face. Three weeks later Mrs Alden noticed thrush in Roberta's mouth (see Fig. 14.4).

Fig. 15.2 Chest radiographs of 5-month-old infants. In the left panel the thymic shadow in a normal healthy child is surrounded by a dashed blue line. In the right panel the heart shadow (surrounded by a dashed red line) is clearly visible in the SCID infant owing to the absence of the thymus. Courtesy of T. Griscom.

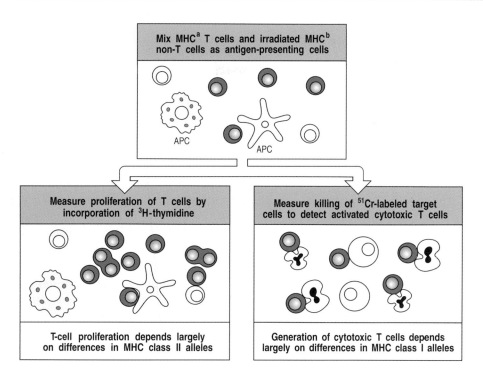

Fig. 15.3 The mixed lymphocyte reaction (MLR) can be used to detect histoincompatibility. Lymphocytes from the two individuals who are to be tested for compatibility are isolated from peripheral blood. The cells from one person (yellow), which also contain antigen-presenting cells, are either irradiated or treated with mitomycin C; they will act as stimulator cells but cannot now respond by DNA synthesis and cell division to antigenic stimulation by the other person's cells. The cells from the two individuals are then mixed (top panel). If the unirradiated lymphocytes (the responders, blue) contain alloreactive T cells, these will be stimulated to proliferate and differentiate to effector cells. Between 3 and 7 days after mixing, the culture is assessed for T-cell proliferation (bottom left panel), which is mainly the result of CD4 T cells recognizing differences in MHC class II molecules, and for the generation of activated cytotoxic T cells (bottom right panel), which respond to differences in MHC class I molecules. When the MLR is used to select a bone marrow donor, the prospective donor's cells are used as the responder cells and the prospective recipient's cells as the stimulator cells.

Aware of the family history, Roberta's pediatrician ordered a chest X-ray and blood studies. No thymic shadow (Fig. 15.2) was seen in the chest X-ray, and the anterior margins of the ribs were flared. Roberta's lymphocyte count was 150 cells μl^{-1} (normal for an infant is >3000 cells μl^{-1}). Her lymphocytes did not respond to the nonspecific T-cell mitogen phytohemagglutinin. A diagnosis of SCID was established.

HLA typing of Roberta, her parents, and her seven siblings revealed that her HLA type was identical with one sister, Ellen, and one brother, John. To test directly for histocompatibility, Ellen's and John's blood cells were tested separately against Roberta's in the mixed lymphocyte reaction (Fig. 15.3). No reaction was seen in either case. John and Roberta also had the same blood type (AB) and so John was chosen as the bone marrow donor. Bone marrow cells (2.8×10^7 cells) were removed from John's iliac crest bone and infused into Roberta (this was calculated to be a dose of 5×10^6 cells kg^{-1} of Roberta's body weight).

Twelve days after the bone marrow transplant, Roberta's lymphocyte count had increased to 500 μl^{-1}, and the response of her blood lymphocytes to phytohemagglutinin had risen to half normal. A karyotype of the responding cells revealed that they had XY sex chromosomes; they were thus of male origin. Two days later Roberta developed a transient graft-versus-host reaction characterized by fever and eosinophilia, which lasted 3 days. She continued to gain weight and was discharged from the hospital. Several weeks later she, her parents, and all her siblings were affected with a severe influenza-like respiratory infection. She recovered from this without problems and her lymphocyte count rose to 1575 μl^{-1}; she also now had a normal response to phytohemagglutinin. Roberta continued to grow and develop normally and remained free of infections. All her lymphocytes continue to have an XY karyotype.

Diagnosis of SCID. Consider bone marrow transplant.

Bone marrow transplant from HLA - and blood group- identical brother.

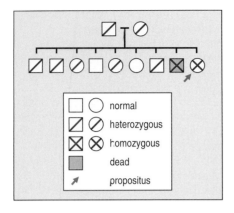

Fig. 15.4 Inheritance of ADA deficiency in Roberta's family. ADA deficiency is inherited as an autosomal recessive condition.

Adenosine deaminase deficiency.

The gene encoding adenosine deaminase (ADA) is located on chromosome 20. SCID due to ADA deficiency appears in homozygotes for defects in the ADA gene that result in an inactive enzyme or no enzyme. As production of ADA from the normal gene in heterozygotes is sufficient to compensate, the disease is inherited as an autosomal recessive condition (Fig. 15.4). Homozygotes for mutations in this gene show the most profoundly lymphopenic form of SCID. The thymus is poorly developed and there is a characteristic abnormality in the rib bones. The pigmented spots that appeared on the trunk and face were the result of a localized graft-versus-host reaction due to maternal cells.

Normally, the adenosine and deoxyadenosine content of cells is limited by ADA, which converts these nucleotides to inosine (and deoxyinosine) and subsequently to urate, which is excreted. The limited amounts of adenosine within cells are converted to adenosine monophosphate (AMP), adenosine diphosphate (ADP), and adenosine triphosphate (ATP). Deoxyadenosine is converted to dAMP, dADP, and dATP (see Fig. 15.1). In the absence of ADA, these metabolites can accumulate in up to 1000-fold excess within cells. Excess dATP in particular inhibits the enzyme ribonucleotide reductase, which is required for the synthesis of all the deoxynucleotides required for DNA synthesis (Fig. 15.5); its inhibition is probably the main culprit in causing the death and nondevelopment of lymphocytes.

The thymus contains 13 times more ADA than any other tissue in the body. As adenosine metabolites are very toxic to lymphocytes, the high level of ADA in the thymus is probably crucial for normal thymocyte development. The reason that lymphocytes are particularly vulnerable to the accumulation of these metabolic poisons is probably because they are relatively deficient in the enzyme 5' nucleotidase. This enzyme degrades AMP and dAMP to adenosine and deoxyadenosine (see Fig. 15.1) and thereby prevents the excessive accumulation of ADP, ATP, dADP, and dATP, even in the absence of ADA.

The administration of ADA, which is commercially available in a form bound to polyethylene glycol (PEG-ADA), clears the noxious metabolites and results in improved immune function. The same beneficial result has been achieved by gene therapy. The patient's white blood cells are transfected *in vitro* with a retroviral vector bearing a normal ADA gene. They are induced to proliferate *in vitro* by treatment with mitogen and then reinfused into the patient. This treatment is very laborious because it has to be repeated every few months to achieve success.

Another enzyme in the purine degradation pathway, purine nucleoside phosphorylase (PNP), degrades guanosine to inosine. Its absence results in excessive accumulation of the guanosine metabolites GMP, GDP, and GTP, as well as the deoxyguanosine metabolites dGMP, dGDP, and dGTP, which, like the adenosine metabolites, are toxic to lymphocytes. Genetic defects in PNP resulting in SCID have also been found.

Fig. 15.5 Role of ribonucleotide reductase in generating deoxynucleotides required for DNA synthesis.

Questions.

1 The absence of a thymic shadow in the chest radiograph of a young infant can be helpful in the diagnosis of SCID but it is not a reliable finding. Why is this so?

2 John and Roberta were histocompatible at the major histocompatibility locus—HLA. Despite this, Roberta sustained a transient graft-versus-host reaction characterized by fever and eosinophilia. How do you explain this?

3 In a mixed lymphocyte reaction, the cells used to stimulate the response are treated with mitomycin, a mitotic inhibitor, to prevent their responding to the responder cells. When Roberta's and John's cells were tested in a mixed lymphocyte reaction to ascertain whether his T cells would respond to Roberta's cells it was not necessary to treat Roberta's cells with mitomycin. Why was this the case?

4 Can you think of a reason for the fact that NK cells are normal in ADA deficiency and are not affected by the metabolic defect?

CASE 16 | Omenn Syndrome

A defect in V(D)J recombination results in severe immunodeficiency.

The development of B cells in the bone marrow and T cells in the thymus is initiated by the assembly of gene segments to make the variable (V) sequence that encodes the V regions of the heavy and light chains of immunoglobulins or of the α and β chains of the T-cell antigen receptors (Fig. 16.1). This process is called V(D)J recombination. A V (variable) and a J (joining) gene segment are joined to make the V-region sequences for the light chains of immunoglobulins or the α chains of T-cell receptors. An additional gene segment, D (diversity), is involved in the rearrangements that produce the V-region sequences for the immunoglobulin heavy chain and the β chain of the T-cell receptor; a D and a J gene segment are joined first, followed by joining of a V gene segment to form VDJ. In all these recombination events, the DNA between the rearranging gene segments is deleted from the chromosome. As there are many different V, D, and J segments in the germline genome, there are several million possible combinations. This is how much of the vast diversity in the antibody and T-cell receptor repertoires is generated. Moreover, small insertions or deletions of nucleotides at the joins between V and D, and D and J segments further contribute to diversity.

The process of V(D)J recombination is initiated by enzymes encoded by the recombinase-activating genes *RAG-1* and *RAG-2*. The RAG-1 and RAG-2 enzymes nick double-stranded DNA. They recognize canonical DNA sequences called recombination signal sequences, which flank the coding gene segments and consist of a heptamer (CACAGTG) followed by a spacer of 12 or

Topics bearing on this case:
V(D)J recombination
RAG enzymes
Severe combined immunodeficiency

Fig. 16.1 Rearrangement of the T-cell receptor genes. The top and bottom rows of the figure show the germline arrangement of the variable (V), diversity (D), joining (J), and constant (C) gene segments at the T-cell receptor α and β loci. During T-cell development, a V-region sequence for each chain is assembled by DNA recombination. For the α chain (top), a V_α gene segment rearranges to a J_α gene segment to create a functional gene encoding the V domain. For the β chain (bottom), rearrangement of a D_β, a J_β, and a V_β gene segment creates the functional V-domain exon. A similar array of gene segments is present at the immunoglobulin loci, and immunoglobulin gene rearrangement follows an essentially similar course.

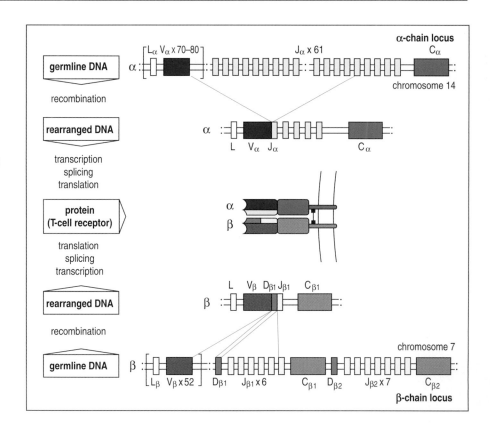

23 bases and then a nonamer (ACAAAAGTG) (Fig. 16.2). RAG-1 binds to the nonamer element followed by binding of RAG-2 to the heptamer. The DNA sequence that forms the border between the heptamer and the coding segment is then nicked, and a break in the double-stranded DNA occurs. A series of DNA-repair proteins carry out the rejoining process that brings a gene segment of each type together.

If either of the *RAG* genes is knocked out by homologous recombination in mice, B-cell and T-cell development is completely abolished and the mice have severe combined immunodeficiency. Mutations in *RAG-1* and *RAG-2* have also been found in cases of human SCID. Certain missense mutations in the *RAG* genes, however, result in partial enzymatic activity, and this gives rise to an interesting immunodeficiency disease called Omenn syndrome.

Fig. 16.2 Each V, D, or J gene segment is flanked by recombination signal sequences (RSSs). This is illustrated here with respect to the immunoglobulin genes. There are two types of RSS. One consists of a nonamer (9 nucleotides, shown in purple) and a heptamer (7 nucleotides, shown in orange) separated by a spacer of 12 nucleotides (white). The other consists of the same 9- and 7-nucleotide sequences separated by a 23-nucleotide spacer (white).

The case of Ricardo Reis: a bright red rash betrays an immunodeficiency.

Fig. 16.3 **Bright red rash on the face and shoulders of an infant with Omenn syndrome.** Photograph kindly provided by Luigi Notarangelo.

At birth, Ricardo seemed to be a normal healthy baby. He gained weight normally and cried vigorously. Soon after birth, however, his mother noticed that he had ten loose bowel movements a day. On the 27th day after birth, he developed a red, raised (papular) rash on his legs, which over the next 7 days spread over his entire body. His parents brought him to the emergency room at Children's Hospital, and also reported that he had had a dry cough for the past week.

On physical examination, Ricardo's weight, length, and head circumference were normal. The diffuse papular scaly rash was worst on his face (Fig. 16.3) but also covered his trunk and extremities (Fig. 16.4). Small blisters were present on his palms and the soles of his feet, which were red. He had purulent conjunctivitis (yellow discharge from his eyes), his eardrums were normal, no lymph nodes could be felt, and his heart and lungs were normal. The liver and spleen were not enlarged.

Ricardo's parents had three normal children, but had had two other children, a boy and a girl, who had died soon after the onset of a similar rash at 1 month old. The parents were first cousins of Portuguese extraction.

Ricardo was admitted to the hospital. Blood tests showed that his hemoglobin was 15.4 g dl^{-1} (normal), platelet count 460,000 (slightly elevated), white blood cell count 8000 μl^{-1} (normal), of which 56% were eosinophils (normal <5%), 23% monocytes (normal 10%), 15% neutrophils and 6% lymphocytes (normal 50%). An examination of his bone marrow revealed a preponderance of eosinophil precursors. Ricardo's serum IgG level was 55 mg dl^{-1} (normal 400 mg), IgA and IgM were undetectable, and IgE was 7200 IU ml^{-1} (normal <50 IU ml^{-1}). A skin biopsy showed that the dermis was infiltrated with large numbers of eosinophils, lymphocytes, and macrophages. Large numbers of cells surrounded the blood vessels. An X-ray of Ricardo's chest showed clear lungs and a normal cardiac shadow; there was no thymic shadow (see Case 15).

In the hospital, a hematology consultant recommended treatment with vinblastine. Despite this therapy, Ricardo's condition rapidly worsened. He developed enlarged lymph nodes in the neck and groin, and pus accumulated in the skin behind his ear. This was drained, and *Staphylococcus aureus* and *Candida albicans* were cultured from the drainage fluid. Thrush (*Candida albicans*) was noticed in his mouth (see Case 14). An immunologist was consulted. He ordered blood tests that revealed an absence of B cells and a paucity of T cells. Ricardo's peripheral blood lymphocytes responded poorly to stimulation with phytohemagglutinin and pokeweed mitogen and with anti-CD3 monoclonal antibody. On FACS analysis, no cells were found that reacted with anti-CD19, which detects B cells (see Case 10). All the lymphocytes were CD3+, of which 90% expressed major histocompatibility complex (MHC) class II molecules, a marker of T-cell activation. Sixty percent of the lymphocytes were CD4+ and 35% were CD8+. Analysis of genomic DNA from Ricardo's peripheral blood lymphocytes by Southern blotting using a J$_\beta$ probe revealed only a small number of different T-cell receptors (oligoclonality). An *in vitro* analysis of V(D)J recombination revealed that it was 20% of normal, suggesting a defect in RAG-1 or RAG-2. The *RAG-1* and *RAG-2* genes were sequenced and a missense mutation was found in the *RAG-1* gene. The T cells were definitively identified as Ricardo's (and not transferred maternal T cells) by the MHC class II type.

While these studies were being carried out, Ricardo developed *Pneumocystis carinii* pneumonia and died of respiratory failure.

Eosinophilia and low lymphocyte count. No thymic shadow.

Lymph nodes enlarged. Opportunistic infections noted. Immunodeficiency?

One-month-old infant with bright red rash and purulent conjunctivitis. Admit to hospital.

Fig. 16.4 Legs and groin of an infant with Omenn syndrome. The skin is bright red and wrinkled from edema and infiltration of inflammatory cells. Photograph kindly provided by Luigi Notarangelo.

Omenn syndrome.

The RAG enzymes essential for V(D)J recombination were first discovered in mice and later identified in humans. Infants with the autosomal recessive form of severe combined immunodeficiency (SCID) were screened for mutations in these genes and several cases were identified in which RAG-1 or RAG-2 was deficient. These infants had a phenotype resembling that seen in adenosine deaminase deficiency (see Case 15) in that they had T⁻B⁻NK⁺ SCID.

Some patients were found with missense mutations in the *RAG* genes such that only partial enzyme activity was expressed. An examination of patients with a form of SCID called Omenn syndrome revealed further missense mutations in *RAG* genes. This syndrome is characterized by early onset of a generalized red rash (erythroderma), failure to thrive, protracted diarrhea, and enlargement of the liver, spleen, and lymph nodes. A high eosinophil count (eosinophilia) is usually encountered, together with a lack of B lymphocytes and a marked decrease in T cells. Immunoglobulins are also markedly decreased, but IgE levels are raised. As only partial ability to execute V(D)J recombination is retained by the mutated enzyme, no mature circulating B cells are detected and the few T cells that are found are oligoclonal; that is, they are the products of a limited number of different clones.

As illustrated by this case, Omenn syndrome is usually rapidly fatal unless it can be treated by bone marrow transplantation; some patients have been rescued with transplants of HLA-identical bone marrow.

Other genetic defects that result in severe impairment of T-cell development can also result in Omenn syndrome. These include IL-7Rα chain deficiency (IL-7 is required for lymphocyte development) and a deficiency of the DNA repair and V(D)J recombination protein Artemis. As with the *RAG* genes, Omenn syndrome occurs when the defect is 'leaky'; that is, due to a missense mutation that severely impairs but does not abolish function, allowing a few T cells to develop. It is likely that in Omenn syndrome the reduced diversity and scarcity of the T regulatory cells (see Case 25) that escape the developmental block prevents them from exerting suppression over the effector cells that escape that block.

Apart from the lymphocyte-specific RAG proteins, V(D)J recombination also involves proteins universally used for DNA repair and recombination in human cells. Cases of SCID associated with mutations in some of these have been found. In addition to T⁻B⁻NK⁺ SCID, the latter syndromes are characterized by sensitivity of cells in general (e.g. fibroblasts) to ionizing radiation, because they are unable to repair radiation-induced DNA damage. These syndromes are often associated with abnormalities in nonlymphoid tissues. Defects in the proteins Artemis, Cernunnos, and DNA ligase IV, which are required both for V(D)J recombination and for DNA repair, are among those found in this category of radiosensitive SCID (RS-SCID).

Questions.

1   How do you explain the high IgE level and eosinophilia in this case?

2   How do you explain the enlargement of the lymph nodes in this patient?

3   Ricardo's T cells were found to be oligoclonal. How was this determined with a J_β probe?

4   What assay was performed to ascertain that Ricardo had only 20% of normal RAG function?

5   How does Ricardo's family history help you determine the mode of inheritance of Omenn syndrome?

6   A bright red rash (erythroderma) is characteristic of Omenn syndrome. What causes this rash?

CASE 17 | MHC Class I Deficiency

A failure of antigen processing.

The class I molecules encoded by the major histocompatibility complex (MHC) are expressed to a greater or lesser extent on the surface of all the cells of the body except the red blood cells. MHC class I molecules bind peptides derived from proteins synthesized in the cytoplasm, and carry them to the cell surface, where they form a complex of peptide and MHC molecule on the cell surface. This complex can then be recognized by antigen-specific CD8 T cells. T cells as a class recognize only peptides presented to them as a complex with an MHC molecule; the T-cell receptors of CD8 T cells recognize only peptides presented by MHC class I molecules, while those of CD4 T cells recognize only peptides presented by MHC class II molecules (see Case 18).

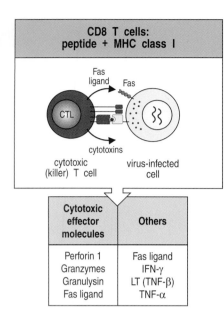

CD8 T cells:
peptide + MHC class I

Fas
ligand   Fas

CTL

cytotoxins

cytotoxic
(killer) T cell   virus-infected
cell

Cytotoxic effector molecules	Others
Perforin 1	Fas ligand
Granzymes	IFN-γ
Granulysin	LT (TNF-β)
Fas ligand	TNF-α

Fig. 17.1 Cytotoxic CD8 T cells recognize and kill virus-infected cells. CD8 T cells recognize virally derived peptides presented by MHC class I antigens at the surface of virus-infected cells. They kill target cells by releasing cytotoxins and cytotoxic cytokines and by binding to Fas on the target cell. LT, lymphotoxin.

MHC class I molecules are involved principally in immune reactions against virus infections. Cytotoxic CD8 T cells specific for viral antigens terminate viral infections by recognizing viral peptides carried by MHC class I molecules on the surface of virus-infected cells, and killing these cells (Fig. 17.1). They release the cytotoxic proteins perforin and granzymes, as well as the cytotoxic cytokines tumor necrosis factors TNF-α and TNF-β. In addition, cytotoxic T cells express the Fas ligand, which engages the cell-surface molecule Fas on target cells. Both processes induce the target cells to undergo programmed cell death (apoptosis).

In humans, the class I and class II MHC molecules are known as the HLA antigens and together they determine the tissue type of an individual. MHC class I molecules are particularly abundant on T and B lymphocytes, and also on macrophages and other cells of the monocyte lineage as well as on neutrophils. Other cells express smaller amounts. Each individual expresses three principal types of class I molecule—HLA-A, HLA-B, and HLA-C. These are heterodimeric glycoproteins, composed of an α chain and a β chain, the latter known also as β$_2$-microglobulin (Fig. 17.2).

The genes encoding the α chains of the human MHC class I molecules are located close together in the major histocompatibility complex on the short arm of chromosome 6 (Fig. 17.3). In humans, the gene encoding β$_2$-microglobulin, the polypeptide chain common to all class I molecules, is located not in the MHC but on the long arm of chromosome 15. The genes encoding the α chains are highly polymorphic, and so there are numerous variants of HLA-A, HLA-B, and HLA-C within the population.

Viral proteins, like cellular proteins, are made in the cytoplasm of the infected cell, and some are soon degraded into peptide fragments by large enzyme complexes called proteasomes. The peptides are then transported from the cytosol into the endoplasmic reticulum by a complex of two transporter proteins called TAP1 and TAP2, which is located in the endoplasmic reticulum membrane (Fig. 17.4). The genes encoding TAP1 and TAP2 are also located in the MHC, in the region containing the class II genes (see Fig. 17.3).

The endoplasmic reticulum contains MHC class I molecules, which enter as separate α and β chains as soon as they have been synthesized and are retained there. After the antigenic peptides enter the endoplasmic reticulum they are loaded onto the complex of α chains and β$_2$-microglobulin. The MHC class I:peptide complex is then released and transported onwards to the cell surface (Fig. 17.5).

This case describes a rare inherited immune deficiency accompanied by the absence of MHC class I molecules on the patients' cells.

peptide-binding
cleft

α$_2$ α$_1$

α$_3$ β$_2$-microglobulin

Fig. 17.2 A schematic representation of a human MHC class I molecule. It is a heterodimeric glycoprotein, composed of one transmembrane α chain bound noncovalently to β$_2$-microglobulin. The α chain is folded into three protein domains, two of which form a cleft into which the peptide antigen binds.

Gene structure of the human MHC

Fig. 17.3 The organization of the major histocompatibility complex (MHC) on chromosome 6 in humans. There are separate regions of class I and class II genes. The class I genes are called HLA-A, HLA-B, and HLA-C. The gene for β_2-microglobulin is located on chromosome 15. The genes for the TAP1:TAP2 transporter are located in the class II region of the MHC.

The children of Sergei and Natasha Islayev: the consequences of a small flaw in the MHC.

Tatiana Islayev was 17 years old when first seen at the Children's Hospital. She had severe bronchiectasis (dilatation of the bronchi from repeated infections) and a persistent cough that produced yellow-green sputum. Tatiana had been chronically ill from the age of 4, when she started to get repeated infections of the sinuses, middle ears, and lungs, apparently due to a variety of respiratory viruses. The bacteria *Haemophilus influenzae* and *Streptococcus pneumoniae* could be cultured from her sputum and she had been prescribed frequent antibiotic treatment to control her persistent fevers and cough. Her brother Alexander, aged 7 years old, also suffered from chronic respiratory infections. Like his sister, he had begun to suffer severe repeated viral infections of the upper and lower respiratory tracts at an early age. He also had severe bronchiectasis and *H. influenzae* could be cultured from his sputum.

Diagram of the TAP transporter complex

Fig. 17.4 The TAP1 and TAP2 transporter proteins form a heterodimer in the endoplasmic reticulum membrane.

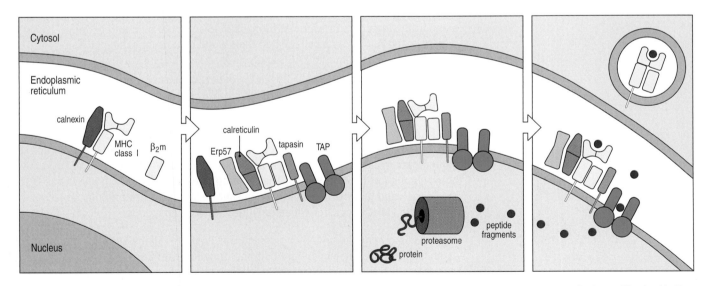

Fig. 17.5 MHC class I molecules do not leave the endoplasmic reticulum unless they bind peptides. MHC class I α chains assemble in the endoplasmic reticulum with a membrane-bound protein, calnexin. When this complex binds β_2-microglobulin (β_2m) it is released from calnexin, and the partly folded MHC class I molecule then binds to the TAP1 subunit of the TAP transporter by interacting with one molecule of the TAP-associated protein, tapasin. It is retained within the endoplasmic reticulum until released by the binding of a peptide, which completes the folding of the MHC class I molecule. Peptides generated by the degradation of proteins in the cytoplasm are transported into the lumen of the endoplasmic reticulum by the TAP transporter. Once peptide has bound to the MHC molecule, the peptide:MHC complex is transported through the Golgi complex to the cell surface.

Brother and sister with symptoms of severe chronic respiratory infection.

Other siblings normal.

Owing to the chronic illness of Tatiana and Alexander, the Islayevs had emigrated recently from Russia to the United States, where they hoped to get better medical treatment. They had three other children, aged 5, 10, and 13 years old, when they came to America, who were all healthy and showed no increased susceptibility to infection. As infants in Moscow, both Tatiana and Alexander had received routine immunizations with oral poliovirus as well as diphtheria, pertussis, and tetanus (DPT) vaccinations. They had also been given BCG as newborn babies for protection against tuberculosis, and had tolerated all these immunizations well.

When they were examined, Tatiana and Alexander both had elevated levels of IgG, at >1500 mg dl^{-1} (normal levels 600–1400 mg dl^{-1}). They had white blood cell counts of 7000 and 6600 cells µl^{-1} respectively. Of these white cells, 25% (1750 and 1650 µl^{-1} respectively) were lymphocytes. Ten per cent of the lymphocytes reacted with an antibody to B cells (anti-CD19) (a normal result) and 4% with an antibody to natural killer (NK) cells (anti-CD16) (normal). The remainder of the lymphocytes reacted with an anti-T-cell antibody (anti-CD3). Over 90% of the T cells were CD4, while 10% were CD8. This represents a profound deficiency of CD8 T cells. Blood tests on their siblings and parents showed no deficiency of CD8 T cells. Both Tatiana and Alexander had normal neutrophil function and complement titers.

There was thus no evidence of a deficiency of humoral immunity. Furthermore, their cell-mediated immunity also seemed normal when tested by delayed hypersensitivity skin tests to tuberculin and antigen from *Candida*, a fungal component of the normal body flora (see Case 37); they developed the normal delayed-type hypersensitivity response of a hard, raised, red swelling some 50 mm in diameter at the site of intradermal injection of these antigens. Both children were found to have high titers of antibodies to herpes virus and cytomegalovirus as well as to mumps, chickenpox, and measles viruses. When asked, the parents recalled that the children had been immunized against influenza several times, and antibodies against five different strains of influenza were found. However the anti-influenza antibodies were present in very low titers. They also had low titers of antibody against the Epstein–Barr virus.

When white blood cells from all family members were typed for HLA antigens by standard typing procedures, no MHC class I molecules at all could be found on Tatiana's and Alexander's cells. When their blood cells were examined using the more sensitive technique of FACS analysis (see Fig. 10.3), it was found that Tatiana and Alexander expressed very small amounts of MHC class I molecules, less than 1% of the amount expressed on the cells of their father (Fig. 17.6). They also expressed MHC class II molecules normally.

The HLA typing revealed that the mother and father shared an MHC haplotype (HLA-A3, -B63, HLA-DR4, -DQ3) and that Tatiana and Alexander had inherited this shared haplotype from both their parents (Fig. 17.7). They were therefore homozygous for the MHC region. The other children were heterozygous. It was thus concluded that the two children's susceptibility to respiratory infections was linked to the MHC locus, for which only they were homozygous.

To try to determine the underlying defect, B cells from Tatiana and Alexander were established as cell lines in culture by transformation with Epstein–Barr virus. The transformed B cells were examined for the presence of messenger RNA for MHC class I molecules: normal levels were found. This eliminated the possibility that they had a structural or regulatory defect in genes encoding the α chain of the MHC class I molecules. Because the gene for β_2-microglobulin maps to chromosome 15, it was highly unlikely that their MHC-linked condition resulted from a defect in that gene. Only one possibility remained to be explored—a mutation in the *TAP1* or *TAP2* genes. When the DNA sequences of these genes were determined, both Tatiana and Alexander were found to have the same nonsense mutation in their *TAP2* genes. Their parents were found to be heterozygous for this mutation.

Fig. 17.6 Fluorescent antibody typing for HLA antigens. The peripheral blood lymphocytes of Tatiana and her father were typed by a fluorescent antibody to the MHC class I molecule HLA-A3. Tatiana's lymphocytes express about 1% of the HLA-A3 expressed on the lymphocytes of her father.

Fig. 17.7 Inheritance of MHC haplotypes in the Islayev family.

MHC class I deficiency.

A recessively inherited immunodeficiency was first suggested by the similarity of Tatiana's and Alexander's condition and its complete absence from other members of the family.

The Islayev children present us with a unique example of a clinical phenotype resulting from MHC class I deficiency. This phenotype has some unexpected features. One would not have expected their apparently normal capacity to fight some types of viral infections successfully, despite their profound deficiency in CD8 T cells and their inability to present viral antigens to CD8 T cells because of the absence of MHC class I antigens. The high levels of antibodies to chickenpox, measles, and mumps viruses in their blood showed that they have been exposed to and had successfully overcome these infections.

They have sustained innumerable respiratory viral infections, however. And their poor antibody responses to a variety of influenza strains show that they may have problems responding to respiratory viruses in general. It may be that some viruses are better able than others to stimulate an increased expression of MHC class I molecules on their cells. Tatiana and Alexander express MHC class I molecules at very low levels and when their isolated B cells were loaded with an antigenic peptide from influenza virus, this stimulated a small increase in the number of MHC class I molecules on these cells (Fig. 17.8). It is therefore possible that other virus infections that they sustained, such as chickenpox, were able to induce sufficient expression of MHC class I molecules to terminate the infection properly.

The repeated respiratory infections have caused anatomic damage to their airways, resulting in the bronchiectasis. The abundant *Haemophilus* and pneumococci in their sputum is characteristic of patients with bronchiectasis, and in their case is not due directly to any deficiency of immunity against these capsulated bacteria (compare with Case 10).

Their profound deficiency in CD8 T cells is a direct consequence of the lack of MHC class I molecules on the cells of the thymus, the organ in which all T cells bearing α:β T-cell receptors mature. An interaction with MHC class I molecules on thymic epithelial cells is crucial for the intrathymic maturation of such T cells as CD8 T cells.

This family reveals that we have much to learn about the role of MHC class I molecules in protection against intracellular pathogens.

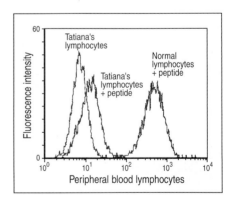

Fig. 17.8 Increased expression of MHC class I molecules in response to virus infection. Peripheral blood lymphocytes from Tatiana were examined with a fluorescent antibody to HLA-A3 before and after loading the cells with an antigenic peptide from influenza virus. There is a small increase in the number of MHC class I molecules expressed on the lymphocytes to approximately 5% of the amount expressed by normal cells loaded with the same peptide.

Questions.

1 | Why was the number of CD8 T cells in Tatiana and Alexander decreased despite normal levels of CD4 T cells?

2 | When the T-cell antigen receptors on their CD8 T cells were examined, it turned out that they were all $\gamma{:}\delta$ receptors and none were $\alpha{:}\beta$ receptors. How do you explain this?

3 | Tatiana and Alexander had normal delayed-type hypersensitivity responses to tuberculin and Candida. Is this surprising in view of their deficiency in CD8 T cells?

4 | Why did Tatiana and Alexander have high levels of serum IgG?

5 | Genetic defects have also been found in the gene encoding TAP1. Do you think that the clinical course in these patients would differ from that observed in Tatiana and Alexander?

CASE 18 | MHC Class II Deficiency

An inherited failure of gene regulation.

The class II molecules of the major histocompatibility complex (MHC) are involved in presenting antigens to CD4 T cells. The peptide antigens that they present are derived from extracellular pathogens and proteins taken up into intracellular vesicles, or from pathogens such as *Mycobacterium* that persist intracellularly inside vesicles. MHC class II molecules are expressed constitutively on antigen-presenting cells, including B lymphocytes, macrophages, and dendritic cells. In humans, together with the MHC class I molecules (see Case 17) they are known as the HLA antigens. They are also expressed on the epithelial cells of the thymus and their expression can be induced on other cells, principally by the cytokine interferon-γ. T cells also express MHC class II molecules when they are activated.

MHC class II molecules are heterodimers consisting of an α chain and a β chain (Fig. 18.1). The genes encoding both chains are located in the MHC on the short arm of chromosome 6 in humans (Fig. 18.2). The principal MHC

Fig. 18.1 Structure of an MHC class II molecule. Panel a shows a computer graphic representation of the MHC class II molecule HLA-DR1. Panel b is a schematic representation of the molecule. It is composed of two transmembrane glycoprotein chains, α and β, each folded into two protein domains. The antigenic peptide binds in a cleft between the two chains. Photograph courtesy of C. Thorpe.

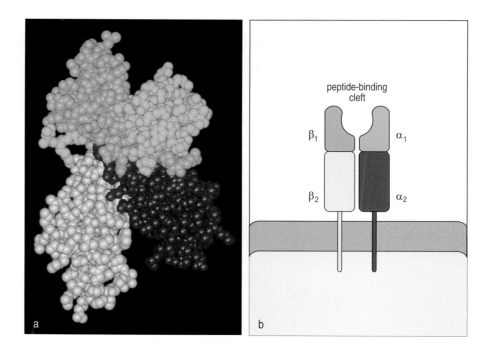

class II molecules are designated DP, DQ, and DR, and like the MHC class I molecules they are highly polymorphic. Peptides bound to MHC class II molecules can be recognized only by the T-cell receptors of CD4 T cells and not by those of CD8 T cells (Fig. 18.3). MHC class II molecules expressed in the thymus also have a vital role in the intrathymic maturation of CD4 T cells.

Expression of the genes encoding the α and β chains of MHC class II molecules must be coordinated strictly and is under complex regulatory control. The regulation of MHC class II gene expression is not fully understood as it involves the action of transcription factors that are defined only in part. The existence of these transcription factors and a means of identifying them were first suggested by the study of patients with MHC class II deficiency.

Fig. 18.2 Detailed map of the MHC class II region. The genes for the α and β chains of the HLA-DP, HLA-DR, and HLA-DQ molecules are shown as DPA, DPB, etc. The situation is complicated as there are two DPA genes, two DPB genes, and several DRB genes. Genes shown in gray and named in italic are pseudogenes. MHC class II genes are shown in yellow. Genes in the MHC region which have immune functions but are not related to the MHC class I and class II genes are shown in purple. Approximate genetic distances given in thousands of base pairs (kb).

The case of Helen Burns: a 6-month-old child with a mild form of severe combined immunodeficiency.

Helen Burns was the second child born to her parents. She thrived until 6 months of age when she developed pneumonia in both lungs, accompanied by a severe cough and fever. Blood and sputum cultures for bacteria were negative but a tracheal aspirate revealed the presence of abundant *Pneumocystis carinii*. She was treated successfully with the anti-*Pneumocystis* drug pentamidine and seemed to recover fully.

As her pneumonia was caused by the opportunistic pathogen *Pneumocystis carinii*, Helen was suspected to have severe combined immunodeficiency. A blood sample was taken and her peripheral blood mononuclear cells were stimulated with phytohemagglutinin (PHA) to test for T-cell function by ^{3}H-thymidine incorporation into DNA. A normal T-cell proliferative response was obtained, with her T cells incorporating 114,050 counts min^{-1} of ^{3}H-thymidine (normal control 75,000 counts min^{-1}). Helen had received routine immunizations with orally administered polio vaccine and DPT (diphtheria, pertussis, and tetanus) vaccine at 2 months old. However, in further tests, her T cells failed to respond to tetanus toxin *in vitro*, although they responded normally in the ^{3}H-thymidine incorporation assay when stimulated with allogeneic B cells (6730 counts min^{-1} incorporated compared with 783 counts min^{-1} for unstimulated cells).

When it was found that Helen's T cells could not respond to a specific antigenic stimulus, her serum immunoglobulins were measured and found to be very low. IgG levels were 96 mg dl^{-1} (normal 600–1400 mg dl^{-1}), IgA was 6 mg dl^{-1} (normal 60–380 mg dl^{-1}), and IgM 30 mg dl^{-1} (normal 40–345 mg dl^{-1}).

Helen's white blood cell count was elevated at 20,000 cells μl^{-1} (normal range 4000–7000 μl^{-1}). Of these, 82% were neutrophils, 10% lymphocytes, 6% monocytes, and 2% eosinophils. The calculated number of 2000 lymphocytes μl^{-1} is low for her age (normal >3000 μl^{-1}). Of her lymphocytes, 7% were B cells as determined by an antibody to CD20 (normal 10–12%) and 57% reacted with antibody to the T-cell

6-month-old girl with pneumonia. SCID? Do lymphocyte function tests.

Low Ig levels, deficiency of CD4 T cells.

Fig. 18.3 Effector CD4 cells recognize antigens bound to MHC class II molecules. CD4 T cells carry the co-receptor molecule CD4, which binds to MHC class II molecules on the antigen-presenting cell and helps to stabilize T-cell receptor:antigen binding. Effector CD4 T cells fall into two main types: T$_H$1 and T$_H$2. T$_H$1 cells are involved mainly in responding to antigens presented by macrophages, whereas T$_H$2 cells respond to antigen presented by B cells, stimulating the differentiation of B cells to plasma cells and the production of antibodies.

Fig. 18.4 Detection of MHC class II molecules by fluorescent antibody. Helen's transformed B-cell line was examined by using a fluorescent antibody to HLA-DQ and -DR. Helen (left panels) expressed approximately 1% of the amount of MHC class II molecules compared with a transformed B-cell line from a normal control (right panels).

Cells from a patient with class II histocompatibility deficiency	Immunofluorescence of normal EBV-transformed cells
HLA-DQ	
HLA-DR	

No HLA-DR type available. Do FACS analysis.

MHC class II deficiency. Bone marrow transplant advisable.

marker CD3. Of these T cells, 34% were positive for CD8, and 20% were positive for CD4. Thus, at 388 cells μl^{-1} her number of CD8 T cells was within the normal range, but the number of CD4 T cells (228 cells μl^{-1}) was much lower than normal (her CD4 T-cell count would be expected to be twice her CD8 T-cell count). The presence of substantial numbers of T cells, and thus a normal response to PHA, ruled out a diagnosis of severe combined immunodeficiency (see Case 14).

Helen's pediatrician referred her to the Children's Hospital for consideration for a bone marrow transplant, despite the lack of a diagnosis. When an attempt was made to HLA type Helen, her parents and her healthy 4-year-old brother, a DR type could not be obtained from Helen's white blood cells. A long-term culture of her B cells was made by transforming them with Epstein–Barr virus and the transformed B cells were then examined for expression of MHC class I and class II molecules with fluorescent-tagged antibodies. It was found that her B cells did not express HLA-DQ or HLA-DR molecules and a diagnosis of MHC class II deficiency was established (Fig. 18.4).

As her brother did not have the same HLA type as Helen, it was decided to use her mother as a bone marrow donor. Helen was given 1 mg kg^{-1} body weight of cytotoxic drug busulfan every 6 hours for 4 days to depress bone marrow function and then 50 mg kg^{-1} of cyclophosphamide to ablate her bone marrow. The maternal bone marrow was depleted of T cells to diminish the chance of graft-versus-host disease developing and was administered to Helen by transfusion. The graft was successful and immune function was restored.

MHC class II deficiency.

MHC class II deficiency is inherited as an autosomal recessive trait. Health problems show up early in infancy. Affected babies present the physician with a mild form of severe combined immunodeficiency (SCID) as they have increased susceptibility to pyogenic and opportunistic infections. However, they differ from SCID patients (see Case 14) in that they have T cells, which can respond to nonspecific T-cell mitogens such as PHA and to allogeneic stimuli. Also unlike SCID patients, they do not sustain graft-versus-host disease when given HLA-mismatched blood transfusions. This is because the host tissue has no MHC class II molecules against which the T cells in the graft can react. SCID patients, in contrast, carry MHC II molecules on some of their tissues, against which T cells in the graft will react to cause graft-versus-host disease. They have no T cells at all and therefore cannot reject the graft. Unlike in some other types of immunodeficiency, progressive infection with the attenuated live vaccine strain BCG has not been observed in MHC class II-deficient patients after BCG vaccination against tuberculosis (most cases of MHC class II deficiency have been observed in North African migrants in Europe, where BCG vaccination is routine). This is because mycobacterial antigens derived from BCG can be presented on MHC class I molecules and infected cells can be destroyed by cyotoxic T cells.

Patients with MHC class II deficiency are deficient in CD4 T cells, in contrast to MHC class I deficiency, in which CD8 T-cell numbers are very low and the levels of CD4 cells are normal (see Case 17). They also have moderate to severe hypogammaglobulinemia.

Genetic linkage analysis in large extended families with MHC class II deficiency has shown that this condition is not linked to the MHC locus on the short arm of chromosome 6 and that the genes encoding the MHC class II molecules at this locus are normal. Interferon-γ induces the expression of MHC class II molecules on antigen-presenting cells from normal people but fails to induce their expression on the antigen-presenting cells of patients with MHC class II deficiency. This suggested that the defect might lie in the regulation of expression of the MHC class II genes.

The search for the cause of the defect was complicated further by the discovery that MHC class II deficiency in different patients seems to have different causes. B-cell lines isolated from class II-deficient patients do not express MHC class II molecules. However, when B cells from two different patients are fused, MHC class II expression is often observed. The fusion of the two cell lines has corrected the defect. This means that one cell must be able to replace whatever is lacking in the other, and thus the two cells must carry different genetic defects causing the MHC class II deficiency. Pairwise fusions were performed on a large number of cell lines from different patients, and at least four complementation groups were found (Fig. 18.5).

These experiments provided clues that led eventually to the identification of the defect. The lack of MHC class II molecules turns out to result from defects in the transcription factors required to regulate their coordinated expression. All four of these transcription factors, which bind to the 5′ regulatory region of the MHC class II genes, have been identified.

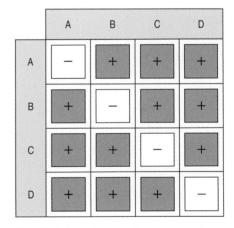

Fig. 18.5 Complementation groups of MHC class II deficiency. B-cell lines isolated from different patients were fused in all pairwise combinations to determine whether they could correct each other's defect. If two cell lines do not correct each other (−), they are in the same complementation group and have the same genetic defect. However, if the defect is corrected (+), the two cell lines belong to two different complementation groups and have two different defects. Four complementation groups, A, B, C, and D, were discovered by this technique.

Questions.

1 | *Why did Helen lack CD4 T cells in her blood?*

2 | *Why did Helen have a low level of immunoglobulins in her blood?*

3 In SCID, lymphocytes fail to respond to mitogenic stimuli. Although Helen was first thought to have SCID, this diagnosis was eliminated by her normal response to PHA and an allogeneic stimulus. How do you explain these findings?

4 If a skin graft were to be placed on Helen's forearm do you think she would reject the graft?

CASE 19 | Multiple Myeloma

A malignancy of terminally differentiated B lymphocytes.

Malignant tumors result from the outgrowth of a single transformed cell. Malignant tumors that result from the clonal outgrowth of B lymphocytes can occur at all stages of B-cell development (Fig. 19.1). These malignant B cells express clonotypic immunoglobulin molecules derived from the same immunoglobulin gene rearrangement, either on their surface or as secreted monoclonal antibody.

Malignancies of plasma cells cause a disease called multiple myeloma. It is a disease of bone because these plasma-cell tumors arise in the bone marrow. As the tumor masses expand, they cause local erosions of the bone, and the appearance on radiographs of multiple bone lesions (Fig. 19.2).

Topics bearing on this case:

The development of B lymphocytes

Monospecificity of B-cell receptors

Structure of the antibody molecule

Immunoglobulin gene rearrangement

B-cell tumors

Measurement of immunoglobulins

Immunoglobulin isotype switching

Fig. 19.1 B-cell tumors represent clonal outgrowths of B cells at various stages of development. Each type of tumor cell has a normal B-cell equivalent, homes to similar sites, and has behavior similar to that cell. Thus, myeloma cells look much like the plasma cells from which they derive, they secrete immunoglobulin, and they are found predominantly in the bone marrow. Many lymphomas and myelomas may go through a preliminary less aggressive lymphoproliferative phase, and some mild lymphoproliferations appear to be benign.

Name of tumor	Normal cell equivalent		Location	Status of Ig V genes
Chronic lymphocytic leukemia	CD5 B-1 cell		Blood	Mutated
Acute lymphoblastic leukemia	Lymphoid progenitor		Bone marrow and blood	Unmutated
Pre-B cell leukemia	Pre-B cell			Unmutated
Mantle cell lymphoma	Resting naive B cell			Unmutated
Follicular center cell lymphoma / Burkitt's lymphoma	Mature B cell		Periphery	Mutated, intraclonal variability
Hodgkin's lymphoma	Germinal center B cell			Mutated +/– intraclonal variability
Waldenström's macroglobulinemia	IgM-secreting B cell			Mutated, no variability within clone
Multiple myeloma	Plasma cell. Various isotypes		Bone marrow	Mutated, no variability within clone

These myelomas secrete a staggering amount of monoclonal immunoglobulin, which may be of the IgG or IgA, or very rarely IgD or IgE, isotype, bearing either kappa (κ) or lambda (λ) light chains (Fig. 19.3). The malignant plasma cells asynchronously synthesize more light chains than heavy chains, so that

Fig. 19.2 Radiographs of the skull and a long bone in a patient with multiple myeloma. Note the 'punched out' lesions in the bones, where the accumulation of malignant plasma cells has eroded the normal calcification. Courtesy of L. Shulman.

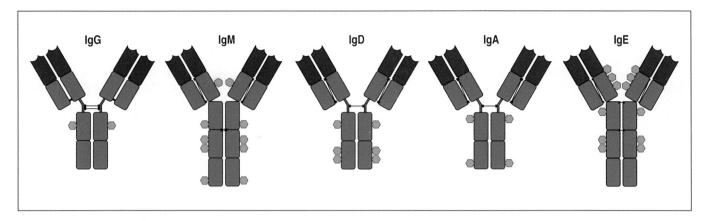

Fig. 19.3 The structural organization of the main human immunoglobulin isotype monomers. The choice of constant-region gene determines the class or isotype of the immunoglobulin made. Both IgM and IgE lack a hinge region but each contains an extra heavy-chain C domain. The isotypes also differ in the distribution of N-linked carbohydrate groups, as shown in turquoise, and in the distribution of disulfide bonds (black line).

immmunoglobulin light chains are excreted in the urine in excessive amounts. In 1846 Dr Charles McIntyre, a physician practicing medicine in London, made a house call on a greengrocer residing in Devonshire Street. Seeing this unfortunate man wasting away with fragile bone disease, it occurred to Dr McIntyre that he might be losing excessive amounts of protein ('animal matter') in his urine. He took a urine specimen back to his consulting rooms and found that, upon heating, a precipitate formed in the urine between 45 and 60°C, and redissolved upon further heating of the urine to boiling point. On the following day, he sent a specimen of this urine to Dr Henry Bence-Jones, Professor of Clinical Chemistry at Guy's Hospital, with a complete description of the bizarre characteristics of the abnormal protein that precipitated, and a question: 'What is this?'. The protein was thenceforth known as a Bence-Jones protein. It took over 100 years to answer the question but, eventually, Bence-Jones proteins were found to be immunoglobulin light chains.

The case of Isabel Archer: the consequences of unrestrained growth of an antibody-producing B-cell clone.

Isabel Archer was a 55-year-old housewife in 1989, when she began to experience excessive fatigue. She had been in good health her entire life. Her 57-year-old husband, a successful lawyer, was also in good health, as were her three sons, all in their 20s. At the time of a routine annual check-up at her physician she reported to him how easily she became fatigued. He found no abnormalities on physical examination.

A blood sample revealed that she had mild anemia; her red blood cell count was 3.5×10^6 μl^{-1} (normal $4.2-5.0 \times 10^6$ μl^{-1}). Her white blood cell count was 3600 μl^{-1} (normal 5000 μl^{-1}). The sedimentation rate of her red blood cells was 32 mm h^{-1} (normal <20 mm h^{-1}). Unclotted whole blood from Mrs Archer was put in a narrow-bore tube to determine how far the red blood cells would sediment in 1 hour.

Mrs Archer, 55 years, previously very healthy but now easily fatigued.

Fig. 19.4 Electrophoresis indicates whether serum immunoglobulins have monoclonal components. An electrophoretic pattern of normal serum run on an agarose gel (lane 1) is shown next to the pattern obtained with a serum sample from Mrs Archer (lane 2). The heterogeneous immunoglobulins from normal serum stained as a smear, whereas the monoclonal component of Mrs Archer's serum ran as a tight protein band. The electrophoresis was performed again with normal serum (lanes 3 + 5) and Mrs Archer's serum (lanes 4 + 6) and this reacted with an antibody to γ chains (lanes 3 + 4) and antibody to κ chains (lanes 5 + 6). The agarose gel was washed to remove all proteins except for antigen:antibody complexes. This shows that Mrs Archer's myeloma protein was IgGκ.

Elevated sedimentation rate, elevated IgG. IgG levels continue to rise.

Sedimentation of the red blood cells is caused by rouleaux formation, in which red blood cells stack on one another, and is hastened when the fibrinogen or IgG content of the blood plasma is elevated. This elevated sedimentation rate prompted the measurement of her serum immunoglobulins. The concentration of IgG was found to be 3790 mg dl^{-1} (normal 600–1500 mg dl^{-1}), that of IgA 14 mg dl^{-1} (normal 150–250 mg dl^{-1}) and that of IgM 53 mg dl^{-1} (normal 75–150 mg dl^{-1}). Electrophoresis of her serum revealed the presence of a monoclonal protein, which on further analysis was found to be IgG with kappa light chains (Fig. 19.4). Radiographs of all of her bones did not show any abnormality. No treatment was advised.

Mrs Archer returned for regular visits to her physician and on each occasion he measured her serum IgG level and noticed that it was gradually increasing. In April 1991 her serum IgG was 4520 mg dl^{-1}, and in January 1992 it was 5100 mg dl^{-1}. By November 1992, her anemia had worsened and her red blood cell count had fallen to 3.0×10^6 μl^{-1}. At the same time her white blood count had fallen to 2600 μl^{-1}.

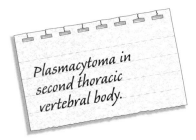

Plasmacytoma in second thoracic vertebral body.

In December 1992, Mrs Archer experienced the sudden onset of upper back pain. She was referred to a radiologist who performed a radiograph of the thoracic spine followed by a magnetic resonance imaging (MRI) scan. He reported to the internist that he found destruction of the second thoracic vertebral body with extrusion of a plasmacytoma (a tumor of plasma cells) from the affected vertebral body compressing the spinal cord. Mrs Archer was treated with the corticosteroid, decadron, and irradiation to her spine. Her symptoms improved. However, her serum IgG level reached 6312 mg dl^{-1} and she required blood transfusions because of her worsening anemia. She was treated with melphalan and prednisone.

IgG level keeps rising.

In April 1993, further chemotherapy was given because of the persisting elevation of her serum IgG. She was treated for 9 months with vincristine, adriamycin, and decadron and her serum IgG fell from 6785 mg dl^{-1} to 5308 mg dl^{-1}. When her serum IgG subsequently rose to 8200 mg dl^{-1} she was treated with a course of cyclophosphamide, etoposide, and decadron, which reduced her serum IgG level to 6000 mg dl^{-1}.

In February 1995, Mrs Archer developed high fever and chest pain. On chest radiograph she was found to have pneumonia of the lower lobe of the left lung. She was treated successfully with antibiotics. She again experienced high fever, shaking chills and chest pains in May 1995.

Because she was hypotensive (low blood pressure) she was admitted to the intensive care unit and given antibiotics intravenously and cardiac pressors to raise her blood pressure. *Streptococcus pyogenes* was cultured from her sputum and blood. She recovered from this episode in the hospital and remains fully active. She requires occasional blood transfusions for her anemia and complains at times of bone pain. Her serum IgG is stable at 6200 mg dl^{-1}.

Multiple myeloma.

Isabel Archer presents us with many, if not most, of the typical features of multiple myeloma, a malignant tumor of plasma cells. A single plasma cell has undergone malignant transformation; its progeny have disseminated to many sites in the bone marrow and are producing prodigious quantities of a monoclonal immunoglobulin.

Multiple myeloma is a very malignant disease that is resistant to most cancer chemotherapy. Methylphenylalanine mustard (melphalan), which Mrs Archer received, is one of the few chemotherapeutic agents that has been effective in the treatment of this disease. Although Mrs Archer was in relatively good health as our case history ended, her outlook for survival is very poor. Recently, bone marrow transplants have been used to cure patients with multiple myeloma.

Myeloma proteins have played an important part in the history of immunology. Subclasses of IgG were first recognized, for example, when a rabbit was immunized with a single human myeloma protein and found to react with 80% of myeloma proteins but not with the other 20%. This led to the conclusion that the 80% that did react belonged to an IgG subclass (IgG1) capable of generating subclass-specific antibodies in rabbits. Four subclasses of IgG were distinguished by immunizing rabbits with single myeloma proteins and testing the antibodies generated for cross-reactivity to other myeloma proteins. Korngold and Lipari had already used this approach to classify Bence-Jones proteins into two groups of proteins, subsequently called kappa and lambda light chains. Later on, a myeloma protein that was available in abundant amounts as a homogenous protein became the first immunoglobulin molecule for which a complete amino acid sequence was obtained.

Questions.

1 The serum IgG from Mrs Archer was assumed to be monoclonal because it migrated as a tight band on electrophoresis in an agarose gel, and because it reacted with antibodies to kappa but not to lambda chains. What other evidence could be brought to bear to prove the monoclonality of this IgG?

2 Mrs Archer became anemic (low red blood cell count) and neutropenic (low white blood cell count). What was the cause of this?

3 As her disease progressed, Isabel Archer became susceptible to pyogenic infections; for example, she had pneumonia twice in a short period. What is the basis of her susceptibilty to these infections?

4 You might conclude that it would be useful to administer gamma globulin intravenously to Mrs Archer to protect her from more pyogenic

infections. Why would this treatment be less successful than in the case of X-linked agammaglobulinemia?

5 A monoclonal immunoglobulin in the serum is called an M-component ('M' for myeloma). Is the presence of an M-component in serum diagnostic of multiple myeloma?

6 Very rarely an individual with multiple myeloma has two M-components in the blood. Although these two M-components derive from different constant-region genes, their antigen-binding regions are both encoded by the same variable-region gene. Can you hypothesize how this happens?

CASE 20 | T-Cell Lymphoma

A malignancy of functional T lymphocytes.

T-cell lymphomas arise when mature T cells or their precursors in the thymus gland undergo malignant transformation and clonal expansion. These tumors seem to represent arrested stages of normal differentiation, and express molecules on their surface that are normally present on T cells at various stages of maturation (Fig. 20.1). Occasionally, T-cell lymphomas even mimic the immune effector functions of their nontransformed normal counterparts.

T cells develop in the thymus gland, where precursors arriving from the bone marrow enter a phase of intense proliferation. As thymocytes proliferate and mature into T cells, they pass through a series of distinct phases (Fig. 20.2; see also Fig. 14.3). These are marked by changes in the status of T-cell receptor genes and the expression of T-cell receptor proteins on the T-cell surface, and by changes in the co-receptor molecules CD4 and CD8. Particular combinations of these cell-surface molecules can thus be used as markers for T cells at different stages of differentiation.

When progenitor T cells first enter the thymus from the bone marrow, their receptor genes are unrearranged and they lack most of the surface molecules characteristic of mature T cells. Interactions with the thymic stroma trigger rapid cell proliferation and expression of the first T-cell specific surface molecule, CD2. At the end of this phase, the immature thymocytes bear

Topics bearing on this case:
Development of T cells
T-cell tumors
Role of cytokines in isotype switching and leukocyte production

Fig. 20.1 T-cell tumors represent monoclonal outgrowths of normal cell populations. Each distinct T-cell tumor has a normal equivalent, as also seen with B-cell tumors, and retains many of the properties of the cell from which it develops. However, unlike B-cell tumors, tumors of T cells lack the intermediates in the T-cell developmental pathway. Some of these tumors represent massive outgrowth of a rare cell type. For example, acute lymphoblastic leukemia is derived from the lymphoid progenitor cell. One T-cell related tumor is also included. Thymomas derive from thymic stromal or epithelial cells. Some characteristic cell-surface markers for each stage are also shown. For example, CD10 (acute lympho-blastic leukemia antigen or CALLA) is a widely used marker for acute lymphoblastic leukemia. Note that T-cell chronic lympho-cytic leukemia (CLL) cells express CD8, whereas the other T-cell tumors mentioned express CD4. Adult T-cell leukemia is caused by the retrovirus HTLV-1.

distinctive markers of the T-cell developmental lineage, such as CD25, but do not express any of the three cell-surface markers that define mature T cells, namely the T-cell receptor complex (detected as CD3), CD4, or CD8. Because they lack CD4 and CD8, such cells are called double-negative thymocytes (CD4⁻CD8⁻).

Some double-negative cells (representing about 20% of all double-negative cells in the thymus) express the genes encoding the rare γ:δ receptor; these cells represent a separate developmental lineage of CD4⁻ CD8⁻ thymocytes. A second double-negative population expresses the pre-T receptor; these cells mature into double-positive (CD4⁺CD8⁺) thymocytes. Once the mature T-cell receptor α chain is expressed, these rapidly proliferating cells phosphorylate RAG-2 protein and degrade it; consequently there is no further gene arrangement.

Small double-positive thymocytes express only low levels of the T-cell receptor, and most (more than 95%) are destined to die. These cells express receptors that cannot recognize self MHC and thus fail positive selection in the thymus. Double-positive cells that do recognize self MHC, however, mature to express high levels of T-cell receptor and subsequently cease to express one or other of the two co-receptor molecules, becoming either CD4 or CD8 single-positive thymocytes. Thymocytes also undergo negative selection during the double-positive stage. Those that survive this dual screening mature into single-positive T cells that are rapidly exported from the thymus to join the peripheral T-cell repertoire (see Figs 20.2 and 14.3).

This case concerns a child with a malignant monoclonal tumor of double-positive immature T cells with abnormal function.

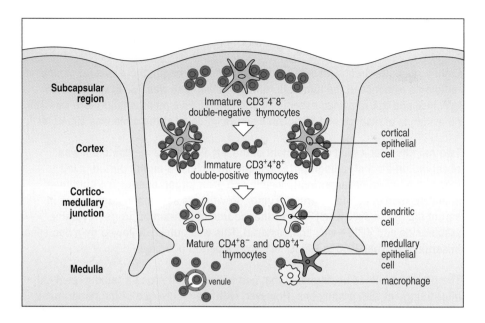

Fig. 20.2 Thymocytes go through several distinct developmental stages, which can be distinguished by the presence or absence of certain cell-surface proteins. The earliest cells to enter the thymus are double-negative thymocytes that lack the T-cell receptor (detected as CD3), and the co-receptors CD4 and CD8. As these cells proliferate and mature into double-positive thymocytes, bearing CD3 and both CD4 and CD8, they move deeper into the thymic cortex. Finally, the medulla contains only mature single-positive T cells, which eventually leave the thymus and enter the bloodstream.

The case of Elizabeth Bennet: the consequence of unrestrained growth of a cytokine-producing T-cell clone.

Elizabeth Bennet was a healthy 8-year-old when she developed chest pain, a cough, and shortness of breath. After 3 months of these symptoms, her parents sought advice from their family doctor.

On physical examination Elizabeth was found to be thin (weighing 28 kg) and 129 cm tall. She had a bulging left chest, no sound of breath over the left chest, and decreased breath sounds over the right chest. Cervical, axillary, and left inguinal lymph nodes were moderately enlarged. The liver and spleen were enlarged; their edges were each 5 cm below the costal margin.

8-year-old female; possible obstruction in left chest; X-ray.

Her chest radiograph revealed a completely opaque left hemithorax and a mediastinal shift to the right. A computed axial tomography scan (CAT scan) showed a solid tumor filling the left hemithorax. Pleural fluid obtained from the left chest contained abundant eosinophils and few lymphocytes but no malignant cells. A biopsy of the tumor showed that 98% of the tumor cells expressed CD3, CD4, and CD8; this suggested that the tumor was derived from the double-positive CD4+CD8+ population of thymocytes.

T-cell tumor with accompanying eosinophilia.

On admission to the Children's Hospital, Elizabeth's leukocyte count was 11,400 cells ml^{-1} (slightly elevated compared with normal), with 67% eosinophils, 31% neutrophils, 7% lymphocytes, and 5% monocytes. Bone marrow aspiration revealed a hypercellular marrow with increased eosinophil precursors. No malignant cells were observed in peripheral blood or bone marrow smears.

Because of the blood and bone marrow eosinophilia, serum IgE level was measured and was found to be >10,000 IU ml^{-1}. There was no personal or family history of allergic disease. A battery of 24 hypersensitivity skin tests for inhalant allergens was negative, as were repeated stool examinations for parasites and their eggs.

Anti-tumor treatment; IgE levels decreased to normal.

The IgE was polyclonal in origin. IgG was 1388 mg dl^{-1} (normal 568–1100), IgA was 123 mg dl^{-1} (normal 57–414), and IgM was 84 mg dl^{-1} (normal 120–274).

Eight weeks after treatment with cytotoxic drugs was begun there was no clinical or radiologic evidence of the tumor. Serum IgE at that time was 767 IU ml^{-1}. Serum IgM, IgG, and IgA remained essentially unchanged. Five months later there was still no evidence of disease and Elizabeth's serum IgE level at that time was 200 IU ml^{-1}.

Two months later a mild enlargement of the left side of the mediastinum was observed on a chest radiograph. Subsequently, over a 3-month period, a left mediastinal mass became clearly visible and grew larger, despite the continued administration of cytotoxic drugs. Tumor growth was accompanied by a rise in serum IgE (up to 3900 IU ml^{-1}). A further course of anti-tumor drugs—cytosine arabinoside and VP16—was administered. This was quickly followed by a decrease in serum IgE, from 3900 to 746 IU ml^{-1} over a period of 8 days.

The peripheral blood eosinophilia that had initially been present rapidly resolved after treatment with prednisone. However, with the recurrence of the tumor mass the blood eosinophil count rose rapidly. Although the neutrophil count rose 2.5-fold at that time, the eosinophil count rose concomitantly from 100 to 6800 ml^{-1}. Supernatants from cultures of the tumor cells contained interleukin (IL)-4 and induced IgE synthesis when added to cultures of normal peripheral blood mononuclear cells. IgE synthesis was blocked by anti-IL-4 neutralizing monoclonal antibody. The same supernatants induced the differentiation of cord blood mononuclear cells into eosinophils after 3 weeks in culture. Differentiation was blocked by anti-IL-5 neutralizing monoclonal antibody.

T-cell lymphoma.

The high division rate of maturing lymphocytes together with the activity of DNA recombination mechanisms responsible for generating antigen receptor diversity provide a fertile ground for malignant transformation and the generation of T-cell lymphomas. Even terminally differentiated T cells retain the capacity to proliferate in response to appropriate stimuli and therefore retain potentially lifelong capability for neoplasia. Once transformed, these T cells or their developmentally arrested precursors frequently retain the pattern of surface antigen expression and function of nontransformed lymphocytes at the same developmental stage.

Neoplastic transformation at various stages of T-cell development gives rise to leukemias and lymphomas with distinct clinical characteristics. Common acute lymphoblastic leukemias (C-ALL or B-ALL) are derived from early lymphoid precursors and do not express CD3, CD4, CD8, or any of the rearranged T-cell receptor genes. T-cell acute lymphoblastic leukemia (T-ALL) and lymphoblastic lymphoma can result from the transformation of any of the stages of T-cell differentiation including double-negative (CD4$^-$ CD8$^-$) and double-positive (CD4$^+$CD8$^+$) precursors or mature single-positive (CD4$^+$or CD8$^+$) T cells. Neoplasms of mature CD4 helper T cells include large-cell lymphoma, adult T-cell leukemia/lymphoma (induced by the retrovirus HTLV-1), cutaneous T-cell lymphoma or Sézary syndrome, and hairy-cell leukemia. Transformation of the mature CD8 single-positive phenotype is observed in some large-cell lymphomas.

A wide variety of immunologic abnormalities can be observed in patients with lymphoma and these reflect the functional effects of the transformed cells. Some T-cell lymphomas retain function: in particular, the cells of cutaneous T-cell lymphomas (Sézary syndrome) can display helper activity and can induce normal B cells to undergo isotype switching and produce immunoglobulin. Unlike normal T cells, however, the transformed Sézary syndrome lymphocytes perform these helper functions without expressing interleukin (IL)-2 receptors and do not require IL-2 for their ongoing proliferation.

Another setting in which the transformation of mature CD4$^+$ helper T cells occurs is in diseases associated with human T-cell leukemia/lymphoma virus (HTLV). These malignancies, also referred to as adult T-cell leukemia (ATL), are particularly prevalent in parts of Japan. They are induced by the HTLV-1 retrovirus; the cells express CD3 and CD4 antigens but not CD8, consistent with a mature helper T cell phenotype. Unlike other T-cell malignancies, these HTLV-1-transformed lymphocytes express high levels of IL-2 receptors and require IL-2 for their proliferation and function.

The lymphoma described in Elizabeth represents a case of transformation of cells with some immunological function. Its pattern of surface antigen expression—CD3$^+$CD4$^+$ and CD8$^+$—is reminiscent of the double-positive cells that constitute most of the thymus and distinguishes this lymphoma from the Sézary syndrome and HTLV-1 neoplasms discussed above. The double-positive thymocytes are in the process of undergoing selection on the basis of their T-cell receptor specificity before maturing to a single-positive CD4$^+$ or CD8$^+$ phenotype and being exported to the peripheral immune organs. Typically, such cells are not found in the periphery and are not known to secrete cytokines or carry out immune effector functions. This double-positive lymphoma was clearly shown to secrete both IL-4 and IL-5, properties usually associated with CD4$^+$ T cells differentiated to the T_H2 phenotype.

Questions.

1 Elizabeth's tumor cells secreted IL-4 and IL-5. What other cytokines could be secreted by the tumor that would also cause eosinophilia and elevation of serum IgE levels?

2 Elizabeth's serum IgG was elevated. How do you explain this?

3 What definitive test will establish that Elizabeth's tumor represents a monoclonal expansion of a T-cell clone?

4 Administration of corticosteroids (e.g. prednisone) results in a rapid decrease in levels of circulating and tissue eosinophils (which is also commonly seen in allergic conditions after corticosteroid treatment). Propose a mechanism for this effect.

5 Besides CD4$^+$ T_H2 cells, what other thymus-derived population of cells represents a potential source of IL-4?

CASE 21

Interferon-γ Receptor Deficiency

The destruction of intracellular microorganisms in macrophages.

Certain pathogens such as mycobacteria, *Listeria*, *Leishmania*, and *Salmonella* take up residence in macrophages and are thereby protected from elimination by antibody or cytotoxic T cells. These microorganisms can be eliminated only when their host macrophages are activated and produce increased amounts of nitric oxide, oxygen radicals, and other microbicidal molecules (Fig. 21.1). The activation of macrophages in an adaptive immune response is masterminded by T_H1 cells (Fig. 21.2); the most important cytokine involved in macrophage activation is interferon (IFN)-γ.

Topics bearing on this case:
Macrophage activation
Microbicidal action of phagocytes
Cytokine receptor signaling pathways

Activated macrophage

Fig. 21.2 T$_H$1 cells activate macrophages to become highly microbicidal. When a T$_H$1 cell specific for a bacterial peptide contacts an infected macrophage, the T cell is induced to secrete the macrophage-activating factor IFN-γ by IL-12 and to express CD40 ligand. Together, these newly synthesized T$_H$1 proteins activate the macrophage.

2½-year-old female child with enlarged lymph nodes. Order blood test.

Fig. 21.1 Activated macrophages undergo changes that greatly increase their antimicrobial effectiveness and amplify the immune response. Once the macrophage is activated, lysosomes fuse with the intracellular vesicles within which the pathogenic bacteria (red) reside, which exposes the microorganisms to degradative enzymes and other microbicidal agents. Activated macrophages also increase their expression of receptors for tumor necrosis factor (TNF), and secrete TNF-α. This autocrine stimulus synergizes with interferon (IFN)-γ secreted by T$_H$1 cells to increase the antimicrobial action of the macrophage, in particular by inducing the production of nitric oxide (NO) and oxygen radicals (O$_2^·$). The macrophage also increases the expression of CD40, which by interaction with the CD40 ligand on T cells upregulates the expression of B7 proteins and increases the expression of class II MHC molecules on the macrophage, thus allowing the further activation of resting CD4 T cells.

When macrophages take up microorganisms by phagocytosis, they secrete interleukin (IL)-12. This cytokine is necessary for the induction of IFN-γ synthesis by T cells and natural killer (NK) cells. IL-12 furthermore favors the maturation of T$_H$1 cells, which activate macrophages, and suppresses the maturation of T$_H$2 cells, which secrete a cytokine, IL-10, involved in the deactivation of macrophages.

IFN-γ acts at a receptor on the macrophage surface. This receptor is composed of two different types of polypeptide chain—IFN-γ receptor 1 and IFN-γ receptor 2, each of which is associated with a tyrosine kinase—Jak1 and Jak2 respectively. When a dimer of IFN-γ binds to two molecules of the IFN-γ receptor 1, it causes them to associate with two IFN-γ receptor 2 chains (Fig. 21.3) and this initiates a signaling pathway that eventually results in changes in gene transcription.

As we shall see in this case, a mutation in the gene encoding the IFN-γ receptor 1 chain has dramatic effects on the capacity to fight certain pathogens.

The case of Clarissa Dalloway: a relentless infection due to mycobacteria.

Clarissa Dalloway was the first child born to a couple who lived in an isolated fishing village on the coast of Maine. Her parents thought that they were distantly related to each other. The fishermen of this village were all descended from English settlers who came there in the late 17th century and there was much intermarriage in the community.

Clarissa was well at birth and developed normally until she was around 2½ years old. Her mother then noticed that she was not eating well, had diarrhea, and was losing weight. She took Clarissa to a pediatrician in the nearest town of Bath, Maine. The pediatrician, Dr Woolf, noted enlarged lymph nodes and ordered an ultrasound and CT scan of the chest and abdomen. These showed enlarged lymph nodes in the mesentery and para-aortic region and Dr Woolf referred Clarissa to the Children's Hospital in Boston.

Blood tests revealed a white blood cell count of 9400 μl^{-1}, of which 55% were neutrophils, 30% lymphocytes, and 15% monocytes (slightly elevated). Her serum IgG was 1750 mg dl^{-1}, IgA 450 mg dl^{-1}, and IgM 175 mg dl^{-1} (these immunoglobulin values are all elevated).

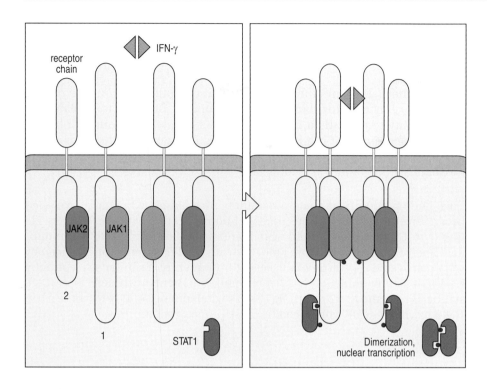

Fig. 21.3 Diagram of the IFN-γ receptor and its activation of the JAK–STAT signaling pathway. A dimer of IFN-γ causes clustering of the four chains of the IFN-γ receptor. The receptor 1 chains become phosphorylated by JAK1, forming a site to which the transcription factor STAT1α can bind. This in turn becomes phosphorylated, dimerizes, and is then transported to the nucleus, where it upregulates the various genes that produce the activated macrophage phenotype. The IFN-γ receptor 1 chain is encoded on chromosome 6, whereas the receptor 2 chain is encoded on chromosome 21.

It was decided to biopsy the enlarged lymph nodes. On histological examination they exhibited marked proliferation of histiocytes, and many neutrophils were seen in the lymph node. There was no granuloma formation and no giant cells were seen. An acid-fast stain for mycobacteria (see, for example, Fig. 30.4) revealed numerous microorganisms, and *Mycobacterium avium intracellulare* was cultured from the lymph nodes as well as from the blood.

Despite appropriate antibiotic treatment for the mycobacterial infection, Clarissa eventually developed infiltrates in the lungs and progressive enlargement of the spleen. At 6 years old she developed sepsis and *Salmonella paratyphi* was cultured from her blood. She was successfully treated with antibiotics for this infection but soon after she developed meningitis and died. *M. avium intracellulare* was cultured from the cerebrospinal fluid.

A detailed family history revealed that Clarissa had three male cousins, two of whom were brothers, who had died of mycobacterial infections. In one case *M. chelonei* had been cultured from lymph nodes; *M. fortuitum* and *M. avium intracellulare* had been cultured from blood and lymph nodes of the two brothers.

Do lymph node biopsy.

Interferon-γ receptor deficiency.

It is estimated by the World Health Organization that 1.7 billion human beings currently alive have been infected with mycobacteria. Yet only a very small fraction of these individuals develop disease due to this infection. The AIDS epidemic has dramatically increased the incidence of mycobacterial disease due to *M. tuberculosis* and to other mycobacterial species such as *M. avium intracellulare*, which are collectively called 'atypical mycobacteria.' In general, atypical mycobacteria do not cause disease in immunologically normal human beings, except for swollen lymph nodes in which these mycobacteria survive. Infection with atypical mycobacteria causes a positive tuberculin reaction to develop.

Although there are many ethnic differences in susceptibility to *M. tuberculosis*, no single genetic factor had ever been found in humans to explain this susceptibility until Clarissa Dalloway and her family were found to have a genetic defect in the IFN-γ receptor 1 gene. The genetic defect was ascertained in this family by mapping the susceptibility gene to chromosome 6q22, the map location of the receptor gene.

This finding prompted the examination of children who had developed progressive mycobacterial infection after immunization with the BCG (Bacille Calmette-Guérin) vaccine, and who did not have severe combined immunodeficiency (see Case 14). More kindreds with the IFN-γ receptor 1 deficiency were discovered as a result of this lead. Subsequently, a similar susceptibility to mycobacterial infection has been found in patients with defects in IL-12 synthesis or in the IL-12 receptor. Thus the dependence of IFN-γ synthesis on IL-12 has been neatly confirmed by these human mutations.

Questions.

1 How do you explain the alarming increase in the incidence of mycobacterial infections in patients with AIDS?

2 Clarissa Dalloway and her cousins had positive tuberculin skin tests. Would you have predicted this?

3 Clarissa and her cousins also did not develop granulomas as a result of infection with atypical mycobacteria. On one occasion mycobacteria were surprisingly cultured from Clarissa's blood. What do these observations tell us?

4 Clarissa also had a problem with salmonella, but she had no problem with pneumococcal infection or with any viruses, such as chickenpox. How do you explain this?

CASE 22 Wiskott–Aldrich Syndrome

Role of the actin cytoskeleton in T-cell function.

Many functions of T cells require the directed reorganization of the cell's cytoskeleton, in particular the actin cytoskeleton. The eukaryotic cell cytoskeleton as a whole consists of actin filaments, microtubules, and intermediate filaments. It provides a framework for the internal structural organization of the cell and is also essential for cell movement, cell division, and many other cell functions. In T cells, as in other animal cells, the actin cytoskeleton is found mainly as a meshwork of actin filaments immediately underlying the plasma membrane (Fig. 22.1). The actin cytoskeleton is a dynamic structure and can undergo rapid reorganization by depolymerization and repolymerization of actin filaments. As we shall see in the case of Wiskott–Aldrich syndrome, an inability of T cells to reorganize their actin cytoskeleton when required has profound effects on their function and thus on immune function as a whole.

Topics bearing on this case:
Methods for measuring T-cell function
Gene knockouts
T-cell help in antibody responses
Isotype switching
Interaction of cytotoxic T cells with their targets

Fig. 22.1 An activated cytotoxic T cell moving over a surface. The actin cytoskeleton is stained green. The red staining indicates the lytic granules containing cytotoxic proteins. Photograph courtesy of Gillian Griffiths.

Reorganization within the T-cell cortical actin cytoskeleton takes place when T cells interact with B cells or other target cells through cell-surface receptors. The functions of T cells in immune defense all involve interactions with other cells that are initiated by direct cell–cell contact via cell-surface receptors. For example, helper T cells interact with B cells through cell-surface receptors to stimulate B-cell proliferation and the subsequent differentiation into antibody-producing plasma cells. T-cell–B-cell interactions are also involved in isotype switching and generation of memory cells, whereas cytotoxic T-cell killing of virus-infected target cells also involves direct contact with the target cell.

The cytoskeleton is linked to cell-surface receptors in the plasma membrane so that events occurring at the membrane can affect cytoskeleton reorganization. For example, cross-linking of T-cell antigen receptors and co-receptors by antigen:MHC complexes leads to their aggregation at one pole of the T cell, with an accompanying concentration of the actin cytoskeleton at that point also. Binding of a helper T cell to a B cell through its T-cell receptors' binding to antigen:MHC complexes on the B-cell surface also leads to a reorganization of the actin cytoskeleton locally in the zone of contact, which in turn causes a microtubule-dependent mechanism to focus the secretory apparatus of the T cell on the point of contact with the B cell (Fig. 22.2); the release of cytokines from the T cell is thus directed to the contact point. Similar cytoskeletal reorganizations occur when a cytotoxic T cell contacts its target cell.

Many other T-cell functions depend on the actin cytoskeleton. Like many other animal cells, T lymphocytes move in a crawling 'ameboid' fashion. The movement of T lymphocytes as they emigrate from the thymus into the blood

Fig. 22.2 Binding of a helper T cell to an antigen-binding B cell causes a reorganization of the cytoskeleton in the T cell. Engagement of the T-cell receptor causes the T cell to express the CD40 ligand (CD40L), which binds to CD40 on the surface of the B cell (top panel). The cross-linking of the receptors at the point of contact leads to a reorganization of the cortical actin cytoskeleton, shown here by the redistribution of the protein talin (shown in red in top left and both center panels), which is associated with the actin cytoskeleton. Subsequent reorientation of the secretory apparatus of the T cell leads to cytokine release at the point of contact with the antigen-binding B cell, as shown by staining for IL-4 (bottom right panel), which shows the IL-4 (green) confined to the space between the B cell and the T cell. Photographs courtesy of A. Kupfer.

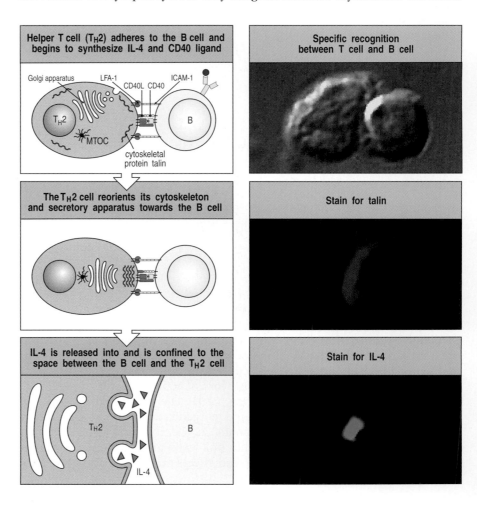

vessels and subsequently 'home' from the bloodstream into lymphoid tissue requires the active participation of the actin cytoskeleton. Cell division induced by the activation of T cells by antigen or by nonspecific mitogens also involves the actin cytoskeleton in that the cell is divided into two by the action of a contractile ring formed of actin filaments and myosin.

T cells from patients with Wiskott–Aldrich syndrome are deficient in all these normal cellular abilities and, in particular, seem unable to interact successfully with B cells and other target cells.

The case of Austin Sloper: the consequences of a failure of T-cell–B-cell interaction.

Austin Sloper was first referred to the Children's Hospital at 2 years of age with a history of recurrent infections, eczema, asthma, and episodes of bloody diarrhea. He had been a full-term baby and appeared quite normal at birth. Routine immunizations of DPT (diphtheria, pertussis, and tetanus) and oral polio vaccine had been given at 3, 4, and 5 months of age without any untoward consequences. At 6 months his mother noticed eczema developing on his arms and legs; this was treated with 1% cortisone ointment. The eczema became infected with *Staphylococcus epidermitidis*, and petechiae (small skin hemorrhages) appeared in the eczematous areas as well as on unaffected areas of skin. By 2 years old he began to have frequent middle ear and respiratory infections, including pneumonia as confirmed by a chest radiograph. Between respiratory infections he started to wheeze and was found to have asthma.

Normally developed male infant; recurrent infections, eczema, asthma.

A blood analysis showed normal levels of hemoglobin of 11.5 g dl^{-1} and a normal white blood cell count of 6750 μl^{-1}. The proportions of leukocytes of different types (for example, granulocytes and lymphocytes) were also normal. The platelet count was, however, low (thrombocytopenia), at 40,000 μl^{-1} (normal 150,000–350,000), and the platelets were abnormally small (Fig. 22.3). Because of the combination of repeated infections, eczema, and thrombocytopenia with small platelets, the Wiskott–Aldrich syndrome (WAS) was diagnosed.

Further immunological investigations at this time revealed levels of IgG of 750 mg dl^{-1} (normal), IgM 25 mg dl^{-1} (decreased), IgA 475 mg dl^{-1} (increased), and IgE 750 IU ml^{-1} (increased). Austin's red blood cells were type O but no anti-A or anti-B isohemagglutinins were present in his serum. He made no immune response to immunizations with a pneumococcal vaccine (composed of capsular polysaccharides) or to polyribose phosphate (PRP) vaccine (composed of the cell-wall carbohydrate of *Haemophilus influenzae*). His titer of antibodies against tetanus toxid (a protein) was 0.1 unit ml^{-1} (a borderline normal result). He was given

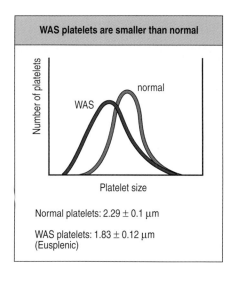

WAS platelets are smaller than normal

y-axis: Number of platelets
x-axis: Platelet size
Labels: WAS, normal

Normal platelets: 2.29 ± 0.1 μm

WAS platelets: 1.83 ± 0.12 μm
(Eusplenic)

Fig. 22.3 Platelet sizing. Blood was drawn from a normal subject and from a patient with WAS into sodium citrate, an anticoagulant. The blood was then centrifuged and allowed to sediment at 1*g* and the platelet-rich plasma was removed. This was diluted in sterile buffer and scanned in a particle sizer. The mean diameter of normal platelets is 2.29 ± 0.1 μm. The mean diameter of platelets from patients with the Wiskott–Aldrich syndrome is 1.83 ± 0.12 μm. Data kindly supplied by Dianne Kenney.

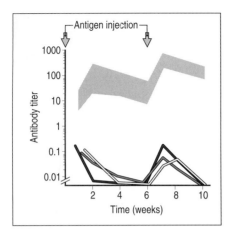

Fig. 22.4 Impaired primary and secondary immune responses in Wiskott–Aldrich syndrome. The figure shows the response of three patients (red, blue, and yellow lines) with Wiskott–Aldrich syndrome to the antigen bacteriophage φX174 given intravenously. Their titer of anti-bacteriophage antibodies was measured at 2, 4, and 6 weeks after the injection. At 6 weeks, a second dose of bacteriophage was administered. The antibody titer was then measured at 1, 2, and 4 weeks after the second injection. The gray shaded area represents the 66% confidence limits of the response of normal children to the same antigen. Data kindly supplied by Hans Ochs.

Thrombocytopenia with small platelets; Wiskott–Aldrich?

Poor immune responses, esp. to carbohydrate antigens; test T-cell function.

Problems with blood clotting continue; splenectomy advisable.

two intravenous doses of the bacteriophage φX174 as a test antigen 6 weeks apart. A primary response was elicited by the first injection but the secondary response was poor (Fig. 22.4).

He had normal numbers of circulating B lymphocytes (11% of peripheral blood lymphocytes) and T lymphocytes (85%) with a normal distribution of CD4+ and CD8+ T cells. However, when his T cells were stimulated with the mitogens concanavalin A and phytohemagglutinin a poor proliferative response was obtained. Nor did his T cells respond to the mitogenic effect of anti-CD3 antibodies.

As he grew older, Austin developed frequent severe nosebleeds. At these times his platelet count was found to have dropped to 10,000 μl^{-1} and he required platelet transfusions to stop the nosebleeds. He frequently developed purpuric lesions (black and blue spots due to hemorrhaging) on the skin.

At 14 years of age his spleen was removed (splenectomy), after which his platelet count immediately rose to 253,000 μl^{-1} and has remained in the normal range, although the platelets continue to be of smaller than normal size (see legend to Fig. 22.3). His absolute T-cell count had progressively decreased with age from 3500 μl^{-1} to 1250 μl^{-1}. He was started on intravenous gamma globulin at a dose of 30 g every 3 weeks (at the time he weighed 60 kg) and was advised to take penicillin before visits to the dentist or during respiratory infections.

Further allergies developed, particularly to nuts. His asthma persisted and he was advised to take theophylline 300 mg twice daily. At age 16 years he developed severe chickenpox with systemic symptoms of varicella hepatitis, which was successfully treated with Acyclovir. As he grew older and continued to be treated with gamma globulin the frequency of infections decreased. Austin is now in his his mid-30s, works full time as a teacher and is in reasonably good health.

After Austin, his mother had three more male children: Jonathan, Thornton, and Matthew. Jonathan and Matthew both had Wiskott–Aldrich syndrome; Thornton was normal. Mrs Sloper had five brothers, all of whom are normal; she had no sisters and no daughters. At age 32 years Jonathan Sloper developed a B-cell lymphoma. Although this was successfully treated with cytotoxic drugs he died of sepsis (bloodstream infection) due to a Gram-negative enterobacillus. Matthew Sloper developed glomerulonephritis due to deposition of immune complexes in his kidney during his 20s and at age 34 years he received a successful kidney transplant from a cadaver donor. His renal function continues to be normal 3 years after the transplant, and he works full time as a business executive. He has not had a splenectomy and his platelet count remains around 45,000 μl^{-1}. He has no problems with abnormal bleeding.

Wiskott–Aldrich syndrome.

Austin Sloper and his two affected brothers illustrate between them all the principal features of Wiskott–Aldrich syndrome (WAS). The syndrome was first described by Wiskott in 1937 in Munich when he observed three male infants with bloody diarrhea, eczema, and thrombocytopenia with small platelets. In 1947 Aldrich and his colleagues studied a large Dutch-American family in Minnesota and established the X-linked inheritance of the disease in five generations of affected males. The immunodeficiency that accompanies WAS was not appreciated until 1968, when it was described by Blaese and Waldmann and by Cooper and Good.

The defective gene responsible for WAS was mapped to the short arm of the X chromosome, and the normal gene has now been cloned. This encodes a protein named the Wiskott–Aldrich syndrome protein or WASP, which was found to have homology with actin-binding cytoskeletal proteins involved in the reorganization of the actin cytoskeleton in white blood cells and platelets. WASP is expressed only in white blood cells and megakaryocytes (from which platelets are derived), which explains the restriction of its effects to immune system and blood-clotting functions. In patients with WAS, T cells and platelets are defective in number and function. T-cell movement, capacity for cell division, capping of antigen receptors, and reorientation of the cytoskeleton on engagement with other cells are all impaired.

Abnormally small platelets are a prerequisite for the diagnosis of WAS. The destruction of blood platelets by the spleen is greatly increased, yet it is not known what the spleen recognizes as abnormal on WAS platelets. The platelets in the circulation become spontaneously activated and they extrude their granules. For this reason they appear to be small and in fact prove to be so when their volume or diameter is measured in a particle sizer (see Fig. 22.3). Splenectomy rapidly alleviates this problem, as we saw in Austin's case. The platelet count rises and the platelet size increases. T cells are also morphologically abnormal in WAS patients as they lose their surface microvilli and assume a characteristically bald appearance (Fig. 22.5).

Fig. 22.5 Loss of microvilli on T cells in Wiskott–Aldrich syndrome. Scanning electron micrographs of normal lymphocytes (panel a) and lymphocytes from a patient with Wiskott–Aldrich syndrome (panel b). Note that the normal lymphocyte surface is covered with abundant microvilli, which are sparse or absent from the patient's lymphocytes. Photographs courtesy of Dianne Kenney.

Patients with WAS have increased susceptibility to both pyogenic bacterial infections and opportunistic infections. Among the latter, severe chickenpox (varicella), herpes simplex and molluscum contagiosum (a viral infection of the skin, particularly skin involved with eczema) are frequently encountered. The increased susceptibility to such viral infections may be at least partly due to the impaired cytotoxic function of CD8 T cells and NK cells in WAS; the impairment seems to be in their inability to attach to target cells. Antibody formation, particularly to carbohydrate antigens, is defective, as we saw for Austin, who could not respond to pneumococcal polysaccharides and the capsular polysaccharide of *H. influenzae*. He also did not make antibodies against the blood group antigens, which are complex carbohydrates.

The defects in antibody production, including the impaired secondary response, reflect a general failure of T-cell–B-cell interactions. The failure to produce antibodies against capsular polysaccharides and blood group antigens in WAS is of particular interest as it confirms that, in humans, the production of antibodies against complex linear polysaccharides is not independent of T-cell help. In humans these anti-polysaccharide antibodies are normally of the IgG2 subclass; they fix complement and protect against pyogenic infections, which are often due to capsulated bacteria.

The importance of anti-polysaccharide antibodies was highlighted by the problems with infection encountered in the case of X-linked agammaglobulinemia (see Case 10), whereas the importance of being able to switch to IgG isotypes is illustrated by the case of hyper IgM syndrome (see Case 11). The fact that the nonproduction of polysaccharide antibodies in WAS patients is due to T-cell defects was confirmed by early attempts at bone marrow transplantation to correct WAS. A few patients were converted into partial T-cell chimeras, which was sufficient to clear their eczema and enable them to make antibodies against carbohydrate antigens. Subsequently, complete success at replacing the T-cell population has been achieved by bone marrow transplantation after irradiation to wipe out the abnormal cells in the host.

Although Austin's T-cell number seemed normal when tested at 2 years old, soon thereafter the total T-cell number in affected males begins to decline. As this happens, the B cells undergo a general polyclonal expansion. Eventually this B-cell expansion can become monoclonal and Epstein–Barr virus associated B-cell lymphocytes appear, as happened in one of Austin's affected brothers. As affected males are now living longer because of improved therapy (WAS used to be fatal in the first decade of life), more are developing B-cell lymphomas in their 30s and 40s. These progressions are reminiscent of what is observed in patients with AIDS.

The structure and function of WASP have been studied intensively during the past few years (Fig. 22.6). The carboxy-terminal portion of the molecule, the so-called VCA domain, binds to the actin-related protein complex (Arp2/3) and this binding initiates the nucleation of actin. Arp2/3 guides the branching of the actin filaments as they polymerize to make new filaments. WASP resides in the cytoplasm in an inactive form because the carboxy-terminal acidic domain (A) is bound to the basic domain so that inactive WASP appears as a hairpin. When the small G protein Cdc42 binds to the G-protein-binding domain (GBD), WASP springs open into a linear activated form and allows the VCA domain to bind to the Arp 2/3 complex.

Mice with 'knockouts' of the gene encoding WASP have defective homing of lymphocytes and defective capping of the T-cell antigen receptor (Fig. 22.7),

Functional domains of WASP

WIP   PIP$_2$ Cdc42   SH3 (Grb2, Nck, Btk)   Arp2/3

WH1   Basic   GBD   Proline-rich   V   C   A

Fig. 22.6 A stick diagram of WASP (501 amino acids) shows the various domains of the molecule. The amino-terminal domain (WH1) binds WIP (the WASP interactive protein, which keeps WASP inactive until it is released from WASP binding). It is followed by a stretch of basic amino acids which bind the carboxy-terminal acid residues (A) and this also keeps WASP in an inactive state. Phosphatidylinositol bisphosphate (PIP$_2$) also binds to this region and moves WASP from the cytoplasm to the lipid rafts of the immunologic synapse in the cell membrane. Then there is a domain that binds the G protein Cdc42, followed by a proline-rich region, which is an SH3 domain that binds many adaptor proteins. Finally, the carboxy-terminal region is composed of verprolin (V)-like, cofilin (C)-like and acidic (A) regions. Verprolin and cofilin are actin-binding proteins.

as well as defective responses of T cells to mitogens and anti-CD3 antibodies. B-cell function seems to be normal. These mice may serve as an animal model for gene therapy for this disease. The high frequency of immune-complex disease and allergy in WAS remains to be explained. It may relate to defective function of regulatory T cells. So does the persistent finding of low IgM and high IgA and IgE, as though these patients make excessive amounts of IL-4. Explanations for these findings are not yet obvious.

Fig. 22.7 Failure of T-cell capping in WASP knockout mice. T cells were obtained from mice in which the WASP gene had been knocked out by homologous recombination (WASP$^{-/-}$), and from normal mice of the same strain. The T cells were treated with anti-CD3ε antibody, which binds to the T-cell receptor:CD3 complex, and then examined with a fluorescent antibody to CD3. Normal resting T cells (panel a) show capping of the T-cell receptor after anti-CD3ε treatment (panel b). Panel c shows resting T cells from WASP$^{-/-}$ mice, and the failure of cap formation (panel d) after stimulation of the T-cell receptor. Photographs courtesy of Scott Snapper.

Questions.

1   All five of Mrs Sloper's brothers were normal. What does this suggest about the source of Mrs Sloper's mutation?

2   When Mrs Sloper and other female heterozygous carriers of WAS are examined for randomness of X-chromosome inactivation, nonrandom inactivation of the WAS-bearing X chromosome is found in all the blood cell lineages—monocytes, eosinophils, basophils, neutrophils, B lymphocytes, and CD4+ and CD8+ T lymphocytes. When hematopoietic stem cells are isolated from these women they also exhibit nonrandom X inactivation. How might this be explained?

3   Human infants do not respond to polysaccharide antigens until they are 18 months old. An examination of the B-cell repertoire for receptors specific for polysaccharide antigens reveals that none can be detected until 18 months of age. Patients with WAS also do not respond to polysaccharide antigens. What do you think you would find if you examined their B-cell receptor repertoire?

4   Can you devise a strategy that might induce isotype switching in WAS B cells to overcome the lack of T-cell—B-cell collaboration?

5   Austin Sloper takes theophylline to improve his asthma symptoms. What is theophylline and how does it work?

CASE 23 X-linked Lymphoproliferative Syndrome

A defect in the immune response to a virus.

Viruses pose a constant challenge to our immune system. Unable to reproduce on their own, they have evolved as parasites, capable of residing within living cells whose biosynthetic machinery they subvert for their own reproduction. An effective two-pronged immune response to these hidden invaders is to kill the host cells within which they reside by means of the cytotoxic cells of the immune system, and to reduce the number of extracellular virus particles by means of antibodies.

Both innate and adaptive immune responses control viral infections. The natural killer (NK) cells of innate immunity (Fig. 23.1) are constantly on surveillance for cells with telltale markers of viral infection. These cells are large granular cells of the lymphocyte lineage, which differ from cytotoxic T lymphocytes in not expressing antigen-specific receptors. Instead they carry a number of receptors that recognize virus-infected cells in other ways and, like cytotoxic T cells, release cytotoxic proteins that induce apoptosis and

Topics bearing on this case:
NK-cell activating receptors and killer function
NK T cells
Cell signaling from tyrosine kinase-associated receptors
Activation and function of cytotoxic T cells

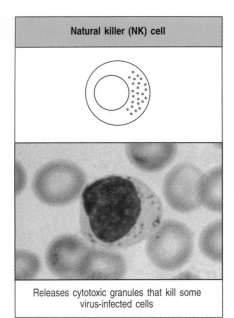

Fig. 23.1 Natural killer (NK) cells. These are large granular lymphocyte-like cells with important functions in innate immunity. Although lacking antigen-specific receptors, they can detect and attack certain virus-infected cells. Photograph courtesy of N. Rooney and B. Smith.

death of the virus-infected cell. These proteins include perforin and granzyme, which are also released by cytotoxic CD8 T cells. This response does not require any previous immunological experience with the virus and is particularly important when an individual first encounters a virus.

In the adaptive immune response, virus-specific cytotoxic T lymphocytes are generated during the primary immune response to the virus and specifically kill infected cells through the release of cytotoxic granules similar to those of NK cells. Naive virus-specific CD8 T cells are activated to their effector status as cytotoxic T cells through engagement of the T-cell antigen receptor, which interacts with a complex of antigen-derived peptide and major histocompatibility complex (MHC) class I molecule at the surface of a cell presenting viral antigens. For naive CD8 T cells, however, engagement of T-cell receptors, although necessary, is not sufficient for activation. Strong co-stimulation by antigen-presenting dendritic cells together with co-stimulatory help from activated CD4 T cells are usually required to activate naive cytotoxic T cells.

During the primary immune response to viruses, memory cytotoxic T cells are also produced. In the event of reexposure to the same virus, either by reinfection from the environment or by reactivation of virus latent in the body, these cytotoxic T cells rapidly recognize and kill infected host cells displaying viral antigens. Equally important to antiviral immunity are virus-specific primary and secondary antibody responses.

In this case we look at the effects of a deficiency in an intracellular signaling molecule found in both NK cells and T cells on the ability of the immune system to control infection by the Epstein–Barr virus (EBV). EBV infects epithelial cells and B cells and is present in small numbers in most humans over the age of 15 years. Its expansion is usually well controlled by cytotoxic T cells and NK cells. A primary EBV infection triggers the activation and cell division of B cells infected by the virus. The infected B cell expresses a number of viral antigens that are targets for specific cytotoxic responses by NK cells and CD8 T cells that keep the proliferation of infected B cells under control. In most people, EBV infection remains asymptomatic, but in a minority of cases (a subset of patients who first encounter the virus in adolescence) it gives rise to acute infectious mononucleosis, which eventually subsides within a span of 6–10 weeks. The normal course of an EBV infection in a healthy teenager is described in Case 27, and you may find it helpful to read that case before embarking on this one. After resolution of the acute infection, the virus persists in a latent form in B cells, salivary glands, and epithelial cells of the nose and throat, and can be shed in saliva. Occasional reactivation of the virus later in life is rapidly brought under control by EBV-specific memory cytotoxic T cells. This cellular immune surveillance is critical in maintaining the balance between host and virus. Primary and acquired deficiencies of T-cell function are associated with a marked susceptibility to lethal EBV infection.

For example, in very rare instances acute EBV infection in boys is not contained, and results in a failure to eliminate the virus that is accompanied by massive overproliferation of lymphocytes (lymphoproliferation), overproduction of cytokines, liver and bone-marrow destruction, B-cell lymphoma and/or dysgammaglobulinemia (selective deficiencies of one or more, but not all, classes of immunoglobulins), in most cases resulting in death. Such susceptibility to overwhelming infection can be inherited through unaffected females, an indication that it is carried by a gene on the X chromosome, and the condition is called X-linked lymphoproliferative syndrome (XLP). The gene involved has been found to encode an intracellular signaling protein called SAP, and defects in SAP are responsible for XLP.

The case of Nicholas Nickleby: inefficient killing of EBV-infected B cells by cytotoxic lymphocytes.

Nicholas was brought to the pediatrician at 5 years old because of several days of fever (38–39°C). He had no cough, runny nose, rash, diarrhea, or any other symptoms of infection. The physical examination revealed only some mildly enlarged lymph nodes in his neck and his parents were advised to treat the fever with acetaminophen. Over the following weeks, the fevers persisted and Nicholas seemed less energetic than usual. He was brought to the doctor several times but the only consistent finding was persistent enlarged, nontender lymph nodes. Finally, after 6 weeks of illness, Nicholas complained of abdominal pain and was referred to the Children's Hospital.

The past medical history was significant, revealing problems with persistent and recurrent middle ear infections (otitis media) as well as several episodes of bacterial pneumonia between the ages of 2 and 3 years. Immunological evaluation at the time had revealed decreased blood levels of IgG of 314 mg dl^{-1} (normal 600–1500 mg dl^{-1}), and normal IgA and IgM. At the time, Nicholas was briefly treated with prophylactic antibiotics with a good response. These were discontinued at the age of 4 years and he had no further infections or follow-up tests before his admission to Children's Hospital. The family history was notable for the presence of a maternal uncle with persistent unexplained low blood platelet count (thrombocytopenia). Nicholas's maternal grandfather had recurrent lymphomas.

On admission to hospital, Nicholas appeared tired but not acutely ill. His temperature was 38.5°C and his heart rate, respiration, and blood pressure were all normal. His height and weight were both in the 25th percentile for age. A few scattered small skin hemorrhages (petechiae) were noted on his legs and feet. Several lymph nodes were palpable in his neck, and these appeared significantly larger than on previous examinations. Supraclavicular, axillary, or inguinal lymph nodes were not enlarged. The tonsils were moderately enlarged but were not red, and there was no evidence of inflammation. The heart sounds were normal. The abdomen was moderately distended but soft, and slightly tender in the right upper quadrant. The liver was enlarged and its edge was palpable 4 cm below the right costal margin.

Laboratory evaluation showed a mild anemia with a hematocrit of 28% (normal 35–40%). The white blood cell count was 6400 μl^{-1} (normal 5–10,000 μl^{-1}) and the platelet count was decreased at 47,000 μl^{-1} (normal 150–200,000 μl^{-1}). The count of different types of white blood cell was remarkable for the very high proportion (22%) of atypical lymphocytes (normal <2%) (see Case 27). Liver function tests indicated liver damage. Tests for antibodies against hepatitis A, B, and C viruses were negative. The titer of IgM antibody against EBV viral capsid antigen (VCA) was positive at >1:40. Anti-VCA IgG antibody was 1:320 and antibodies against Epstein–Barr nuclear antigen (EBNA) and early antigen (EA) were undetectable, consistent with an acute EBV infection (see Fig. 27.4). Circulating EBV genome was detected in Nicholas's blood cells by the polymerase chain reaction (PCR).

A chest X-ray showed clear lungs and a normal-sized heart, but the lymph nodes in the mediastinum were enlarged. Ultrasound examination of the abdomen revealed a significant amount of free fluid in the abdominal cavity (ascites) and an enlarged liver. An abdominal CT scan revealed marked enlargement of lymph nodes in the retroperitoneum.

5-year-old child with persistent unexplained fever and abdominal pain.

History of recurrent otitis media.

High proportion of atypical lymphocytes in blood count. Test for EBV infection.

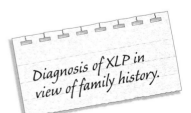

Diagnosis of XLP in view of family history.

Fig. 23.2 Northern blot analysis of SAP mRNA in peripheral blood mononuclear cells obtained from Nicholas. RNA was prepared from the Jurkat T-cell line (Jk), and from the peripheral blood mononuclear cells of a control (C) and the patient (P). A ^{32}P-labeled probe specific for SAP was used to probe the Northern blot, and a β-actin probe was used as a loading control. Courtesy of Hans Oettgen.

In the light of the family history and laboratory evidence of acute EBV infection, a diagnosis was made of X-linked lymphoproliferative syndrome (XLP) with fulminant infectious mononucleosis. Nicholas was kept in hospital and initially treated with an antiviral agent, acyclovir, and intravenous immune globulin (IVIG) in an attempt to control the EBV infection. However, the fever persisted and his liver dysfunction and ascites rapidly worsened. The glucocorticoid dexamethasone was added to his therapy. Despite these aggressive interventions, Nicholas developed severe shock symptoms, resembling those after bloodstream infection (sepsis), with diffuse vascular leakage, a fall in blood pressure (hypotension), poor circulation, and multiorgan failure. All cultures for bacterial pathogens were negative. He died 10 days after admission to hospital.

A post-mortem examination showed that the liver was markedly enlarged. Fluid had accumulated in the abdomen and around the lungs (pleural effusions). EBV was identified by culture and PCR in the liver and bone marrow. There was a striking infiltration of the liver, spleen, and lymph nodes by a mixed population of mononuclear cells including small lymphocytes, plasma cells, and lymphoblasts. In the liver, these infiltrates were associated with extensive tissue death (necrosis). Examination of the bone marrow revealed a decreased number of erythroid, megakaryocytic, and myeloid cells, along with increased numbers of histiocytic cells, lymphocytes, and plasma cells.

To confirm a diagnosis of X-linked lymphoproliferative syndrome, the function of the SAP gene (*SH2D1A*) was analyzed by Northern blotting, revealing a complete absence of *SH2D1A* mRNA (Fig. 23.2). None of the four exons encoding SAP could be isolated by PCR, consistent with a complete deletion of the gene. Such complete deletions have now been shown to be common in XLP.

X-linked lymphoproliferative disease (XLP).

The gene responsible for many, if not all, cases of familial XLP has been mapped to the X chromosome at position Xq25 and has been identified as *SH2D1A*, which encodes the intracellular signaling protein SAP. Patients with this defect show uncontrolled T-cell activation, especially in response to an EBV infection, but a reduced capacity to kill EBV-infected B cells. In a minority of cases, symptoms of XLP occur without evident past or current EBV infection. Boys presenting with EBV-induced fulminant infectious mononucleosis, and who have a family history of affected male relatives, will have XLP as a result of mutations in the SAP gene. The fulminant infectious mononucleosis after their initial encounter with Epstein–Barr virus commonly proves lethal; among 161 boys with XLP, 57% died of this disease. Of those who survived, half developed lymphomas, as did Nicholas's maternal grandfather, and the other half became agammaglobulinemic as a result of the destruction of their B cells. In rare instances, the bone marrow of affected males may be destroyed, resulting in the fatal disease of aplastic anemia or in thrombocytopenia, as in Nicholas's maternal uncle.

Fig. 23.3 Receptor–ligand interactions in the activation of NK cells. NK cells bear both activating and inhibitory receptors. For example, the NKR-P1 family of receptors binds to polysaccharide ligands and provides an activating signal. Killer-cell immunoglobulin receptors (KIRs) bind to MHC molecules on target cells and give a negative signal. As a result, cells expressing low levels of MHC molecules are susceptible to killing by NK cells. The SLAM-family member CD244 interacts with CD48, a protein that is upregulated on B cells after EBV infection. This interaction may enhance NK cell activity by signaling via SAP and Fyn.

In normal individuals, EBV-infected B cells are attractive targets for killing by both NK cells and virus-specific effector cytotoxic T cells. NK cells express receptors that activate or inhibit the cell's killer activity. The fate of potential target cells is determined by the balance of activating and inhibitory signals they deliver to NK cells through these receptors (Fig. 23.3). Inhibitory receptors interact with MHC class I molecules and help prevent NK cells from attacking healthy, uninfected cells. NK cells also carry a receptor called CD244

(2B4), a member of the SLAM (signaling lymphocytic activation molecule) family, which can act as an activating or an inhibitory receptor, depending on which intracellular signaling proteins it is associated with.

The cytoplasmic portions of SLAM receptors have tyrosine-containing motifs that provide potential docking sites for the intracellular adaptors SAP (SLAM-associated protein; encoded by *SH2D1A*) or an alternative adaptor protein EAT-2. SAP is composed of a single SH2 domain and is found in all T cells, germinal center B cells, NK cells, and thymic NK T cells, which are a subpopulation of double-negative lymphocytes with features of both NK and T cells that carry relatively invariant T-cell receptors recognizing glycolipids. SAP is unusual among intracellular adaptor proteins in that its single SH2 domain can both bind to the tails of activated SLAM receptors and recruit the cytoplasmic Src kinase Fyn to the receptor complex by interacting with the SH3 domain of Fyn.

B cells infected with the Epstein–Barr virus (EBV) increase expression of the SLAM family member CD48 on their surface, and in normal individuals CD48 interacts with CD244 on NK cells, providing an activating signal that enables them to kill the infected B cells (see Fig. 23.3). Signaling via CD244 is critical in driving NK killing of EBV-infected target cells. CD244 contains several tyrosines in its cytoplasmic tail, which after the receptor is activated become phosphorylated and constitute docking sites for cytosolic proteins containing SH2 domains. Under normal conditions, SAP binds to these cytoplasmic tyrosines, and recruits the Src-family tyrosine kinase Fyn to the receptor complex, allowing propagation of the activating signal onward (Fig. 23.4). In the absence of SAP, active Fyn is not recruited to CD244, and alternative signaling pathways are activated, which result in inhibition of killing by the NK cells.

The SLAM family is also implicated in the function of CD8 cytotoxic T cells, with receptors such as SLAM itself (CD150) and CD244 on T cells interacting with SLAM-family molecules on antigen-presenting dendritic cells, B cells, and monocytes. With the exception of the CD48/CD244 pair, SLAM-family molecules are homotypic, interacting with an identical molecule on another cell. In SAP knockout mice and in patients with XLP, the CD8 T-cell response is exaggerated, suggesting that signaling via SAP normally downregulates this response. In the absence of SAP, SLAM-family receptor signaling is dysregulated, leading to unchecked proliferation of cytokine-producing CD8 cells but impaired cytotoxic function.

Fig. 23.4 The effect of defects in SAP on NK-cell killer function. In normal NK cells (left panel), interaction with EBV-infected B cells provides positive signals that lead to activation of the NK cytotoxic mechanism. These positive signals are provided by ligands that bind activating NK receptors (e.g. carbohydrate ligands binding NKR-P1A) and by the binding of NK-cell CD244 to CD48 on the B cell. The cytosolic tail of CD244 contains tyrosine residues, which interact with the SH2 domain of SAP. SAP couples CD244 to activation of the tyrosine kinase Fyn, thus enhancing NK-cell activation. In the absence of SAP (right panel), this activating signal is interrupted by the binding of the phosphatase SHP-1 to the receptor.

The overproduction of cytokines as a result of uncontrolled T-cell proliferation seems to be important in causing the life-threatening tissue damage that occurs in XLP. On infection with EBV, patients with XLP suffer a destruction of the liver that is most probably due to uncontrolled cytokine-mediated injury. Indeed, plasma levels of a number of T-cell-derived cytokines, including interferon (IFN)-γ, interleukin (IL)-2, and tumor necrosis factor (TNF)-α, are all markedly elevated. The same cytokines are present at much lower (or undetectable) levels in normal EBV-infected individuals. In patients with fulminant infectious mononucleosis, the uncontrolled lymphocyte proliferation and cytokine secretion lead to a syndrome of severe inflammation of the liver (hepatitis), destruction of bone marrow cells, and systemic shock. T-cell derived cytokines, particularly TNF-α, as well as monocyte-derived IL-1, result in enhanced vascular permeability and loss of intravascular volume. This is analogous to the clinical scenario observed in toxic shock syndrome (see Case 26) in which uncontrolled T-cell activation and cytokine secretion can also lead to multiorgan damage and a clinical picture resembling sepsis.

In a second potential mechanism of tissue injury, particularly in the liver, T-cell derived cytokines may promote the expression of Fas on hepatocytes. The ligand, FasL, is expressed on the surface of activated T cells and can induce hepatocyte apoptosis (see Case 24). T-cell derived cytokines, particularly IFN-γ, also trigger the activation of monocytes/macrophages, which engage in indiscriminate phagocytosis of surrounding cells. Histologic analysis of tissues from patients with fulminant infectious mononucleosis often reveals erythrophagocytosis, a phenomenon in which macrophages appear to engulf entire red blood cells and others are heavily laden with cellular debris. Erythrocyte phagocytosis is further enhanced because the polyclonal B-cell activation induced by EBV infection (see Case 27) produces complement-fixing antibodies that interact with red-cell antigens. Finally, numbers of NK T cells are severely diminished in XLP, possibly because activation of the kinase Fyn is essential for their development.

While more than half of the patients with XLP suffer from an extreme susceptibility to EBV, one-third of the patients develop dysgammaglobulinemia without an episode of severe mononucleosis. Experiments with SAP-deficient mice show that SAP is essential for the functional integrity of both T- and B-cell responses to soluble T-dependent antigens, and pivotal for IgM and IgG responses to viruses, well-defined protein antigens, and haptens, and for immunoglobulin class switching and germinal center formation. Thus, when SAP is absent, both helper T cells and B cells are defective in function, which may explain the progressive dysgammaglobulinemia in a subset of patients with XLP without involvement of EBV.

Patients with XLP are also predisposed to the development of B-cell lymphomas. Most lymphomas occur because of B-cell transformation by EBV, and the classic 8;14 chromosomal translocation of EBV-induced Burkitt's lymphoma has been observed in some patients with XLP. In some cases there is no direct evidence for a role for EBV in the B-cell transformation. In these cases, altered (or absent) SAP function may interfere with tumor surveillance by interfering with the function of tumor-specific NK cells. In some families, males with the identical mutation in *SH2D1A* may be present with fulminant infectious mononucleosis, dysgammaglobulinemia, or B-cell lymphoma.

Questions.

1 Did Nicholas's history provide any clues to his diagnosis before his presentation with fulminant infectious mononucleosis?

2 How can the diagnosis of X-linked lymphoproliferative syndrome be made?

3 If a girl presents with fulminant infectious mononucleosis and evidence of acute EBV infection, what might be the basis of her disease?

4 Nicholas has a 3-year-old brother Alexander with SAP deficiency. What can be done to prevent the occurrence of disease in this boy?

5 Some individuals with common variable immunodeficiency (see Case 13) have turned out to have mutations in SAP. What is the basis of this observation?

CASE 24 Autoimmune Lymphoproliferative Syndrome (ALPS)

Increased survival of lymphocytes as a result of a mutation in Fas.

When antigen-specific lymphocytes are activated through their antigen receptors, they undergo blast transformation and then begin to increase their numbers exponentially by cell division. This clonal expansion can continue for up to 7 or 8 days, so that lymphocytes specific for the infecting antigen increase vastly in number and can come to predominate in the population. In the response to certain viruses, 50% or more of the CD8 T cells at the peak of the response are specific for a single virus-derived peptide:MHC class I complex. After clonal expansion, the activated lymphocytes undergo their final differentiation into effector cells; these remove the pathogen from the body and so terminate the antigenic stimulus.

When an infection has been overcome, activated effector T cells are no longer needed, and cessation of the antigenic stimulus prompts them to undergo programmed cell death, or apoptosis (Fig. 24.1). Apoptosis is widespread in the immune system and can be induced by several mechanisms; for example, the granule proteins released by cytotoxic T cells kill their target cells by inducing apoptosis. Another well defined apoptotic pathway is that triggered by the interaction of the receptor molecule Fas with its ligand, called Fas ligand (FasL) (Fig. 24.2), which induces apoptosis in the Fas-bearing cell. FasL is a member of the tumor necrosis factor (TNF) family of membrane-associated cytokines, and Fas is a member of the TNF receptor (TNFR) family. Both Fas and FasL are normally induced on lymphocytes and other cell types during the course of an adaptive immune response. Apoptosis induced by cytotoxic T cells bearing FasL is a minor mechanism of cytotoxicity, whereas apoptosis in lymphocytes themselves, induced through Fas, appears to be an important mechanism of lymphocyte homeostasis, as this case shows.

Topics bearing on this case:

Apoptosis

Fas–Fas ligand interactions

TUNEL staining

Autoimmune disease

Lymphocyte activation

Lymphocyte survival

Fig. 24.1 Apoptosis. Apoptosis is a form of induced 'cell suicide' in which the cell undergoes chromatin compaction and DNA fragmentation, followed by cell shrinkage and internal degradation. Panel (a) shows a healthy cell with a normal nucleus. Early in apoptosis (panel b), the chromatin in the nucleus becomes condensed (red) and, although the cell sheds membrane vesicles, the integrity of the cell membrane is retained, in contrast to the necrotic cell in the upper part of the same field. In late stages of apoptosis (panel c), the cell nucleus (middle cell) is very condensed, no mitochondria are visible, and the cell has lost much of its cytoplasm and membrane through the shedding of vesicles. Photographs ($\times$ 3500) courtesy of R. Windsor and E. Hirst.

The case of Ellen O'Hara: uncontrolled lymphocyte proliferation in the absence of infection or malignancy.

18-month-old girl, enlarged spleen. Order blood tests.

Ellen O'Hara was born after a normal and uncomplicated pregnancy, was breast fed, and received her routine immunizations without any adverse reactions. At 18 months old, during a routine check-up by her pediatrician, she was found to have an enlarged spleen (splenomegaly) and extensive enlargement of her lymph nodes (lymphadenopathy) (Fig. 24.3). According to her parents, she had had no unusual infections and appeared to be growing and developing normally.

Fig. 24.2 Binding of FasL to Fas initiates the process of apoptosis in the Fas-bearing cell. Binding of trimeric FasL to trimeric Fas brings the death domains in the Fas cytoplasmic tails together. A number of adaptor proteins containing death domains bind to the death domains of Fas, in particular the protein FADD, which in turn interacts through a second death domain with the protease caspase-8. Clustered caspase-8 can transactivate, cleaving caspase-8 itself to release an active caspase domain that in turn can activate other caspases. The ensuing caspase cascade culminates in the activation of the caspase-activatable DNase (CAD), which is present in all cells in an inactive cytoplasmic form bound to an inhibitory protein called I-CAD. When I-CAD is broken down by caspases, CAD can enter the nucleus where it cleaves DNA into the 200 base pair fragments that are characteristic of apoptosis.

Fig. 24.3 Lymphadenopathy in ALPS.
Young girl with ALPS with very enlarged lymph nodes in her neck. Photograph courtesy of Jennifer Puck.

Laboratory tests revealed that Ellen's white blood cell count was 12,500 μl^{-1}, of which 9175 were lymphocytes (normal 3000–7500). Her serum immunoglobulins were all elevated: IgG, 4000 mg dl^{-1} (normal 520–1500); IgM, 400 mg dl^{-1} (normal 40–200) and IgA, 1660 mg dl^{-1}. Flow cytometry analysis of her lymphocytes revealed that 29% were CD19-positive B cells (normal 5–15%; CD19 is a component of the B-cell co-receptor complex) and 65% were CD3-positive T cells (normal 61–84%; CD3 is a component of the T-cell receptor complex). Of the CD3-positive T cells, 14% carried the co-receptor protein CD4 and 18% the co-receptor CD8. She thus had many $CD3^{+}4^{-}8^{-}$ T cells—so-called double-negative (DN) T cells (normally, DN T cells are either absent or less than 2% of circulating T cells). A biopsy of a lymph node from Ellen's neck showed extensive enlargement of the follicles (hyperplasia) and a marked increase in the numbers of immunoblasts and plasma cells in the paracortical area. No infectious agents were cultured from the lymph node, despite the fact that the observed changes resembled those caused by a viral infection. Although more than 50% of the T cells in the lymph node were double negatives, no chromosomal abnormality was found on karyotyping, and there was no evidence of oligoclonality of the T-cell receptor, thus ruling out a malignancy (see Case 20).

In the absence of evidence of infection or malignant disease, autoimmune lymphoproliferative disease was diagnosed and Ellen received the anti-inflammatory steroid prednisone and the immunosuppressant drug cyclosporin A. Her lymph nodes rapidly reduced in size following this therapy but enlarged again when therapy was discontinued.

Ellen continued to grow and develop normally, and when she reached adolescence the size of her lymph nodes spontaneously decreased. At 18 years of age, repeat blood counts revealed that her platelet count was 75,000 μl^{-1} (normal 150,000–250,000). An autoantibody to platelets was found in her serum. A diagnosis of idiopathic thrombocytopenic purpura (low platelet numbers accompanied by red or purplish-red spotty skin discoloration due to local hemorrhages) was made. She was treated with the steroid dexamethasone, and the condition resolved. At age 32, Ellen's blood neutrophil count fell to <1000 μl^{-1} (normal 2500–5000). She was found to have developed an autoantibody against granulocytes.

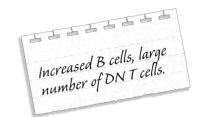

Increased B cells, large number of DN T cells.

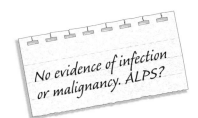

No evidence of infection or malignancy. ALPS?

TUNEL assay, defective apoptosis.

Ellen's family history was informative in that her paternal grandfather had splenomegaly and generalized lymphadenopathy as a child, and his spleen was removed at age 25. At age 60 he developed a B-cell lymphoma. Ellen's father also had splenomegaly and lymphadenopathy but no clinical symptoms. When blood lymphocytes from Ellen's father and paternal grandfather were examined by flow cytometry, a large number of double-negative T cells were found. On the other hand, her brother, mother, and maternal grandparents had normal T cells. The TUNEL assay for apoptotic cells (Fig. 24.4) was performed on blood mononuclear cells from Ellen, her parents, and her paternal grandfather. The cells were first stimulated *in vitro* with phytohemagglutinin for 3 days, and growth of the resulting T-cell blasts was continued for 3 weeks by adding IL-2 to the cultures. The cultures were then divided and half were exposed to an antibody to Fas, which mimics the function of FasL. The percentage of cells undergoing apoptosis was then counted. Sixty percent of her mother's T cells underwent apoptosis whereas only 2% of Ellen's cells, <1% of her father's cells and 1.4% of her paternal grandfather's cells demonstrated programmed cell death (normal controls 35–70%). The Fas and FasL genes were examined in DNA samples from Ellen, her father, and her paternal grandfather. An identical single-base transversion, causing a premature termination codon, was found in one of the alleles of the Fas gene in these DNA samples. The Fas genes in Ellen's mother and brother were normal.

Autoimmune lymphoproliferative syndrome.

Autoimmune lymphoproliferative syndrome (ALPS) is characterized by splenomegaly and lymphadenopathy from early childhood, and, frequently, autoimmunity. Affected individuals can develop autoimmune hemolytic anemia, neutropenia, thrombocytopenia, and hepatitis (inflammation of the liver). Most patients with ALPS are heterozygous for a dominant mutation in the Fas gene, and their activated T cells do not undergo Fas-mediated apoptosis *in vitro*, as is the case for Ellen and her father and grandfather. Other patients with ALPS have been found to have mutations in the genes encoding FasL or caspase-10, an enzyme involved in triggering apoptosis via the Fas pathway.

| The DNA in apoptotic cells is extensively nicked | The enzyme TdT adds biotinylated nucleotides to the free 3' ends of the nicked DNA | Enzymes coupled to streptavidin bind to the labeled bases; the enzyme generates a colored reaction product |

Fig. 24.4 The TUNEL assay. When cells undergo apoptosis their DNA becomes fragmented and they can be revealed by labeling the fragmented DNA using the enzyme terminal deoxynucleotidyltransferase (TdT). TdT adds nucleotides to the ends of DNA fragments; biotin-labeled nucleotides (usually dUTP) are most commonly added in this assay (second panel). The biotinylated DNA can be detected by using streptavidin, which binds to biotin, coupled to enzymes that convert a colorless substrate into a colored insoluble product (third panel). Cells stained in this way can be detected by light microscopy, as shown in the photograph of apoptotic cells (stained red) in the thymic cortex. Photograph courtesy of R. Budd and J. Russell.

The clinical and immunologic features of ALPS bear a striking resemblance to a lymphoproliferative disease observed in mice with *lpr* or *gld* mutations. The *lpr* phenotype results from the absence of Fas, whereas the *gld* phenotype is caused by a mutation in FasL. A progressive accumulation of DN T cells is observed in both these strains of mice (note that these circulating CD3$^+$ DN T cells should not be confused with the immature CD3$^-$ CD4$^-$ CD8$^-$ 'double-negative' thymocytes that are a normal stage of T-cell development in the thymus). The mice make antibodies to double-stranded DNA, similar to the situation in human systemic lupus erythematosus (see Case 43). Consistent with these findings in mice, patients with ALPS have defective T-cell apoptosis and abnormal accumulations of DN T cells. When B cells are activated, they also express Fas and become susceptible to Fas-mediated apoptosis. Thus, activated B cells in ALPS are not properly eliminated. The serum concentrations of immunoglobulins rise (hypergammaglobulinemia), the number of B cells is increased (B-cell lymphocytosis), and pathological autoantibody production ensues. Because T and B cells are not eliminated normally, patients with ALPS are predisposed to develop lymphomas. Autoimmunity may result because Fas-mediated killing is a mechanism for removing autoreactive B cells.

Questions.

1 Patients with ALPS are heterozygous for the mutation in Fas or FasL; they have one normal allele and one mutant allele. How do you explain the dominant inheritance?

2 Ellen's great-aunt (her paternal grandfather's sister) was found to have the same Fas mutation as Ellen, yet she had no symptoms. How can this be explained?

3 It is advantageous for viruses to inhibit apoptosis so that the host cells in which they thrive do not get eliminated by apoptosis induced by recognition by cytotoxic T cells. How might a virus accomplish this?

4 When Fas is activated by FasL it associates with and activates caspase-8 (see Fig. 24.2). When the gene encoding caspase-8 is knocked out in mice, this proves to be lethal at the fetal stage. Would it be worthwhile to search for caspase-8 mutations in patients with ALPS when there is no mutation in Fas or FasL?

CASE 25 | Immune Dysregulation, Polyendocrinopathy, Enteropathy X-linked Disease

A failure of peripheral tolerance due to defective regulatory T cells.

The primary role of the immune system is to recognize pathogens and eliminate them from the body. However, an equally important task is to distinguish potentially dangerous antigens from those that are harmless. Countless innocuous foreign antigens are encountered every day by the lungs, gut, and skin, the interfaces between the body and the environment. Similarly, the body contains numerous self antigens that might bind to the specific antigen receptors on B and T cells. Activation of the immune system by such innocuous antigens is unnecessary and may lead to unwanted inflammation. Allergic and autoimmune diseases are well-known examples of such unwanted and potentially destructive responses.

Topics bearing on this case:

Central tolerance
Peripheral tolerance
Regulatory T cells
Autoimmunity

Fig. 25.1 Mechanisms of peripheral immunologic tolerance. T cells that are physically separated from their specific antigen—for example, by the blood–brain barrier—cannot become activated, a circumstance referred to as immunologic ignorance. T cells that express Fas (CD95) on their surface can receive signals from cells that express Fas ligand, leading to their deletion. The activation of naive T cells can be inhibited if the cell-surface protein CTLA-4 (CD152) binds B7.1 (CD80) on antigen-presenting cells (APCs). Regulatory T cells (mainly CD4 CD25 Foxp3-expressing) can inhibit, or suppress, other T cells, most probably through the production of inhibitory cytokines such as IL-10 and TGF-β.

Fortunately, unwanted immune responses are normally prevented or regulated by the phenomenon of immunologic tolerance. This is defined as nonresponsiveness of the lymphocyte population to the specific antigen, and arises at two stages of lymphocyte development. Central tolerance is the result of the removal of self-reactive lymphocytes in the central organs; an autoimmune disease due to a defect in central tolerance is described in Case 38. Peripheral tolerance, in contrast, inactivates those T and B cells that escape central tolerance and exit to the periphery. Defects in either central or peripheral tolerance can result in unwanted or excessive immune responses.

Several mechanisms of peripheral tolerance exist (Fig. 25.1). One whose importance is increasingly being recognized is the network of regulatory cells that prevent or limit the activation of T cells, including self-reactive T cells, and the consequent destructive inflammatory processes. When these regulatory cells do not function properly, problems can arise. A key cell type responsible for the maintenance of peripheral tolerance is the CD4 CD25 regulatory T cell (T$_{reg}$), also known as the natural regulatory T cell, which seems to become committed to a regulatory fate while still in the thymus and represents a small subset of circulating CD4 T cells (5–10%).

Although these cells were first characterized by their cell-surface CD25 (the α chain of the IL-2 receptor), this protein also appears on other T cells after activation. Natural T$_{reg}$ cells are better characterized by their expression of the transcription factor Foxp3, which is essential for their specification and function as regulatory cells. Over the past decade they have emerged as crucial to the maintenance of peripheral tolerance. Neonatal thymectomy in mice and thymic hypoplasia in humans (DiGeorge syndrome) result in impaired generation of natural T$_{reg}$ cells and the development of organ-specific autoimmune disease. The generation of natural T$_{reg}$ cells in the thymus requires interaction with self-peptide:MHC class II complexes on cortical epithelial cells.

A second group of regulatory T cells seem to be induced from naive CD4 T cells in the periphery. They are CD4$^+$ CD25$^-$ and are heterogeneous, including subsets known as T$_H$3, T$_R$1, and a CD4$^+$ CD25$^-$ Foxp3-positive subset. NK cells and NK T cells have also been shown to be able to regulate immune responses. As a group, regulatory cells represent just one mechanism in a complex system of immunologic tolerance, acting to prevent or rein in unwanted immune responses.

The following case illustrates how a breakdown in peripheral tolerance as a result of a defect in regulatory T cells leads to a constellation of allergic symptoms, gastrointestinal symptoms, and autoimmune disease in infancy.

The case of Billy Shepherd: a defect in peripheral tolerance leading to dermatitis, diarrhea, and diabetes.

Billy was born at full term and developed atopic dermatitis (see Case 34) shortly after birth. This was treated by skin hydration, local application of hydrocortisone and antihistamines to control itching, the treatment being only partly successful. At 4 months of age Billy developed an intractable watery diarrhea. Although he had initially gained weight well, by now his weight had fallen below the third centile for his age. At 6 months old, Billy started to develop high blood glucose levels and glucose in the urine. He was diagnosed with type 1 diabetes (insulin-dependent diabetes mellitus; see Case 46) and was referred by his pediatrician to the endocrine clinic at the Children's Hospital.

Infant with dermatitis, diarrhea, and diabetes. Underlying cause?

When first seen at the clinic, Billy weighed 5 kg (the third centile for age is 6.3 kg). He had diffuse eczema and sparse hair (Fig. 25.2). His cervical and axillary lymph nodes and spleen were enlarged. Laboratory tests revealed a normal white blood cell count of 7300 μl^{-1}, normal hemoglobin of 11.3 g dl^{-1}, and a normal platelet count of 435,000 μl^{-1}. The percentage of eosinophils in the blood was high at 15% (normal <5%), and IgE was also elevated, at 1345 IU ml^{-1} (normal <50 IU ml^{-1}). Autoantibodies were found against glutamic acid decarboxylase (the GAD65 antigen) and to pancreatic islet cells. Billy was started on insulin therapy, which controlled his blood glucose level.

Take family history.

Because of the persistent diarrhea and failure to thrive, Billy required parenteral (intravenous) nutrition to maintain his weight. An endoscopy was ordered to ascertain the cause of his persistent diarrhea, and a duodenal biopsy revealed almost total villous atrophy—an absence of villi in the lining of the duodenum—with a dense infiltrate of plasma cells and T cells (Fig. 25.3).

When Billy's mother was questioned, she revealed that there had been another son, who had died in infancy with severe diarrhea and a low platelet count. On the basis of Billy's symptoms and the family history, IPEX (immune dysregulation, polyendocrinopathy, enteropathy X-linked) was suspected. A FACS analysis of Billy's peripheral blood mononuclear cells revealed a lack of CD4 CD25 cells as well as a lack of CD4 Foxp3$^+$ cells. Sequencing of Billy's FOXP3 gene revealed a missense mutation, confirming the diagnosis.

With the diagnosis established, immunosuppressive therapy, including cyclosporin and tacrolimus, was started. Billy's diarrhea, glucose control, and eczema all improved markedly. After several months, however, his symptoms began to return

Fig. 25.2 **Eczematous rash on the face of a baby boy with IPEX.** Photograph courtesy of Talal Chatila, UCLA.

Bone marrow transplant advisable.

and he stopped gaining weight. Shortly afterwards, he developed thrombocytopenia (a deficiency of blood platelets) and anti-platelet antibodies were detected.

The decision was made for Billy to be given a bone marrow transplant from his 5-year-old HLA-identical sister. In the weeks of conditioning leading up to the transplant, the diarrhea and eczema resolved. After transplantation, full engraftment of his sister's stem cells was established. Two weeks after transplantation, anti-platelet, anti-GAD65, and anti-islet cell antibodies could not be detected. A year after the transplant, Billy continues to be symptom-free, although analysis for chimerism reveals that only 30% of his T cells are derived from his sister's cells.

Fig. 25.3 Photomicrograph of duodenal biopsy from a child with IPEX. The section was stained with hematoxylin and eosin. Note the dense mononuclear cellular infiltrate. Photograph courtesy of Talal Chatila, UCLA.

Immune dysregulation, polyendocrinopathy, enteropathy X-linked disease (IPEX).

IPEX is a very rare disease caused by mutations in the gene for the forkhead transcription factor Foxp3, which is essential for the function of CD4 CD25 T_{reg} cells. Foxp3 expression is restricted to a small subset of TCRα:β T cells and defines two pools of regulatory T cells: CD4+ CD25^{high} T cells and a minor population of CD4+ CD25$^{lo/neg}$ T cells. Ectopic expression of *FOXP3 in vitro* and *in vivo* is sufficient to convert naive murine CD4 T cells to T_{reg} cells. In contrast, overexpression of *FOXP3* in naive human CD4+ CD25− T cells *in vitro* will not generate potent suppressor activity, suggesting that additional factors are required. Foxp3 expression and suppressor function can, however, be induced in human CD4+ CD25− Foxp3− cells by cross-linking of the T-cell receptor and stimulation via the co-stimulatory receptor CD28, or after antigen-specific stimulation. This suggests that *de novo* generation of T_{reg} cells in the periphery may be a natural consequence of the human immune response.

T_{reg} cells are anergic *in vitro*. They fail to secrete IL-2 or proliferate in response to ligation of their T-cell receptors, and depend on the IL-2 generated by activated CD4 T cells to survive and exert their function. An *in vitro* assay that measures the ability of CD4 CD25 T cells to suppress CD4 T-cell proliferation is commonly used to test for T_{reg} function (Fig. 25.4). How T_{reg} cells suppress immune responses *in vivo* is still unclear. There is some evidence for contact-dependent inhibition, whereas other studies suggest that regulatory T cells exert their function by secreting immunosuppressive cytokines such as IL-10 or transforming growth factor-β (TGF-β), or by directly killing their target cells in a perforin-dependent manner.

Several lines of evidence show that Foxp3 is both necessary and sufficient for the development and function of CD4 CD25 T_{reg} cells in mice. A mutation in *Foxp3* is responsible for an X-linked recessive inflammatory disease in the *Scurfy* mutant mouse. Male mice hemizygous for the mutation succumb to a CD4 T-cell-mediated lymphoproliferative disease characterized by wasting and multi-organ lymphocytic infiltration. T_{reg} cells are absent in *Scurfy* mice and in mice that have another spontaneous mutation in the *Foxp3* gene. In addition, specific ablation of Foxp3 in T cells only is sufficient to induce the full lymphoproliferative autoimmune syndrome observed in the *Foxp3*-knockout mice. The *Scurfy* phenotype can be rescued by the introduction of a Foxp3 transgene or by bone marrow reconstitution, demonstrating the

Fig. 25.4 Results of a functional assay for regulatory T cells from a normal individual. CD4 responder cells and CD4 CD25 suppressor cells were mixed, together with antigen-presenting cells, at the ratios shown on the horizontal axis. The cells were stimulated with immobilized plate-bound anti-CD3 and soluble anti-CD28 for 3 days, then assessed for proliferation as measured by the incorporation of ^{3}H-labeled thymidine into DNA.

causative role of Foxp3 in pathogenesis. Thus, the lack of Foxp3-expressing T_{reg} cells alone is sufficient to break self-tolerance and induce autoimmune disease.

In humans, missense or frameshift mutations in *FOXP3* result in loss of function of T_{reg} cells and uninhibited T-cell activation. As seen in Billy's case, the most common symptoms are an intractable watery diarrhea, leading to failure to thrive, dermatitis, and autoimmune diabetes developing in infancy. The diarrhea is due to widespread inflammation of the gut, including the colon (colitis), that results in villous atrophy, which reduces the absorptive capacity of the intestinal lining and thus contributes to wasting. Other diseases of immune dysregulation also seen include autoimmune thrombocytopenia, neutropenia, anemia, hepatitis, nephritis, hyperthyroidism or hypothyroidism, and food allergies. Autoantibodies also accompany these autoimmune diseases. Affected patients may also suffer more frequent infections, including sepsis, meningitis, or pneumonia, although the reason for the increased susceptibility to infection is unclear. Patients generally have normal immunoglobulin levels (except for the elevated IgE), and their ability to make specific antibody is intact.

Questions.

1 Billy continued to remain asymptomatic even though he eventually only had around 30% engraftment of his sister's bone marrow stem cells (as judged from the proportions of T cells). Why is full engraftment not necessary in patients with IPEX?

2 Why did Billy's diarrhea improve while he was being prepared for transplantation?

3 Intravenous immunoglobulin (IVIG) has been used to treat IPEX. How might this be an effective therapy?

4 The occurrence of colitis in IPEX suggests that T_{reg} cells may be implicated in its pathogenesis and that they might be used therapeutically in the more common forms of colitis. Is there experimental data to support this claim?

5 What other gene mutations can give rise to a clinical picture similar to IPEX?

CASE 26

Toxic Shock Syndrome

Superantigens cause excessive stimulation of T cells and macrophages.

A conventional protein antigen activates only those T lymphocytes that bear a T-cell receptor specific for peptides derived from the antigen, which are a tiny subset of the total T-cell pool. For example, only about 1 in 10,000 circulating T cells from donors immunized with tetanus toxoid subsequently proliferate when stimulated by the toxoid. In contrast, a special class of antigens known as superantigens directly activate large numbers of T cells. Superantigens are bacterial and viral proteins that bind as the whole unprocessed protein to MHC class II molecules outside the peptide-binding groove and also to the lateral face of certain subsets of V_β chains of the T-cell receptor.

Superantigen therefore cross-links MHC class II molecules on antigen-presenting cells to T-cell receptors with the appropriate V_β specificity (Fig. 26.1). Cross-linking results in the activation of both the T cell and the antigen-presenting cell as a result of signals initiated by the T-cell receptor and the MHC class II molecule respectively. Each superantigen is capable of binding to a limited group of V_β regions. For example, the toxic shock syndrome toxin-1 (TSST-1), which is produced by certain strains of the Gram-positive bacterium *Staphylococcus aureus*, stimulates all those T cells that express the $V_\beta2$ gene segment. As there is a limited repertoire of V_β gene segments, any superantigen will stimulate between 2% and 20% of all T cells.

This mode of stimulation is not specific for the pathogen and thus does not lead to adaptive immunity. Instead, it causes excessive production of cytokines by the large number of activated CD4 T cells, the predominant responding population. This massive cytokine release has two effects on the host: systemic toxicity, which manifests itself as toxic shock syndrome, and suppression of the adaptive immune response. Both these effects contribute to the pathogenicity of microbes that produce superantigens.

Topics bearing on this case:
T-cell activation
Superantigens
Macrophage activation
Toxic shock
Cytokines

Bacterial superantigen e.g. SE, TSST-1	Viral superantigen

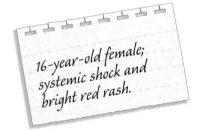

16-year-old female; systemic shock and bright red rash.

Started period day before. Toxic shock syndrome?

Fig. 26.1 Superantigens bind directly to T-cell receptors and MHC molecules. Superantigens interact with MHC class II molecules and T-cell receptors in a way that is quite distinct from the way that normal peptide antigens bind. Superantigens bind independently to MHC class II molecules and to T-cell receptors, binding to the V_β domain of the T-cell receptor, away from the peptide-binding site. The T-cell receptor α chain is not directly involved in binding superantigen.

The case of Claire Bourbon: life-threatening shock from a superantigen.

Claire Bourbon was a healthy 16-year-old with a history of mild asthma and allergic rhinitis who suddenly became ill with a fever, general muscle aches, and dizziness. She felt nauseous and vomited. Claire's temperature rose to 39.8°C and her mother rushed her to the family physician. En route she briefly lost consciousness, and a red rash developed over her arms and spread rapidly to most of the body.

Upon arrival at the physician's she appeared quite ill and was immediately referred to the emergency department. She was alert but listless and her general condition gave cause for concern. On examination, her temperature was 37.8°C, and heart rate and respiration rate were markedly elevated, at 140 beats per minute and 24 breaths per minute respectively. Blood pressure was depressed—98/67 lying supine, 83/49 when seated, and 67/25 when standing—and showed evidence of significant volume depletion.

A bright red rash of flat and raised lesions was apparent on her trunk and extremities, but there were no petechiae (small subcutaneous hemorrhages) and no signs of localized infection.

Questioning revealed that Claire had not taken alcohol or drugs and had not been exposed to other ill individuals. Her last menstrual period had been 6 weeks before, and she had developed vaginal bleeding on the day previous to the onset of her illness. She had not used a tampon overnight, but had inserted one that morning, before she became ill.

Given her critical status, extensive laboratory tests were carried out. Her white blood cell count was raised, at 21,000 cells μl^{-1}, with a predominance of neutrophils and band forms (immature neutrophils), indicating increased mobilization of neutrophils from the bone marrow. Serum electrolyte levels were within normal limits. The blood coagulation time was slightly prolonged and serum transaminase levels were raised; both of these signs are consistent with abnormal liver function.

Cerebrospinal fluid (CSF) was normal and showed no evidence of infection. Cultures of blood, urine, CSF, and vaginal fluid were made and Claire was given a cephalosporin antibiotic (ceftriaxone) along with 2 liters of fluid intravenously. Her blood pressure improved and she was immediately admitted to the intensive care unit, where she developed petechiae. She was treated with intravenous fluids, two anti-staphylococcal antibiotics (oxacillin and clindamycin) as well as a cephalosporin (cefotaxime) with gradual improvement in her overall condition.

Her blood, urine, and CSF cultures remained sterile, but her vaginal culture was positive for abundant *S. aureus*. She was subsequently transferred to the regular in-patient ward and treated for 7 days with anti-staphylococcal antibiotics. The rash gradually faded.

Toxic shock syndrome.

Claire suffered from staphylococcal toxic shock syndrome (TSS), a striking example of the dramatic physiologic alterations caused by superantigens. TSS is a serious disease characterized by rapid onset of fever, a rash, organ failure, and shock. The majority of cases occur in menstruating women, typically in their teenage years, but cases do occur in all age groups. TSS is typically associated with a localized *S. aureus* infection (for example, subcutaneous abscesses, osteomyelitis, and infected wounds), staphylococcal food poisoning, or local colonization, as occurred in the vagina in this case. When kept in the vagina for a long time (>12 hours) tampons soaked with menstrual fluids can enhance the growth of the bacteria which are the source of superantigens. It is unlikely that the tampon played a part in Claire's disease as it was inserted less than 6 hours before the onset of symptoms. As well as TSST-1, toxigenic strains of *S. aureus* can produce other enterotoxins (such as enterotoxin B) that act as superantigens, with similar clinical consequences. In addition, microorganisms other than *S. aureus* secrete superantigens that can cause disease (Fig. 26.2).

Disease	Superantigen	TCR V_β
Definite role for superantigen		
Toxic shock syndrome	TSST-1	$V_\beta 2$
Staphylococcal food poisoning	SEA	$V_\beta 3$, $V_\beta 11$
	SEB	$V_\beta 3$, $V_\beta 12$, $V_\beta 14$, $V_\beta 15$, $V_\beta 17$, $V_\beta 20$
	SEC	$V_\beta 5$, $V_\beta 12$, $V_\beta 13.1$–2, $V_\beta 14$, $V_\beta 15$, $V_\beta 17$, $V_\beta 20$
	SED	$V_\beta 5$, $V_\beta 12$
	SEE	$V_\beta 5.1$, $V_\beta 6.1$–3, $V_\beta 8$, $V_\beta 18$
Streptococcal toxic shock syndrome	SPE-A	$V_\beta 8$, $V_\beta 12$, $V_\beta 14$
Scarlet fever	SPE-B	$V_\beta 2$, $V_\beta 8$
Mycoplasma arthritidis (rodent)	MAM	$V_\beta 17$
Clostridium perfringens	Enterotoxin	$V_\beta 6.9$, $V_\beta 22$
Suspected role for superantigen		
HIV	CMV	$V_\beta 12$
Type I diabetes mellitus	MMTV-like	$V_\beta 7$
Rabies virus	Nucleocapsid	$V_\beta 8$
Toxoplasmosis	?	$V_\beta 5$
Mycobacterium tuberculosis	?	$V_\beta 8$
Yersinia enterocolitica	?	$V_\beta 3$, $V_\beta 6$, $V_\beta 11$
Kawasaki disease	?	$V_\beta 2$, $V_\beta 8$

Fig. 26.2 Superantigens, V_β usage and disease.

Fig. 26.3 Expansion in numbers of V$_{\beta}$2 T cells in toxic shock syndrome (TSS). Panel a, FACS analysis. Peripheral blood T cells from a normal control (left) and a patient with acute TSS (right). Cells are stained with anti-V$_{\beta}$2 monoclonal antibody with a fluorescein tag. There is an increased percentage of V$_{\beta}$2 T cells in the patient. The horizontal axis represents the mean fluorescence intensity. Panel b, time course of persistence of high numbers and return to normal of V$_{\beta}$2 T cells in a patient with TSS.

Consistent with the V$_{\beta}$2 specificity of TSST-1, examination of the circulating lymphocytes from patients in the acute phase of TSS typically reveals a much higher proportion of circulating V$_{\beta}$2 T cells compared with cells using other V$_{\beta}$ segments. As the illness resolves there is a gradual return to near normal proportions. The expansion in the numbers of V$_{\beta}$2 cells can be measured by examining the surface expression of T-cell receptors containing a V$_{\beta}$2 region using immunofluorescence (Fig. 26.3) or by semiquantitative measurement of mRNA transcripts encoding V$_{\beta}$2 T-cell receptor chains using reverse transcription and the polymerase chain reaction (RT-PCR) (Fig. 26.4).

Although all the T cells activated by a given superantigen share a common V$_{\beta}$ region, they will differ in their specificity for conventional peptide antigens. Sequencing of the T-cell receptor from superantigen-activated T cells reveals a different use of D and J gene segments by the β chains and a wide diversity of α chains. These receptors will encompass a wide variety of antigen specificities. In contrast to superantigen, conventional antigen induces the clonotypic expansion of T cells. All the T cells in any given clone will have identical D and J gene segments in their β chains and identical α chains. Because the pool of V$_{\beta}$-restricted T cells activated by superantigen may contain autoreactive T cells, it has been postulated that superantigens may trigger autoimmune disease.

Fig. 26.4 RT-PCR analysis of T-cell receptor V$_{\beta}$ mRNA. Autoradiograms of T-cell receptor chain transcripts amplified by reverse transcription followed by polymerase chain reaction (RT-PCR). T cells from a patient with toxic shock syndrome and a control individual were stimulated with anti-CD3 antibody and IL-2 before the extraction of RNA and generation of cDNA. Each reaction contained specific oligonucleotide primers to expand the particular V$_{\beta}$ gene segment indicated (170–220 base pairs), as well as a C$_{\alpha}$ gene segment (approx. 600 bp) as a control to ascertain that equivalent amounts of mRNA were used. Photograph courtesy of Y. Choy.

Many of the manifestations of TSS are the result of massive and unregulated cytokine production triggered by the activation of immune system cells. TSST-1 is more effective than bacterial lipopolysaccharide in inducing the synthesis and secretion of interleukin (IL)-1 and tumor necrosis factor (TNF)-α by monocytes. TSST-1 is also a potent T-cell mitogen for those T cells whose receptors it engages; it also induces them to produce large amounts of cytokines, including IL-2 and interferon (IFN)-γ.

IL-1 and TNF-α are critical in the induction of the acute-phase response and induce fever and the production of IL-6. IL-1 and TNF-α also activate vascular endothelium and, together with IL-2, increase vascular permeability with the subsequent leakage of fluid from the intravascular space into the perivasculature. It is these effects of the massive overproduction of TNF-α that result in the toxic shock: edema and intravascular volume depletion leading to hypotension and shock with multiple organ failure.

Susceptibility to TSS seems to correlate with a poor antibody response to TSST-1. While the majority of healthy individuals have protective antibody titers to TSST-1, more than 80% of patients with TSS lack anti-TSST-1 antibodies during the acute illness, and most fail to develop anti-TSST-1 antibodies following convalescence. Possible explanations include an inability on the patient's part to mount an antibody response against TSST-1 and staphylococcal enterotoxins, or a specific inhibition of such a response by the toxins.

Questions.

1 How do you determine whether a protein behaves as a superantigen?

2 Explain the rapid progression of clinical symptoms following introduction of superantigen compared with the delay in apparent responses to conventional antigen.

3 What are the potential mechanisms of liver injury in TSS?

4 Is Claire susceptible to another bout of TSS?

CASE 27   Acute Infectious Mononucleosis

Cytotoxic T cells terminate viral infection.

All viruses, and some bacteria, multiply inside infected cells; indeed, viruses are highly sophisticated parasites that do not have a complete biosynthetic or metabolic apparatus of their own and, in consequence, must replicate inside a living cell. Once inside a cell, a pathogen is not accessible to antibodies and has to be eliminated by other means.

Some intracellular bacteria live and multiply in membrane-bound phagosomes within macrophages and are killed by antibacterial agents released into these vacuoles after macrophage activation by CD4 T_H1 cells (see for example Case 30). Viruses, in contrast, together with those bacteria that live in the cytosol, can be eliminated only by destruction of the infected cell itself. This role in host defense is fulfilled by the cytotoxic CD8 T cells of adaptive immunity and the natural killer (NK) cells of innate immunity.

CD8 cytotoxic T cells kill infected cells by recognizing foreign, pathogen-derived peptides that are transported to the cell surface bound to MHC class I molecules (see Fig. 17.1). The peptides carried by MHC class I molecules come from the degradation of proteins in the cytosol and so cytotoxic T cells act against pathogens whose proteins are found in the cytosol of the host

Topics bearing on this case:

Activation of cytotoxic T cells

Cell killing by cytotoxic T cells

Processing and presentation of cytosolic antigens

Fig. 27.1 MHC class I molecules present antigen derived from proteins in the cytosol. In cells infected with viruses, viral proteins are synthesized in the cytosol. Peptide fragments of viral proteins are transported into the endoplasmic reticulum, where they are bound by MHC class I molecules, which then deliver the peptides to the cell surface.

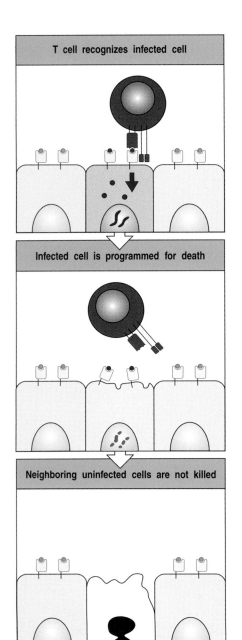

cell at some stage in their life-cycle (Fig. 27.1). The critical role of cytotoxic CD8 T cells in host defense is seen in the increased susceptibility of animals artificially depleted of cytotoxic T cells to many viral and intracytosolic bacterial infections. Mice and humans lacking the MHC class I molecules that present antigen to CD8 cells are also more susceptible to such infections.

Cytotoxic T cells kill their infected targets with great precision and neatness, by inducing apoptosis in the infected cell while sparing adjacent normal cells; this strategy minimizes tissue damage (Fig. 27.2). CD8 cytotoxic T cells release two types of preformed cytotoxin—the fragmentins or granzmes, which seem able to induce apoptosis in any type of target cell, and the protein perforin, which is thought to act as a translocator protein to enable granzymes to cross the membrane of the target cell (Fig. 27.3). A membrane-bound molecule, the Fas ligand, which is expressed on CD8 T cells as well as on some CD4 T cells can also induce apoptosis by binding to Fas on a limited range of target cells. Together, these properties allow the cytotoxic T cell to attack and destroy virtually any infected cell. Cytotoxic CD8 T cells also produce the cytokine interferon (IFN)-γ; this cytokine inhibits viral replication, induces MHC class I expression, and also activates macrophages. As well as combating infection by viruses and intracytosolic bacteria, CD8 T cells are important in controlling some protozoal infections; they are crucial, for example, in host defense against *Toxoplasma gondii*, an intracellular protozoan.

The importance of cytotoxic T cells in the control of viral replication is highlighted by many aspects of Epstein–Barr virus (EBV) infection, which is described in this case study. EBV (also known as human herpesvirus 4) is a member of the virus family Herpesviridae. It has a double-stranded linear DNA genome enclosed in an icosahedral capsid and a lipid envelope and replicates its DNA genome in the host cell nucleus. EBV infects only humans and is one of the most successful infective agents on its obligate host. It can even be thought of as a commensal that only seldom causes injury to the host; anywhere from 60–98% of healthy adults show serological evidence of infection with EBV. The virus infects mainly B cells and epithelial cells.

Fig. 27.2 Cytotoxic T cells kill target cells bearing specific antigen while sparing neighboring uninfected cells. All the cells in a tissue are susceptible to the induction of apoptosis by the cytotoxins of armed effector CD8 T cells but only infected cells are killed. Specific recognition by the T-cell receptor identifies which target cell to kill, and the polarized release of cytotoxic granules (see Fig. 27.3) ensures that neighboring cells are spared.

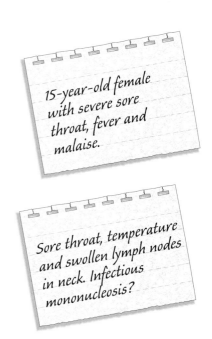

Fig. 27.3 The cytotoxic granules of cytotoxic T cells are released in a polarized fashion. Perforin molecules, as well as several other effector molecules, are contained in the granules of cytotoxic T cells (panel a: G, granules; N, nucleus; M, mitochondria; Go, Golgi apparatus). When a CD8 cytotoxic T cell recognizes its target, the granules are released onto the target cell (panel b, bottom right quadrant). Photographs courtesy of E. Podack.

The case of Emma Bovary: a bad sore throat from a B-cell infection.

Emma Bovary was a healthy 15-year-old when she suddenly developed a very sore throat accompanied by fever and malaise. Her throat was so swollen she had difficulty swallowing. Over the next few days the fever waxed and waned, her sore throat became worse and she became progressively more tired and anorectic (lost her appetite). On the third day of illness her pediatrician noted severe pharyngitis and took a throat culture for β-hemolytic streptococci; the culture proved negative.

Emma's symptoms persisted, and she was unable to eat as she could hardly swallow. She said she had no difficulty breathing but that her left upper abdomen felt slightly uncomfortable. Emma's 1-year-old brother became ill at the same time, but did not have such severe symptoms. He was merely listless and felt warm. He had no particular physical symptoms, and seemed to recover completely after a few days.

On physical examination on the tenth day of illness, Emma appeared very ill. She had a high temperature (38.2 °C), pulse rate of 84, respiratory rate of 18, and blood pressure 85/55. Her mouth was dry and her tonsils were red and enlarged. They met in the midline, leaving a passage of only 2 × 2 cm approximately. Palatal petechiae (very small hemorrhages under the mucosa) could be seen. Her anterior and posterior cervical lymph nodes were swollen and tender (lymphadenopathy); the largest nodes were 2 × 2 cm. Her abdomen felt soft and the liver was enlarged, the edge being palpable 2 cm below the right costal margin. The spleen was also enlarged; the tip was easily palpable under the left costal margin.

A blood test gave a white blood cell count of 18,590 μl⁻¹ with 39% neutrophils, 27% lymphocytes, 22% atypical lymphocytes (very high), and 11% monocytes (high); her

15-year-old female with severe sore throat, fever and malaise.

Sore throat, temperature and swollen lymph nodes in neck. Infectious mononucleosis?

hematocrit was 45% and the platelet count 397,000 μl^{-1}. Serum electrolytes were normal. Another throat culture was obtained and blood tests for Epstein–Barr virus (EBV) were ordered.

In the meantime a presumptive diagnosis of acute infectious mononucleosis was made with complications including partial pharyngeal obstruction and mild dehydration. Emma was admitted to the hospital and received 1 liter normal saline intravenously followed by 20 mg methylprednisolone (a corticosteroid) intravenously every 12 hours.

Her throat culture again proved negative for streptococcus but her blood serum was positive for IgM and IgG antibodies against EBV capsid antigen. Emma improved quickly with the symptomatic treatment and was discharged on the second day after admission to complete her recovery at home.

Acute infectious mononucleosis.

Emma shows many of the clinical features characteristic of acute infectious mononucleosis (IM) induced by the Epstein–Barr virus, a disease also known as glandular fever in some countries. She had severe pharyngitis with petechiae on the palate, swollen lymph nodes in the neck, enlarged liver and spleen, and large numbers of atypical lymphocytes (the mononucleosis after which the disease is named) in her blood. These cells, also known as Downey–McKinlay cells, are large cells with foamy basophilic cytoplasm and fenestrated nuclei. They are mostly T cells, with a preponderance of CD8 cytotoxic T cells, and are present in 90% of patients with IM, where they sometimes constitute the majority of blood leukocytes.

It is these cells that control the acute infection by destroying EBV-infected B cells. The atypical lymphocytosis in IM is a reflection of the increased CD8 T cell cytotoxic activity. In the vast majority of individuals, the infection is brought under control but not eradicated, because the viral genome persists latently in many B cells. In the latent phase some viral antigens are produced and peptides derived from them are presented by MHC I molecules at the surface of the infected cell, thus enabling latently infected cells to be recognized and destroyed by EBV-specific cytotoxic T cells. Some of the latently infected B lymphocytes become transformed; they are able to propagate themselves indefinitely if removed from the presence of EBV-specific cytotoxic T cells, and are potentially malignant. In healthy people after infection, approximately 1 in 10^6 B cells is transformed. In patients with immune deficiency or who are immunosuppressed, infected B cells can grow unchecked. In immuno-deficient patients, EBV infection can cause immunoblastic lymphoma and B-cell lymphoma.

EBV has a long incubation period: the time between primary EBV infection and the onset of illness is 30–50 days. Infection in infancy or early childhood is almost always asymptomatic, or results in mild disease, as evidenced by Emma's younger brother, who probably also had a primary EBV infection. In developed countries, primary infection is delayed in about half of the population to adolescence and early adulthood, so that 30–50% of primary EBV infection results in acute IM.

EBV enters B cells by binding to the B-cell surface molecule CD21 (also called complement receptor 2 (CR2) because it acts as a receptor for the complement fragment C3dg). This receptor is also present on a small subpopulation of T cells and on various types of epithelial cell in the nasopharynx, parotid gland duct, female cervix, and male urethra. After an active phase of viral multiplication, latency is established in B cells and, in some cases, epithelial cells. The EBV DNA is maintained during latency as an extrachromosomal DNA within the nucleus. Virus production is reactivated from time to time and periodic shedding of infectious virus in oral secretions of healthy infected people is common and lifelong. Adolescents like Emma often catch the disease through kissing; although Emma was not yet sexually active, she had dated several boys in her class.

Definitive diagnosis of EBV infection is best made by serological or molecular biological tests. Infected B cells are stimulated to secrete immunoglobulin, producing, among other antibodies, a so-called heterophile IgM antibody whose detection is one of the most widely used diagnostic tests for EBV infection. This heterophile antibody is not specific for EBV antigens but binds to antigens present on heterologous red blood cells (that is, those of other animals, such as sheep or goat) and agglutinates them. In addition, specific antibody responses are generated against several EBV-specific antigens, and the appearance of different antibodies is informative as to the time of infection and the pattern of virus replication (Fig. 27.4). The presence of IgM antibody against the EBV capsid antigen (VCA) indicates that the infection is acute; this antibody declines gradually in the convalescent phase. Because of the long incubation period of the virus, anti-VCA IgG antibody is also detectable at the onset of illness. The continued presence of EBV antigens in the host maintains antibody production throughout life.

Antibodies to so-called early antigens (EA) are produced mainly in the convalescent phase (1–6 months after disease onset); they then disappear. The absence of EA antibody indicates that the virus is mostly quiescent, and is not undergoing replication on a large scale. If the infection is reactivated, EA antibody titers rise again. Antigens expressed later in the viral life cycle, during its latent phase, are the EBNAs (Epstein–Barr nuclear antigens). Appearance of EBNA antibody indicates that the virus has been present in the body for at least a few months.

In very young, immunosuppressed, or immunodeficient patients, antibody formation can be so impaired that serologic diagnosis is impossible. In these cases EBV antigens can be detected by immunofluorescence microscopy on

	Viral capsid antigen; IgM	Viral capsid antigen; IgG	Early antigen	Epstein–Barr virus nuclear antigen
Never exposed	–	–	–	–
Acute infection	+	+	+/–	–
Recent infection	+/–	+	+/–	+/–
Past infection	–	+	–	+
Reactivated or chronic infection	–	+	+/–	+/–

Fig. 27.4 Serologic diagnosis of EBV infection.

blood or tissue (e.g., lymph node) specimens or by *in situ* hybridization for small EBV RNAs (EBERs). Alternatively, viral DNA can be amplified from infected cells and tissue by the polymerase chain reaction with oligonucleotide primers specific for EBV DNA.

EBV infection is mitogenic for B cells, overriding the normal regulatory mechanisms preventing them from dividing. Activation of B cells by virus infection also leads to immunoglobulin production, as noted above. Several EBV genes are critical for B-cell activation. One is the viral gene *BCRF-1,* which encodes a protein, also called VIL-10, very similar to human interleukin-10 (IL-10). The BCRF-1 protein enhances the activation and proliferation of EBV-infected cells. EBV infection also stimulates endogenous synthesis of IL-10 and IL-6, with further autocrine stimulatory effects on B cells. IL-10 also inhibits the T-cell production of IL-2 and IFN-γ, and enhances the production of B-cell stimulatory cytokines such as IL-4. IL-6 may inhibit the ability of NK cells to destroy EBV-infected cells.

In most patients IM is a self-limited disease for which supportive therapy suffices. Nucleoside analogues such as acyclovir or ganciclovir, or the DNA polymerase inhibitor foscarnet, have limited ability to inhibit the replication of EBV *in vitro*. The clinical usefulness of these drugs is so far unproven. They are frequently administered to patients with fulminant disease, or to immunosuppressed patients. Corticosteroids are often prescribed as a palliative measure, especially when airway obstruction is a potential concern. In the most extreme cases, when respiratory distress is present, tonsillectomy can be required. Corticosteroids reduce virus shedding and provide some symptomatic relief due to their anti-inflammatory effects. They do not significantly alter the course of the disease.

The ability of EBV to transform B cells is an extremely useful laboratory tool. When peripheral blood B cells are cultured with EBV, they become immortalized at a relatively high frequency and can be propagated indefinitely *in vitro*. This allows the general study of various aspects of B-cell biology, as well as providing material for the study of individual patients.

The virus-encoded Epstein–Barr virus nuclear antigen 2 (EBNA-2) is critical for transformation. EBNA-2 protein interacts with transcription factors leading to the activation of several host genes such as those encoding B-cell activating molecules such as CD21 (the EBV receptor) and CD23 (FcϵRII, the low-affinity receptor for IgE), and viral genes such as LMP-1 (latent membrane protein-1, the primary oncogene of EBV). EBNA-3, -4, -5, and -6 also have a role in B-cell transformation.

Acute IM is only one of the possible outcomes of EBV infection. The Epstein–Barr virus is in fact named after two workers who studied Burkitt's lymphoma in Africa in the 1960s and first cultured the virus from these patients. This B-cell lymphoma is strongly associated with EBV infection in Africa, but not in other parts of the world; the high rate of EBV infection in early infancy together with the high incidence of malaria in Africa seem to be the predisposing factors. EBV infection is also strongly associated with nasopharyngeal carcinoma in Southeast Asia. This may perhaps be due to a particular strain of EBV that no longer possesses the epitope provoking an immunodominant response in people with a certain HLA class I allele that is present in a high proportion of people in Southeast Asia. Such people are therefore not so efficient at clearing cells infected with this EBV strain, making transformation and eventual malignancy more likely.

Questions.

1 Patients with humoral immunodeficiency (an impaired antibody response) are susceptible to infection with some viruses such as poliomyelitis or enteric viruses but they have no problems with EBV. Why?

2 There is a high risk of EBV-induced lymphoproliferative malignancy after T-cell depleted bone marrow transplantation. Why?

3 In vitro transformation by EBV of B cells from umbilical cord blood rarely fails. Transformation of B cells in blood cultures from some adults is difficult. Why?

4 Why is heterophile antibody produced during EBV infection?

5 Males with X-linked agammaglobulinemia (XLA) never get infected with EBV. How do you explain this?

CASE 28 | Mixed Essential Cryoglobulinemia

Chronic infection can lead to immune-complex disease.

The extent and duration of antigenic stimulation is the major determinant of immunoglobulin levels in the blood and body fluids. Animals reared under germ-free conditions have low levels of immunoglobulins in the blood (hypogammaglobulinemia), whereas individuals with chronic infections have high levels (hypergammaglobulinemia). There are many chronic infectious diseases in which infection persists because of a failure of the immune response to eliminate the causative agent; examples are malaria, leishmaniasis, and hepatitis. There are other, mostly viral, infections in which the immune response fails to clear the infection because of the relative invisibility of the infectious agent to the immune system (as in herpes simplex and hepatitis C virus infection).

The persistent immune response to microbial antigens can lead to the formation of immune complexes. The immune complexes may themselves become antigenic and IgM antibodies are then formed to the IgG in the immune complexes. These anti-IgG IgM antibodies are called rheumatoid factor, because they were first detected in patients with rheumatoid arthritis (see Case 42).

Topic bearing on this case:

Immune-complex disease

Fig. 28.1 Classification of cryoglobulinemia.

Category	Immunological characterization	Associated diseases	Percentage of patients with cryoglobulinemia
I	Monoclonal immunoglobulin	Lymphoproliferative disorders (e.g., multiple myeloma, Waldenström's macroglobulinemia)	10–15
II	Polyclonal IgG and monoclonal IgM rheumatoid factor	Virtually all have chronic hepatitis C virus infection	35–40
III	Polyclonal IgG and polyclonal IgM rheumatoid factor	Chronic hepatitis C virus infection Autoimmune disorders (e.g., systemic lupus erythematosus, Sjögren's syndrome)	50

The serum proteins and immunoglobulins in these immune complexes may have the peculiar property of precipitating in the cold (< 37°C). Such proteins are called cryoglobulins and are not present in normal blood, or are very difficult to detect. The detectable presence of cryoglobulins in the blood (cryoglobulinemia) is associated with the deposition of immune complexes in the kidneys, small blood vessels and other organs, and can cause fatal immune-complex disease (Fig. 28.1).

The case of Billy Budd: immunological consequences of an occult infection with hepatitis C virus.

Billy Budd was a truck driver who had been well all his life until he was 42 years old, when he complained to his wife of chronic fatigue and aching joints for some months. Raised red spots appeared on his shins and his wife urged him to visit their family physician, Dr Melville.

On a physical examination Dr Melville found that Billy's liver was enlarged and that its edge was palpable several centimeters below the rib cage (normally the liver edge is not palpable); the spleen could not be felt. Dr Melville also noted that Billy's lower legs had multiple purpuric spots (raised lesions caused by blood leaking from damaged blood vessels; Fig. 28.2).

Various blood tests, including liver function tests, were ordered and a week later Billy returned to see Dr Melville. His blood tests had revealed a very high level of serum alanine aminotransferase (ALT), at 831 U l^{-1} (normal < 40 U l^{-1}). Other liver function tests, such as those for serum albumin and prothrombin levels, were normal. White blood cell and platelet counts were normal.

When questioned, Billy denied alcohol abuse and intravenous drug abuse, and said that his wife had been his only sexual partner for 25 years. When asked if he had ever received a blood transfusion, he said that he had had one 20 years ago, when he was 22 years old, after he fractured his tibia in a road accident. Dr Melville then ordered tests for antibodies to hepatitis B and hepatitis C viruses. Billy had a high titer of antibody to hepatitis C virus and his serum contained large amounts of

Fig. 28.2 Purpura on the leg of a patient with chronic infection with hepatitis C and cryoglobulinemia. Photograph courtesy of Sanjiv Chopra.

hepatitis C RNA, as determined by the polymerase chain reaction (PCR). A test for blood cryoglobulins was also positive (Fig. 28.3). A urine sample contained 4 + protein (normal 0) and innumerable red blood cells per high-power field, indicating damage due to immune complexes in the glomeruli of the kidney. Dr Melville referred Billy to a gastroenterologist.

Before he was due to see the gastroenterologist, Billy appeared in the emergency room of the New England Deaconess Hospital with fever, severe abdominal pain, and vomiting blood. An arterial perforation was found in the small intestine and intestinal contents had leaked from the perforation into the peritoneum, causing peritonitis. On microscopic examination of a sample of gut wall tissue, the intestinal blood vessels revealed inflammation (vasculitis), with deposition of immune complexes in the vessel walls.

Fig. 28.3 Cryoprecipitate in Billy's blood. Billy's plasma was refrigerated overnight and then placed in a tube that is used to measure the hematocrit. The tube was centrifuged in the cold and the amount of cryoprecipitate was measured. It was 4% of the plasma volume. Photograph courtesy of Sanjiv Chopra.

Mixed essential cryoglobulinemia.

So-called cryoprecipitates—precipitates that appear when sera are stored at low temperatures and disappear at 37°C—were first noted many years ago, before it was appreciated that they were composed of immunoglobulins and other serum proteins. Cryoprecipitates are made by proteins that come out of solution, or at least partly gel, at cold temperatures and dissolve again on rewarming. This definition is based on *in vitro* behavior, however, and the true mechanism of cryoprecipitation remains obscure. It was eventually realized that the appearance of cryoprecipitates in the blood was associated with aching joints, fatigue, and purpura, especially in patients with liver disease due to chronic hepatitis C infection.

When the circulating cryoglobulin complexes are deposited in the walls of blood vessels they activate complement. The ensuing release of the complement fragment C5a, a potent chemoattractant, causes leukocytes to infiltrate the blood vessel walls and cause vasculitis (Fig. 28.4). The inflammation causes the small blood vessels to burst and release blood into the skin, causing purpura. If the complexes are deposited in the synovia of joints, arthritis ensues. Vasculitis in a critical area, such as the brain or intestine, is rare, but can result in fatal bleeding.

Some monoclonal immunoglobulins produced by malignant cells in diseases such as multiple myeloma (IgG) or Waldenström's macroglobulinemia (IgM) may also behave as cryglobulins and precipitate in the cold. The classification of the cryoglobulins is given in Fig. 28.1.

Questions.

1   What treatment do you think the gastroenterologist will suggest to Billy?

2   Rituximab (Rituxan), a monoclonal antibody directed against CD20, has shown some benefit in pilot clinical studies as a novel therapy for cryoglobulinemia. Why might rituximab be expected to help Billy's condition? What might be some of the risks of this therapy?

Fig. 28.4 Comparison of a healthy intestine with an intestine from a patient with cryoglobulinemia. Panel a: healthy intestine. Panel b: intestinal vasculitis from a patient with cryoglobulinemia. The photomicrographs of the gut wall were obtained by endoscopy. Photographs courtesy of Sanjiv Chopra.

CASE 29 | Rheumatic Fever

A streptococcal infection triggers an autoimmune disease of the heart.

Many clinical observations indicate that infection can trigger autoimmune disease, the state in which the body attacks its own tissues, but in most cases the link between the infection and the disease is tenuous. One exception is the childhood disease rheumatic fever, in which the heart becomes inflamed, sometimes leading to serious damage. This disease is a sequel to a sore throat or scarlet fever caused by *Streptococcus pyogenes* group A β-hemolytic streptococci. Rheumatic fever is now far less common than it used to be in industrialized countries, having markedly declined since the 1930s and 1940s, but it is still relatively frequent in Eastern Europe and in developing countries.

Rheumatic fever is an example of an autoimmune disease resulting from 'molecular mimicry', in which an antigen on the infecting agent is similar to an antigen on the host tissues (Fig. 29.1). Antibodies made by the body against the antigen on the pathogen help to clear the infection, but also cause a subsequent autoimmune attack on the patient's own tissues, leading to inflammation and tissue damage.

Topics bearing on this case:

Autoimmune disease

Molecular mimicry

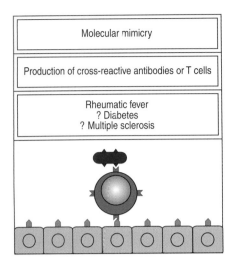

Molecular mimicry

Production of cross-reactive antibodies or T cells

Rheumatic fever ? Diabetes ? Multiple sclerosis

Fig. 29.1 Molecular mimicry is one way in which infectious agents can break self-tolerance. An immune response to bacterial antigens with similarities to self tissue antigens can lead to an autoimmune reaction against the self tissue.

The connection between rheumatic fever and a preceding streptococcal throat infection was first made a century ago. Because neither streptococci nor their degradation products could be found in the heart tissue of patients with rheumatic fever, it was realized that the disease, which follows about 2–4 weeks after a sore throat or scarlet fever, was a consequence of the immune response to the infection. The infection can be quite mild; in many instances patients have no recollection of a recent throat infection, but serological tests show that antibodies against group A streptococcal antigens are present. Rheumatic fever only follows a sore throat (pharyngitis) or scarlet fever caused by *S. pyogenes*, whereas other types of infections caused by *S. pyogenes*, such as those of the skin (impetigo) or the bloodstream do not lead to it.

The inflammation of acute rheumatic heart disease is distinctive, in that foci of myocardial cell death (necrosis) are found surrounded by T lymphocytes, plasma cells, and large macrophages. These inflammatory foci, which resemble granulomas, are called Aschoff nodules after the German pathologist Ludwig Aschoff, who first described them in the early part of the twentieth century. The demonstration that IgG and complement are deposited in cardiac tissue in patients with rheumatic fever was among the first applications of immunofluorescence to immunopathology in the 1950s.

About the time that Aschoff described the distinctive pathological lesions in rheumatic hearts, Rebecca Lancefield started her lifelong work at the Rockefeller Institute in New York to define the Lancefield groupings of pathogenic streptococci, and the virulence factors (M proteins) of group A streptococci. She and her colleagues identified over 80 different types of the M protein against which distinct protective antibodies are made. This antigenic variation among *S. pyogenes* strains means that antibodies made against one strain do not protect against a subsequent infection with another strain. We now know that M proteins contain stretches of amino-acid sequence that are identical with sequences in the myosin and tropomyosin of human cardiac muscle. Thus, the immune response to the M protein produces antibodies that cross-react with cardiac myosin and tropomyosin (Fig. 29.2), and this so-called molecular mimicry is thought to be the major cause of the heart inflammation.

There are many other examples of molecular mimicry. Perhaps the best known is the identity of antigenic determinants on the cell wall of *Escherichia coli* and other normal gut flora with the A and B blood-group antigens present on some people's red blood cells. As a consequence, anyone not possessing these antigens will produce anti-A and anti-B antibodies (isohemagglutinins) once the gut has become colonized with microorganisms, and will react adversely if later given a transfusion with A, AB, or B blood (see Case 45, where this is described).

Only 3% of patients with untreated streptococcal pharyngitis develop rheumatic fever, even though it seems that most M protein serotypes of *S. pyogenes* have the potential to cause it. This raises an interesting question as to whether such patients are genetically or otherwise predisposed to develop rheumatic fever. There is no clear answer at present, although such genetic susceptibility markers have been tentatively identified. It would be an oversimplification to assume that rheumatic fever results simply from the immune response to M proteins. Patients with rheumatic fever have an abnormally brisk immune response to several other streptococcal antigens that cross-react with mammalian tissue.

Fig. 29.2 Immunofluorescent staining of heart muscle with serum obtained from an acute rheumatic fever patient. Photograph courtesy of John Zabriskie.

The case of Ethan Frome: the sudden appearance of a heart murmur after a sore throat.

Ethan, a 4-year-old boy who had always been in good health, complained one morning that he was unable to stand up. He had a fever and his knees were swollen and painful. His mother took him to the pediatrician, who ordered an X-ray of his knees, but no abnormality was found. However, he had an erythrocyte sedimentation rate of 98 mm h^{-1} (normal < 20). He did not improve, and on a return visit a week later the pediatrician detected a heart murmur and referred Ethan to Dr Wharton, a cardiologist at the Children's Hospital. She learned from Ethan's mother that he and two of his siblings had had 'strep throats' 2 months ago, which had been treated with amoxicillin.

On physical examination Ethan weighed 16.8 kg (normal) and was 107 cm tall (normal). His heart rate was 107 (rapid), respiratory rate 22 (normal) and blood pressure 100/47. He had a systolic murmur heard best at the apex of the heart (due to mitral valve insufficiency), and a second murmur heard during diastole at the left upper sternal border (due to insufficiency of the aortic valve). Because of the previous strep infection the cardiologist suspected rheumatic fever and ordered an echocardiogram. This revealed that Ethan had incompetence of his mitral and aortic heart valves and a dilated left atrium and ventricle. At the end of diastole his left ventricle was 3.86 standard deviations above the mean in volume.

Ethan's white blood count was 12.6 × 10^3 μl^{-1}, his platelet count was 440 × 10^3 μl^{-1} (both elevated), and his sedimentation rate was 64 mm h^{-1}. His anti-streptolysin O titer was positive, at 1:1600, and his anti-strep DNase B titer was also positive, at 1:1920. IgG was 1520 mg dl^{-1}, IgA 213 mg dl^{-1}, and IgM 141 mg dl^{-1}, which are all slightly higher than normal for his age.

Ethan's mother was told to keep him in bed and to give him 80 mg of aspirin four times a day. He was also given oral digoxin to increase heart muscle contractility and Lasix (a diuretic) to decrease blood volume and thus place less stress on the heart. He also took 250 mg of penicillin twice a day.

Two weeks later Ethan was much improved; the fever had disappeared and the heart murmurs were fainter. The cardiologist advised his mother to stop the aspirin and the digoxin. One month later, Ethan's diastolic murmur had disappeared. His sedimentation rate had fallen to normal (16 mm h^{-1}) and his anti-streptolysin O titer to 1:400. He was able to resume normal activity, but continued the prophylactic penicillin.

Four-year-old boy with fever and painful knee joints.

Recent strep throat. Rheumatic fever? Do echocardiogram.

Rheumatic fever.

It was not realized until the 1880s that a constellation of symptoms such as migratory arthritis, carditis (inflammation of the heart), and subcutaneous nodules were all part of the disease that we now call rheumatic fever. At that time rheumatic fever was rampant among children and teenagers living in overcrowded slums. In every large city hospitals were built especially for the care of children with this chronic disease, for which there was no satisfactory treatment. In one of these institutions, the House of the Good Samaritan in Boston, Massachusetts, T. Duckett Jones observed more than 1000 patients with rheumatic fever in the 1930s and 1940s. Because there was no laboratory test to establish a diagnosis (there is still no single diagnostic test today) he developed what are still known as the Jones criteria. There are five major and four minor criteria; diagnosis of rheumatic fever depends on the presence of two major criteria or one major and two minor criteria (Fig. 29.3). Ethan showed the two major criteria of carditis and arthritis, and all the minor criteria.

During the Second World War there was an alarming incidence of rheumatic fever among young US Air Force recruits stationed at the Warren air force base in Wyoming. The study of this episode by Charles Rammelkamp and colleagues is a classic of the epidemiological literature and established some fundamentally important facts about how the disease might be prevented. Recruits treated with penicillin within 10 days of acquiring a 'strep throat' did not get rheumatic fever. Furthermore, the strep infection could be acquired only from a 'strep carrier' (someone with a positive throat culture for *S. pyogenes*) who slept no more than two beds away. The infection could not be acquired in any other way, such as through infected bedding. As a result of this study, penicillin prophylaxis, as given to Ethan, became routine for rheumatic fever patients, to prevent further strep infections that might aggravate the heart disease.

A migratory arthritis is one major criterion for a diagnosis of rheumatic fever. It involves the ankles, knees, wrists and elbows, which become hot, red, and swollen in succession. Typically, as the swelling in one joint subsides it flares up in another. The fact that the arthritis responds quickly to aspirin was known even a hundred years ago.

Another major criterion is inflammation of the heart. As in Ethan's case, this usually becomes manifest as previously unheard heart murmurs. The ring of tissue from which the heart valves are suspended becomes inflamed and dilates so that the valves no longer approximate each other when the heart contracts. Thus, the valves become incompetent and this can lead to retrograde flow from the ventricle to the atrium—called mitral valve insufficiency when the left side of the heart is involved. The inflammation also leads to a delay in conduction of contraction signals (prolonged PR interval) that can be discerned on the electrocardiogram and is one of the minor criteria for diagnosis. The echocardiography used in Ethan's diagnosis uses ultrasound (high-frequency sound waves) to create images of cardiac structures. By

Fig. 29.3 Jones criteria for diagnosis of acute rheumatic fever, updated in 1992 by the American Heart Association.

Revised Jones criteria for diagnosis of acute rheumatic fever	
Major criteria	**Minor criteria**
Carditis (inflammation of the heart)	Arthralgia (aching joints without apparent swelling)
Polyarthritis (arthritis successively in several of the large joints)	Fever
Sydenham's chorea (see text)	Increased acute-phase response as judged by a raised erythrocyte sedimentation rate (ESR) or raised levels of C-reactive protein (CRP)
Erythema marginatum (rash)	Prolonged PR interval (delay in conduction of contraction signals in heart, which can be discerned in an electrocardiogram)
Subcutaneous nodules	

combining this with Doppler tests that use sound waves to measure blood flow, cardiologists can assess the structure and function of the heart and its valves. The carditis is best treated with corticosteroids because aspirin is not effective for this purpose.

The third major criterion for diagnosis, Sydenham's chorea, was first described by the English physician Thomas Sydenham in the seventeenth century. This frightening manifestation of rheumatic fever includes sudden uncontrollable jerking of the limbs, facial twitching, and slurred speech, among other signs. These bizarre movements disappear when the patient is asleep, and patients with Sydenham's chorea benefit from sedation. They have antibodies against the cells of the caudate nucleus of the basal ganglia, which interfere with its normal function of controlling body movement.

Another major criterion is erythema marginatum. This distinctive painless rash may appear anywhere on the body as faint red lines, forming circles and other patterns. The fifth major criterion is the occurrence of hard subcutaneous nodules on the extensor surfaces of the limbs, particularly near the elbow and knee joints, but also under the scalp.

Questions.

1   About 20% of strains of group A streptococci do not produce streptolysin O. What would you do to ascertain a recent streptococcal infection in a patient when a diagnosis of rheumatic fever is suspected?

2   Erythema marginatum is a major criterion for the diagnosis of rheumatic fever. Patients with hereditary angioneurotic edema (see Case 4) also develop intermittent erythema marginatum. What does this suggest about the cause of this rash?

3   The migratory arthritis of rheumatic fever responds rapidly to aspirin. How does aspirin work?

4   Aschoff nodules seen in the hearts of patients with rheumatic fever are a form of granuloma. What does this imply about the pathogenesis of rheumatic fever?

5   M proteins bind factor H of the alternative pathway of complement activation. How does this enhance the virulence of Group A streptococci?

CASE 30 Lepromatous Leprosy

T_H1 versus T_H2 responses in the outcome of infection.

Mature naive CD4 T cells emerging from the thymus can differentiate into effector CD4 T cells of several different phenotypes, each with different functions in the immune response. Two of these subsets are called T_H1 and T_H2 (Fig. 30.1). They develop from naive T cells activated by pathogen antigens in the presence of different signals provided by antigen-presenting cells and the local environment. These two types of effector T cell are distinguished chiefly by the cytokines that they secrete when they encounter their target cell. T_H1 cells secrete interleukin-1 (IL-2), interferon-γ (IFN-γ), and lymphotoxin (LT, formerly known as TNF-β); T_H2 cells, often called helper T cells, secrete IL-4, IL-5, and IL-10 (Fig. 30.2).

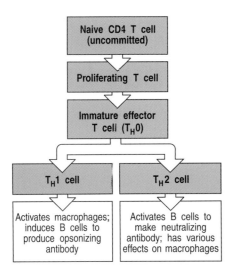

Fig. 30.1 Immature effector CD4 T cells can differentiate into either T_H1 or T_H2 cells. **Fig. 30.1 Immature effector CD4 T cells can differentiate into either T$_H$1 or T$_H$2 cells.** Naive CD4 T cells first respond to peptide:MHC class II complexes by making IL-2 and proliferating. These cells then differentiate into a cell type known as T$_H$0, which has some of the effector functions characteristic of T$_H$1 and T$_H$2 cells. The T$_H$0 cell has the potential to become either a T$_H$1 or a T$_H$2 cell. T$_H$0 cells might also have some effector actions in their own right.

The consequences of the decision to differentiate into T$_H$1 or T$_H$2 cells are profound, as selective production of T$_H$1 cells enables the immune response to activate macrophages and cell-mediated immunity, whereas selective production of T$_H$2 biases the response towards antibody production only. The decision as to which pathway a naive T cell will follow is made during its first encounter with antigen. An antigen that interacts strongly with the T-cell receptor causes the cell to mature into a T$_H$1 cell, whereas a weak interaction leads to T$_H$2 development. Differentiation is also dependent on cytokines. T$_H$1 differentiation is dependent on IL-12 and IFN-γ, whereas T$_H$2 differentiation is dependent on IL-4 (see Fig. 30.2). These cytokines trigger pathways of signal transduction; for example, mice deficient in the intracellular signaling molecule STAT6, which is induced by IL-4, fail to develop T$_H$2 cells.

Other factors influencing the T-cell phenotype are the amount of antigen present and thus which cells are most likely to present it. Large amounts of antigen are usually presented by dendritic cells, which produce IL-12 and therefore favor T$_H$1 differentiation. A limited amount of antigen leads to preferential presentation by antigen-specific B cells that take up antigen more avidly; they induce T$_H$2 differentiation. The co-stimulatory molecules (B7.1 versus B7.2) expressed by the antigen-presenting cells also influence the maturation process in that B7.1 (CD80) is more likely to provoke T$_H$1 development, and B7.2 (CD86) to provoke T$_H$2 development.

Because the decision to differentiate into T$_H$1 versus T$_H$2 cells occurs early in an adaptive immune response, the ability of pathogens to stimulate cytokine production by cells of the innate, nonadaptive immune system has an important role in determining the subsequent course of the response. Infectious agents that invade or nonspecifically activate macrophages and NK cells, as do most viruses and intracellular bacteria such as mycobacteria, induce cells to secrete IL-12, thus favoring the differentiation of T$_H$1 cells, which secrete IFN-γ. This loop is amplified because IFN-γ in turn favors T$_H$1 development and blocks the development of T$_H$2 cells (see Fig. 30.2). IL-12 also enhances the proliferation of T$_H$1 cells but has no effect on T$_H$2 cells because they lack the β chain of the IL-12 receptor and therefore do not respond to the mitogenic effects of IL-12.

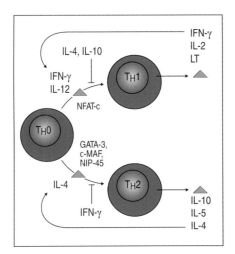

Fig. 30.2 Differentiation of T$_H$0 cells into either T$_H$1 or T$_H$2 cells depends on the cytokines present. IL-4 induces T$_H$2 differentiation, whereas IL-12 and IFN-γ induce differentiation into T$_H$1 cells. IL-4, which is secreted by T$_H$2 cells, also inhibits T$_H$1 development. Similarly, cytokines produced by T$_H$1 cells, for example IFN-γ, inhibit T$_H$2 differentiation. T$_H$1 or T$_H$2 development is driven by transcription factors induced by the cytokines. Induction of the transcription factors GATA-3, c-MAF and NIP-45 in naive T cells leads to T$_H$2 development, whereas induction of NFAT-c leads to T$_H$1 development.

The differentiation of T$_H$2 cells is favored by pathogens, such as parasites, that elicit IL-4 production from specialized subsets of cells that include mast cells, eosinophils, and thymic-derived T cells expressing both the NK1.1 marker and a T-cell receptor of restricted V$_\beta$ and invariant V$_\alpha$ chain usage. This loop is amplified by the cytokines produced by T$_H$2 cells—IL-4 and IL-10. IL-4 promotes the development of T$_H$2 cells and IL-10 blocks T$_H$1 development (see Fig. 30.2).

Once one T$_H$ phenotype becomes dominant in the course of a response, it is difficult to shift the antigen-specific response to the other. One reason for this is that the cytokine products of T$_H$1 and T$_H$2 cells are reciprocally inhibitory (see Fig. 30.2). The outcome of certain infections is greatly influenced by the type of T-cell response elicited. As we see in this case, infection with *Mycobacterium leprae*, the leprosy bacillus, is a good example.

The case of Ursula Iguaran: a T$_H$2 response to the leprosy bacillus has severe consequences.

Ursula first sought medical advice when she was 18 years old, having left her home in Colombia to attend Harvard University on a scholarship. From the age of 16, she had started to notice a gradual loss of sensation on the backs of her hands and had developed hypopigmented lesions over both arms. The lesions progressively became worse and she noticed that she was losing her eyelashes and hairs from her eyebrows. She also experienced recurrent nose bleeds. A month after first noticing the hair loss she decided to seek medical help.

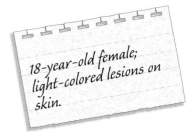

18-year-old female; light-colored lesions on skin.

On examination at the physician's office, Ursula seemed to be well apart from her immediate symptoms. She reported a history of mild asthma, which required treatment with inhaled β$_2$-adrenergic agents on an as-needed basis. Multiple hypopigmented macules (coin-sized raised lesions with ill-defined borders) were evident on her skin, along with cutaneous nodules 1 cm in diameter. These lesions were predominantly on her elbows, wrists, and hands (Fig. 30.3) and showed traces of dried blood; she also had similar lesions on her knees, ears, and buttocks. The absence of eyelashes and the ends of her eyebrows was obvious.

Acid-fast bacilli (? mycobacteria) in skin lesions; leprosy?

Cardiovascular and abdominal examinations were normal; a neurological examination was negative except for a decreased response to pinprick on the outer edges of the right and left hand and the right fourth and fifth fingers. There was a flexion contracture of the fourth and fifth fingers of both hands so that she could not straighten these fingers completely.

On blood test, her hematocrit was 35.1%; white blood count was 7100 μl^{-1}, with 68% neutrophils, 23% lymphocytes, 5% monocytes, and 4% eosinophils (all normal values). Serum electrolytes were normal. Because her symptoms were becoming more severe, she was referred to a dermatologist. She told him that she had grown up in a small village on the Caribbean coast of Colombia where many people, including her mother, had leprosy. The dermatologist performed a biopsy of the lesions on her left arm and right forearm, which revealed numerous acid-fast bacilli in clumps. A routine hematoxylin and eosin stain of lesion tissue showed up numerous Virchow's cells (highly vacuolated cells of the macrophage lineage also known as foam cells) and few lymphocytes (Fig. 30.4). Cultures for acid-fast bacilli were negative.

The suspected diagnosis of lepromatous leprosy led to a more extensive immunologic work-up. Delayed hypersensitivity skin tests with intradermal injections of candida, mumps, and tuberculin antigens showed no reactions when the injection sites were inspected 48 and 72 hours later. Ursula's serum IgG was mildly elevated at 1800 mg dl^{-1} (normal 600–1100 mg dl^{-1}); her IgA and IgM levels were normal.

A diagnosis of lepromatous leprosy was made on the basis of the presence of acid-fast bacilli in the biopsy and Ursula's progressive neurologic symptoms. She was placed on a multiple drug regime consisting of dapsone, clofazamine, and rifampin, drugs that kill *M. leprae*. Her skin lesions gradually flattened and improved.

Fig. 30.3 Cutaneous nodules in lepromatous leprosy. Patients with lepromatous leprosy have multiple skin lesions. This photograph shows subcutaneous nodules on the hand. Photograph courtesy of E. Gonzalez.

Fig. 30.4 Responses to *M. leprae* are sharply differentiated in lepromatous and tuberculoid leprosy. The photographs show sections of lesion biopsies stained with hematoxylin and eosin. Infection with *M. leprae* bacilli, which can be seen in the right-hand photograph as numerous small dark red dots inside macrophages, can lead to two very different forms of the disease. In tuberculoid leprosy (left), growth of the microorganism is well controlled by T_H1-like cells that activate infected macrophages. The tuberculoid lesion contains granulomas (see Case 2) and is inflamed, but the inflammation is localized and causes only local peripheral nerve damage. In lepromatous leprosy (right), infection is widely disseminated and the bacilli grow uncontrolled in macrophages. In the late stages there is severe damage to connective tissues and to the peripheral nervous system. There are several intermediate stages between these two polar forms. Photographs courtesy of G. Kaplan.

Infection with *Mycobacterium leprae* can result in different clinical forms of leprosy	
There are two polar forms, tuberculoid and lepromatous leprosy, but several intermediate forms also exist	
Tuberculoid leprosy	**Lepromatous leprosy**

Tuberculoid leprosy	Lepromatous leprosy
Organisms present at low to undetectable levels	Organisms show florid growth in macrophages
Low infectivity	High infectivity
Granulomas and local inflammation. Peripheral nerve damage	Disseminated infection. Bone, cartilage, and diffuse nerve damage
Normal serum immunoglobulin levels	Hypergammaglobulinemia
Normal T-cell responsiveness. Specific response to *M. leprae* antigens	Low or absent T-cell responsiveness. No response to *M. leprae* antigens

Lepromatous leprosy.

The classical clinical feature in patients with leprosy is the association of cutaneous lesions, neuropathologic changes, and deformities. Leprosy is caused by *Mycobacterium leprae*, which colonizes macrophages and other host cells and multiplies within them. Mycobacteria within macrophages are protected from attack by antibody and can be eliminated only when their host macrophages are activated and produce increased amounts of nitric oxide, oxygen radicals, and other microbicidal molecules. *M. leprae* grows best at 30°C and therefore lesions tend to appear on the extremities—the colder areas of the body—for example the hands, ears, and buttocks as in Ursula's case. Unlike *M. tuberculosis*, *M. leprae* does not grow in culture.

The clinical symptoms of leprosy vary depending on the type of immune response to the mycobacteria. The clinical spectrum is typically divided into two polar forms, tuberculoid and lepromatous leprosy, although intermediate forms exist. Tuberculoid leprosy is associated with a vigorous cell-mediated (T_H1) response against the bacillus. This results in macrophage activation with efficient killing of intracellular mycobacteria, localized tissue damage, and usually a milder clinical picture. In the lepromatous form, the T_H1-cell mediated response is defective and a T_H2 response predominates; this leads to a vigorous but ineffective antibody response against *M. leprae* and dissemination of the bacilli to other sites in the body, which results in further tissue destruction and aggravation of the symptoms. The importance of T_H1-derived IFN-γ in containing mycobacterial infections is further illustrated by the observation that infants with genetic defects in the IFN-γ receptor die from disseminated mycobacterial infections (see Case 21).

Infection with *M. leprae* illustrates a situation in which the same microorganism, in different individuals, can trigger either a T_H1 or a T_H2 response. A T_H1 response predominates in tuberculoid leprosy, in which the mycobacteria are contained within well-circumscribed granulomas and propagate poorly, usually accompanied by subsequent minimal tissue damage. In contrast, a T_H2 response predominates in lepromatous leprosy, in which the mycobacteria propagate rapidly, with resulting extensive tissue damage. Analysis of mRNA isolated from lesions of patients with lepromatous and tuberculoid leprosy illustrates the cytokine patterns in the two forms of the disease. T_H2 cytokines (IL-4, IL-5, and IL-10) dominate in the lepromatous form, whereas T_H1 cytokines (IL-2, IFN-γ, and LT) dominate in the tuberculoid form (Fig. 30.5).

The neurologic damage in leprosy has two main causes. It can arise from bacterial multiplication within Schwann cells—the cells that form the insulating myelin sheath around some nerve cell axons. Disruption of the myelin sheath interferes with the normal conduction of nerve impulses along the axon. In the tuberculoid form, nerve damage also arises from the formation of granulomas and inflammation of the tissue surrounding the nerve. The nerve damage results in dysfunctional nerve terminals, resulting in decreased sensation and eventually a loss of motor function.

The nose bleeds that Ursula experienced are common in leprosy. They are due to large numbers of *M. leprae* in the nasal tissue with extensive involvement of the nasal mucosa, leading to congestion and breakage of blood vessels.

The T_H2 response can influence the course of the infection in various other ways. By binding to mycobacterial antigens displayed on the surface of infected cells, antibodies to the leprosy bacillus can interfere with the action of

Fig. 30.5 Cytokine patterns in leprosy lesions. The cytokine patterns in the two polar forms of leprosy are distinctly different, as shown by Northern blot analysis of the mRNA from lesions of three patients with lepromatous leprosy and three patients with tuberculoid leprosy. Cytokine mRNAs typically produced by T_H2 cells predominate in the lepromatous form whereas cytokines produced by T_H1 cells predominate in the tuberculoid form. Cytokine blots courtesy of R.L. Modlin.

cytotoxic CD8 T cells. CD8 T cells can, in addition to their cytolytic function, also respond to antigen by secreting cytokines. Patients with lepromatous leprosy have CD8 T cells that suppress the T_H1 response by making IL-10 and LT. IL-10 inhibits the development of T_H1 cells and inhibits both cytokine release from macrophages and their capacity to kill internalized microorganisms. LT also inhibits macrophage intracellular killing capacity. Inhibition of macrophages leads to decreased production of IL-12, fewer T_H1 cells, and more T_H2 cells. In contrast, patients with the less destructive tuberculoid leprosy lack suppressor CD8 T cells and thus make a vigorous T_H1 response, leading to macrophage activation and the destruction of the leprosy bacilli.

Questions.

1 Ursula did not respond to candida and mumps antigens, which are common recall antigens, with a delayed-type hypersensitivity reaction. Give a possible explanation.

2 Which cytokine might be beneficial to a patient with lepromatous leprosy?

3 Describe the mechanism for Ursula's hypergammaglobulinemia.

4 Why would Ursula be prone to asthma?

CASE 31

Acquired Immune Deficiency Syndrome (AIDS)

Infection can suppress adaptive immunity.

Certain infectious microorganisms can suppress or subvert the immune system. For example, in lepromatous leprosy, *Mycobacterium leprae* induces T cells to produce lymphokines that stimulate a humoral response but suppress the development of a successful inflammatory response to contain the leprosy bacillus. The leprosy bacillus multiplies and there is a persistent depression of cell-mediated immune responses to a wide range of antigens (see Case 30). Another example of immunosuppression is provided by bacterial superantigens, such as toxic shock syndrome toxin-1. Super-antigens bind and stimulate large numbers of T cells by binding to certain V_β chains of the T-cell receptor, inducing massive production of cytokines by the responding T cells (see Case 26). This, in turn, causes a temporary suppression of adaptive immunity.

Topics bearing on this case:
Failure of cell-mediated immunity
Infection with the human immuno-deficiency virus (HIV)
Control of HIV infection
Drug therapy for HIV infection
ELISA test

Fig 31.1 HIV binds to CD4 T cells through its coat glycoprotein gp120. The gp120 molecule on the surface of the virus binds CD4 on T cells and macrophages; the viral protein gp41 then mediates fusion of the enveloped virus with the target cells, allowing the viral genome to enter the cell.

At the beginning of this century, when tuberculosis was the leading cause of death and fully half the population was tuberculin-positive, it was well known that an intercurrent measles infection would cause a well-contained tuberculosis infection to run rampant and result in death. The mechanism responsible is now known to be the suppression of IL-2 synthesis after binding of measles virus to CD46 on macrophages.

Some of the microorganisms that suppress immunity act by infecting lymphocytes. Infectious mononucleosis or glandular fever is caused by a virus (Epstein–Barr virus) that infects B lymphocytes. The infection activates cytotoxic CD8 T cells, which destroy the B cells in which the Epstein–Barr virus is replicating. In the third week of infection, at the height of activation of CD8 T cells, all adaptive immunity is suppressed. The cytokines responsible for the immunosuppression are not well defined but probably include IL-10 and TGF-β (see Case 27).

The human immunodeficiency virus (HIV) presents a chilling example of the consequences of infection and destruction of immune cells by a microorganism. The T-cell surface CD4 molecule acts as the receptor for HIV (Fig. 31.1). CD4 is also expressed on the surface of cells of the macrophage lineage and they too can be infected by this virus. The chemokine receptors CCR5 and CXCR4 act as obligatory co-receptors for HIV. As we shall see, the primary infection with HIV may go unnoticed, and the virus may replicate in the host for many years before symptoms of immunodeficiency can be seen. During this period of clinical latency, the level of virus in the blood and the number of circulating CD4 cells remains fairly steady but in fact both virus particles and CD4 cells are being rapidly destroyed and replenished, as rounds of virus replication take place in newly infected cells. When the rate at which CD4 cells are being destroyed exceeds the capacity of the host to replenish them, their number decreases to a point where cell-mediated immunity falters. As we have seen in other cases, such as severe combined immunodeficiency (see Case 14), the failure of cell-mediated immunity renders the host susceptible to fatal opportunistic infections.

42-year-old man with a cat scratch that will not heal, and fever.

Lymphocyte count very low. Test for HIV. HIV test positive.

The case of Martin Thomas: a police officer whose past comes back to haunt him.

Martin Thomas is a 48-year-old African-American police officer who has always been in good health. He has been married for 10 years and has one child, an 8-year-old daughter. Six months ago he went to the emergency room at the local hospital complaining of a fever and a swollen right hand. He was admitted to hospital for the hand infection, which was assumed to be the result of a cat scratch. His blood lymphocyte count was found to be very low, so a blood sample was sent to be tested for antibodies against the human immunodeficiency virus (HIV). Both an ELISA (enzyme-linked immunosorbent assay) (Fig. 31.2) and a Western blot (Fig. 31.3) revealed the presence of anti-HIV antibodies. Officer Thomas was referred to Dr Wright, an AIDS specialist, at the Massachusetts General Hospital.

Martin Thomas told Dr Wright that he had had several homosexual encounters before his marriage 10 years ago. He had always been in good health until 6 months before the present consultation, when he began to have drenching night sweats several times a week. Over this period his body weight had gone down from 94.5 kg to 90 kg. He could not remember having any infections other than the one

Fig. 31.2 Use of the enzyme-linked immunosorbent assay (ELISA) to detect the presence of antibodies to the HIV coat protein gp120. Purified recombinant gp120 is coated onto the surface of plastic wells to which the protein binds nonspecifically; residual sticky sites on the plastic are blocked by adding irrelevant proteins (not shown). Serum samples from the individuals being tested are then added to the wells under conditions where nonspecific binding is prevented, so that only binding to gp120 causes antibodies to be retained on the surface. Unbound antibody is removed from all wells by washing, and anti-human immunoglobulin that has been chemically linked to an enzyme is added, again under conditions that favour specific binding alone. After further washing, the colorless substrate of the enzyme is added, and colored material is deposited in the wells in which the enzyme-linked anti-human immunoglobulin is found. This assay allows arrays of wells known as microtiter plates to be read in fiberoptic multichannel spectrometers, greatly speeding the assay.

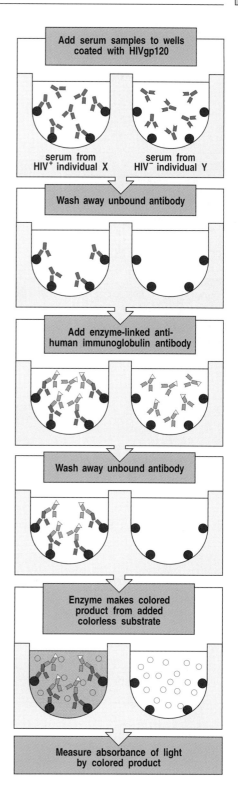

in his hand, nor any rashes, gastrointestinal problems, cough, shortness of breath or any other symptoms. His mother had been 84 years old when she died of a heart attack, and his father had died at age 87 from cirrhosis of the liver, cause unknown. His wife and child were both in good health and his wife had recently tested negative for anti-HIV antibodies. Mr Thomas told Dr Wright that he did not smoke or use intravenous drugs. He drank large amounts of beer at weekends. A cat and a dog were the only pets in the house.

On physical examination his blood pressure was 130/90, his pulse rate 92, and temperature 37.5°C (all normal). Nothing abnormal was found during the physical examination. His white blood cell count was 5800 μl^{-1} (normal), his hematocrit was 31.3 and his platelet count 278,000 μl^{-1} (both normal). His CD4 T-cell count was very low at 170 μl^{-1} (normal 500–1600) and his load of HIV-1 RNA was 67,000 copies ml^{-1}.

Mr Thomas was prescribed trimethoprim sulfamethoxazole for prophylaxis against *Pneumocystis carinii* pneumonia (see Case 14). He was also given a combination antiretroviral therapy consisting of zidovudine (Retrovir, AZT), lamivudine (Epivir, 3TC), and efavirenz (Sustiva). He was counseled about safe sex with his wife.

After 5 weeks of this therapy his HIV-1 viral load declined to 400 copies of RNA ml^{-1} and after 8 weeks to < 50 copies of RNA ml^{-1}, in other words to undetectable levels. In the meantime his CD4 T-cell count rose to 416 μl^{-1}. The prophylaxis for *Pneumocystis* was discontinued. Mr Thomas remains well and active and works full time.

Acquired Immune Deficiency Syndrome (AIDS).

AIDS is caused by the human immunodeficiency virus (HIV) of which there are two known types, HIV-1 and HIV-2. HIV-2 was largely confined to West Africa but now seems to be spreading into Southeast Asia. HIV infections in North and South America and in Europe are exclusively from HIV-1. HIV can be transmitted by homosexual and heterosexual intercourse, by infusion of contaminated blood or blood products, or by contaminated needles, which are the major source of infection among drug addicts. The infection can also be passed from mother to child during pregnancy, during delivery or, more uncommonly, by breastfeeding. Somewhere between 25% and 35% of infants born to HIV-positive mothers are infected. The rate of infection of infants has been decreased fourfold by giving HIV-positive pregnant women the anti-viral agent zidovudine.

Fig 31.3 Western blotting is used to identify antibodies to the human immunodeficiency virus (HIV) in serum from infected individuals. The virus is dissociated into its constituent proteins by treatment with the detergent SDS, and its proteins are separated by SDS-PAGE. The separated proteins are transferred to a nitrocellulose sheet and reacted with the test serum. Anti-HIV antibodies in the serum bind to the various HIV proteins and are detected by using enzyme-linked anti-human immunoglobulin, which deposits colored material from a colorless substrate. This general methodology will detect any combination of antibody and antigen and is used widely, although the denaturing effect of SDS means that the technique works most reliably with antibodies that recognize the antigen when it is denatured.

Contact with the virus does not necessarily result in infection. The standard indicator of infection is the presence of antibodies to the virus coat protein gp120. The initial infection, as in Mr Thomas's case, may pass unnoticed and without symptoms. More often a mild viral illness within 6 weeks of infection is sustained, with fever, swollen lymph nodes, and a rash. It subsides at about the time that seroconversion occurs, and although virus and antibody persist, the patient feels well. A period of clinical latency lasting years, and perhaps even decades, may ensue during which the infected person feels perfectly well. Then he/she begins to experience low-grade fever and night sweats, excessive fatigue, and perhaps oral candidiasis (thrush) in the mouth. Lymph nodes in the neck or axillae or groin may swell. Weight loss may become very marked. These are the prodromal symptoms of impending AIDS. (A prodrome is a concatenation of signs and symptoms that predict the onset of a syndrome.) The number of CD4 T cells in the blood may have been normal up to this time but, with the onset of the prodrome, the CD4 T-cell count begins to fall (Fig. 31.4). When the number of CD4 T cells decreases to the range of 200–400 cells μl^{-1}, the final phase of the illness, which is called AIDS, starts. At this time serious, eventually fatal, opportunistic infections as well as certain unusual malignancies occur (Fig. 31.5).

At any time after the infection, HIV may infect megakaryocytes, which have some surface CD4. Because megakaryocytes are the bone marrow progenitors of blood platelets, extensive infection of megakaryocytes causes the platelet count to fall (thrombocytopenia) and bleeding to occur. HIV may also infect the glial cells of the brain. Glial cells are of monocyte–macrophage lineage and have some CD4 on their surface. The infection of glial cells may cause dementia and other neurological symptoms.

Questions.

1   When Mr Thomas was first seen by Dr Wright, a chest X-ray revealed some enlarged hilar lymph nodes. If a lymph-node biopsy had been obtained, in what way would its histopathology have differed from that of lymph nodes from patients with severe combined immunodeficiency (SCID; see Case 14) or X-linked agammaglobulinemia (XLA; see Case 10)?

2   The course of an HIV infection in adults is very different from that in an infant infected in utero or intrapartum (during birth). What are the

Fig. 31.4 The typical course of infection with HIV. The first few weeks are typified by an acute influenza-like viral illness, sometimes called seroconversion disease, with high titers of virus in the blood. An adaptive immune response follows, which controls the acute illness and largely restores CD4 T cell levels but does not eradicate the virus. Opportunistic infections and other symptoms become more frequent as the CD4 T-cell count falls, starting at around 500 cells μl^{-1}. The disease then enters the symptomatic phase.

major differences between pediatric and adult AIDS and how do you account for them?

3   What are the mechanisms of resistance to the progression of HIV infection?

4   A few individuals, mostly hemophiliacs, are known to have been infected with HIV as long as 20 years ago and yet they remain asymptomatic. What factors may contribute to long-term survival with this infection?

5   Dr Wright told Mr Thomas to take zidovudine, lamivudine, and efavirenz. What are these drugs?

6   What do HIV protease inhibitors do? Does Mr Thomas need one of these drugs?

7   What is the mechanism of CD4 T-cell depletion in HIV infection?

8   What is the most important known determinant of the progression of HIV infection?

9   Which cytokine released during an HIV infection causes weight loss?

Fig. 31.5 A variety of opportunistic pathogens and cancers can kill AIDS patients. Infections are the major cause of death in AIDS, with respiratory infection with *Pneumocystis carinii* being the most prominent. Most of these pathogens require effective macrophage activation by CD4 T cells or effective cytotoxic T cells for host defense. Opportunistic pathogens are present in the normal environment but cause severe disease primarily in immuno-compromised hosts, such as AIDS patients and cancer patients. AIDS patients are also susceptible to several rare cancers, such as Kaposi's sarcoma and lymphomas, suggesting that immune surveillance by T cells may normally prevent such tumors. EBV, Epstein–Barr virus.

Infections	
Parasites	*Toxoplasma* spp. *Cryptosporidium* spp. *Leishmania* spp. *Microsporidium* spp.
Bacteria	*Mycobacterium tuberculosis* *Mycobacterium avium* *intracellulare* *Salmonella* spp.
Fungi	*Pneumocystis carinii* *Cryptococcus neoformans* *Candida* spp. *Histoplasma capsulatum* *Coccidioides immitis*
Viruses	Herpes simplex Cytomegalovirus Varicella zoster

Malignancies
Kaposi's sarcoma (invasive) Non-Hodgkin's lymphoma, including   EBV-positive Burkitt's lymphoma Primary lymphoma of the brain

CASE 32 | Acute Systemic Anaphylaxis

A life-threatening immediate hypersensitivity reaction to peanuts.

Adaptive immune responses can be elicited by antigens that are not associated with infectious agents. Inappropriate immune responses to otherwise innocuous foreign antigens result in allergic or hypersensitivity reactions and these unwanted responses can be serious. Allergic reactions occur when an already sensitized individual is reexposed to the same innocuous foreign substance, or allergen. The first exposure generates allergen-specific antibodies and/or T cells; reexposure to the same allergen, usually by the same route, leads to an allergic reaction.

Acute systemic anaphylaxis is a type I IgE-mediated hypersensitivity reaction (Fig. 32.1) in which the response is so rapid and overwhelming as to be life-threatening. As with any type I hypersensitivity reaction, the first exposure to the allergen generates allergen-specific IgE antibodies, which become bound to Fc receptors (FcεRI) on the surface of mast cells. Reexposure to the same allergen, usually by the same route, leads to an allergic reaction due to cross-linking of the IgE antibodies by the allergen. Antibody cross-linking induces the mast cells to release a variety of chemical mediators, particularly histamine, which increases the permeability of blood vessels, and leukotrienes, which affect smooth muscle, causing bronchospasm (see Fig. 33.4).

Topics bearing on this case:
Class I hypersensitivity reactions
Allergic reactions to food
Mast-cell activation via IgE

Type I immune-mediated tissue damage	
Immune reactant	IgE antibody
Antigen	Soluble antigen
Effector mechanism	Mast-cell activation
	antigen
Example of hyper-sensitivity reaction	Allergic rhinitis, asthma, systemic anaphylaxis

Fig. 32.1 Type I hypersensitivity reactions.

Allergens introduced systemically are most likely to cause a serious anaphylactic reaction through activation of sensitized connective tissue mast cells. The disseminated effects on the circulation and on the respiratory system are the most dangerous, while localized swelling of the throat can cause suffocation. Ingested antigens cause a variety of symptoms through their action on mucosal mast cells (Fig. 32.2).

Any allergen can provoke an anaphylactic reaction but those that most commonly cause acute systemic anaphylaxis are antibiotics, such as penicillin, and other therapeutic drugs (Fig. 32.3). These act as haptens, binding to host proteins. Proteins in food, most commonly milk, eggs, shellfish, legumes, and nuts, can also cause systemic anaphylaxis. Contact with protein antigens found in latex, a common constituent of rubber gloves, is also known to cause anaphylaxis.

Type I allergic responses are characterized by activation of CD4 helper cells (T_H2 cells) and IgE antibody formation. The allergen is captured by B cells through their antigen-specific surface IgM and is processed so that its peptides are presented by MHC class II molecules to T-cell receptors of antigen-specific T_H2 cells. The interleukins IL-4 and/or IL-13 produced by the activated T_H2 cells induce a switch to production of IgE, rather than IgG, by the B cell (see Fig. 11.4).

This case concerns a child who suffered from life-threatening systemic anaphylaxis caused by an allergy to peanuts.

The case of John Mason: a life-threatening immune reaction.

22-month-old child, unconscious, swollen face, difficulty breathing. Give epinephrine immediately.

John was healthy until the age of 22 months, when he developed swollen lips while eating cookies containing peanut butter. The symptoms disappeared in about an hour. A month later, while eating the same type of cookies, he started to vomit, became hoarse, had great difficulty in breathing, started to wheeze and developed a swollen face. He was taken immediately to the emergency room of the Children's Hospital, but on the way there he became lethargic and lost consciousness.

On arrival at hospital, his blood pressure was catastrophically low at 40/0 mmHg (normal 80/60 mmHg). Pulse was 185 beats min^{-1} (normal 80–90 beats min^{-1}), and respiratory rate 76 min^{-1} (normal 20 min^{-1}). His breathing was labored. An anaphylactic reaction was diagnosed. John was immediately given a subcutaneous injection of 0.15 ml of a 1:1000 dilution epinephrine (adrenaline). An intravenous solution of physiologic (0.15 M) saline was started at the rate of 20 ml kg^{-1} body weight h^{-1}. Also 25 mg of the anti-histamine Benadryl (diphenhydramine hydrochloride) and 25 mg of the anti-inflammatory corticosteroid Solu-Medrol (methylprednisolone) were administered intravenously. A blood sample was taken to test for histamine and the enzyme tryptase.

Within minutes of the epinephrine injection, John's hoarseness improved, the wheezing diminished and his breathing became less labored. His blood pressure rose to 50/30 mmHg, the pulse decreased to 145 beats min^{-1} and his breathing to

Connective tissue mast cells		Mucosal mast cells	
Route of allergen entry			
Intravenous: high dose	Subcutaneous: low dose	Inhalation: low dose	Ingestion

Mast-cell activation

| General release of histamine, systemic anaphylaxis | Local release of histamine, wheal-and-flare reaction | Allergic rhinitis (upper airway), caused by increased mucus production and nasal irritation. Asthma (lower airway) due to contraction of bronchial smooth muscle and increased mucus secretion | Contraction of intestinal smooth muscle induces vomiting. Outflow of fluid into gut causes diarrhea. Antigen diffuses into blood vessels and is widely disseminated causing urticaria (hives) or anaphylaxis. |

Fig. 32.2 The dose and route of allergen administration determines the type of IgE-mediated allergic reaction that results. There are two main classes of mast cell, those associated with blood vessels, called connective tissue mast cells, and those found in submucosal layers, called mucosal mast cells. In an allergic individual, all these are loaded with IgE directed against a particular antigen. Allergen in the bloodstream activates connective tissue mast cells throughout the body, resulting in the systemic release of histamine and other mediators. If very large numbers of mast cells are activated, this leads to widespread vasodilation, leading to a catastrophic loss of blood pressure, constriction of airways and swelling of the epiglottis—the anaphylactic shock syndrome. Ingested allergen penetrates across gut epithelia, causing vomiting due to smooth muscle contraction; the food allergen is also disseminated in the bloodstream, causing urticaria and other manifestations of a systemic allergic response.

61 min^{-1}. Thirty minutes later, the hoarseness and wheezing got worse again and his blood pressure dropped to 40/20 mmHg, pulse increased to 170 beats min^{-1} and respiratory rate to 70 min^{-1}.

John was given another injection of epinephrine and was made to inhale an aerosolized solution of 0.15 ml (a 5 mg ml^{-1} solution) albuterol (a β_2-adrenergic agent) in 2 ml physiologic saline. This treatment was repeated once more after 30 minutes. One hour later, the child was fully responsive, his blood pressure was

Blood pressure very low. Anaphylactic reaction.

Fig. 32.3 IgE-mediated reactions to extrinsic antigens. All IgE-mediated responses involve mast-cell degranulation, but the symptoms experienced by the patient can be very different depending on whether the allergen is injected, inhaled, or eaten, and depending on the dose of the allergen.

IgE-mediated allergic reactions			
Syndrome	Common allergens	Route of entry	Response
Systemic anaphylaxis	Drugs Serum Venoms	Intravenous (either directly or following oral absorption into the blood)	Edema Vasodilation Tracheal occlusion Circulatory collapse Death
Acute urticaria (wheal-and-flare)	Insect bites Allergy testing	Subcutaneous	Local increase in blood flow and vascular permeability
Allergic rhinitis (hay fever)	Pollens (ragweed, timothy, birch) Dust-mite feces	Inhaled	Edema of nasal mucosa Irritation of nasal mucosa
Asthma	Danders (cat) Pollens Dust-mite feces	Inhaled	Bronchial constriction Increased mucus production Airway inflammation
Food allergy	Shellfish Milk Eggs Fish Wheat	Oral	Vomiting Diarrhea Pruritis itching Urticaria (hives) Anaphylaxis (rarely)

70/50 mmHg, pulse was 116 beats min^{-1} and his respiratory rate had fallen to 46 min^{-1}. An injection of Susphrine (long-acting epinephrine) was administered, and the child was admitted to the hospital for further observation.

Treatment with Benadryl (25 mg) and methylprednisolone (1 mg kg^{-1} body weight) intravenously every 6 hours was continued for 24 hours, by which time the facial swelling had subsided, and John's blood pressure, respiratory rate, and pulse were normal. He had stopped wheezing and when the doctor listened to his chest with a stethoscope it was clear.

John was taken off all medication and observed for another 12 hours. He remained well and was discharged home. His parents were instructed to avoid giving him foods containing peanuts in any form, and were asked to bring him to the Allergy Clinic for further tests in a few days' time.

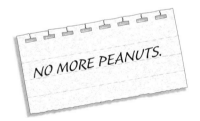

NO MORE PEANUTS.

Acute systemic anaphylaxis.

Anaphylaxis presents a medical emergency and is the most urgent of clinical immunologic events; it requires immediate therapy. It results from the generation and release of a variety of potent biologically active mediators and their concerted effects on a number of target organs. John showed classic rapid-onset symptoms of anaphylaxis, starting with vomiting and swelling of the face and throat, and constriction of the bronchial smooth muscle, which led to his difficulty in breathing. This was soon followed by a catastrophic loss of blood pressure, due to leakage of fluid from the blood vessels. Anaphylaxis can also cause urticaria (hives), heart arrhythmias, and myocardial ischemia, and gastrointestinal symptoms such as nausea, pain, and diarrhea. All these signs and symptoms can occur singly or in combination (Fig. 32.4).

Fatal allergic reactions to the venoms in bee and wasp stings have been recognized for at least 4500 years and account today for approximately 40 deaths each year in the USA. In 1902, Portier and Richet reported that a second injection of a protein from a sea anemone caused a fatal systemic reaction in dogs that had been injected previously with this protein. Since this form of immunity was fatal rather than protective, it was termed 'anaphylaxis' to distinguish it from the prophylaxis (protection) generated by immunization.

Anaphylaxis requires a latent period for sensitization after the first introduction of antigen followed by reexposure to the sensitizing agent, which can be any foreign protein or a hapten. In the early part of the twentieth century the most frequent cause of systemic anaphylaxis was horse serum, which was used as a source of antibodies to treat infectious diseases. The widespread use of penicillin now makes it a major culprit in anaphylactic death in the USA, with a current estimated mortality of around 100 cases each year. Newly introduced chemicals and drugs add continually to the list of agents capable of causing anaphylaxis.

The rate of fatal anaphylaxis from any cause is estimated at 0.4 cases per million individuals per year. The risk of nonfatal anaphylaxis is difficult to assess and has been estimated to range from 0.1% to 1% for people to whom penicillin is administered or who are stung by bees or wasps. Although, in John's case, the reaction was brought on by eating a food, an antigen administered by subcutaneous, intramuscular, or intravenous injection is more likely to induce a clinical anaphylactic reaction than one that enters by the oral or respiratory route. Atopic individuals (see Case 33) are no more susceptible to anaphylaxis than nonatopic individuals.

Fig. 32.4 Mean arterial pressure and epinephrine levels in a representative patient with insect-sting anaphylactic shock. Time 0 indicates the onset of the anaphylactic reaction as reported by the patient. The arrows indicate administration of antihistamines and epinephrine.

Questions.

1 Anaphylaxis results in the release of a variety of chemical mediators from mast cells, such as histamine and leukotrienes. Angioedema (localized swelling caused by an increase in vascular permeability and leakage of fluid into tissues) is one of the symptoms of anaphylaxis. With the above in mind, why did John get hoarse and why did he wheeze?

2 When his parents brought John back to the Allergy Clinic, a nurse performed several skin tests by pricking the epidermis of his forearm with a shallow plastic needle containing peanut antigens. John was also tested in a similar fashion with antigens from nuts as well as from eggs, milk, corn, and wheat. Within 5 minutes John developed a wheal, 10 x 12 mm in size, surrounded by a red flare, 25 x 30 mm (see Fig. 33.7), at the site of application of the peanut antigen. No reactions were noted to the other antigens. A radioallergosorbent test (RAST) was performed on a blood sample to examine for the presence of IgE antibodies to peanut antigens. It was positive. What would you advise John's parents to do?

3   Why was John treated first with epinephrine in the emergency room?

4   What other drug in the epinephrine family was John given?

5   Why was John given a blood test for histamine and the enzyme tryptase?

6   Why was the skin testing for peanuts not done in the hospital immediately after John had recovered, rather than at a later visit?

7   The incidence of peanut allergy is increasing. Why?

8   Can you think of a therapeutic modality that would prevent peanut allergy?

CASE 33 | Allergic Asthma

Chronic allergic disease caused by an adaptive immune response to inhaled antigen.

Chronic allergic reactions are much more common than the acute systemic anaphylaxis reaction discussed in Case 32. Among these are allergic reactions to inhaled antigens (Fig. 33.1), which range in severity from a mild hay fever to potentially life-threatening allergic asthma, the disease discussed in this case. Once an individual is sensitized, the allergic reaction becomes worse with each subsequent exposure to allergen, which not only produces allergic symptoms but also increases the levels of antibody and T cells reactive to the allergen.

Allergic asthma is an example of a type I hypersensitivity reaction. Type I reactions involve the activation of helper CD4 T_H2 cells, IgE antibody formation, mast-cell sensitization, and recruitment of eosinophils. The IgE antibodies formed in sensitized individuals bind to and occupy high-affinity $Fc\varepsilon$ receptors ($Fc\varepsilon RI$) on the surface of tissue mast cells and basophils (Fig. 33.2). When the antigen is encountered again, it cross-links these bound IgE molecules, which triggers the immediate release of mast cell granule contents, particularly histamine and various enzymes that increase blood flow and vascular permeability. This is the early phase of an immediate allergic reaction.

Topics bearing on this case:
Inflammatory reactions
NK T cells
Differential activation of T_H1 and T_H2 cells
IgE-mediated hypersensitivity
Skin tests for hypersensitivity
Radioimmunoassay
Tests for immune function

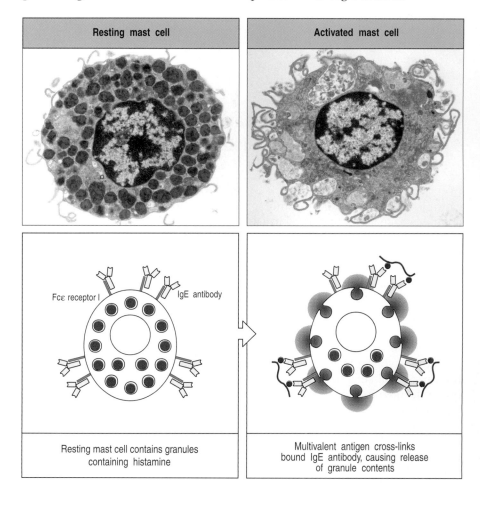

Fig. 33.1 Allergic reactions require previous exposure to the allergen. In this example, the first exposure to pollen induces the production of IgE anti-pollen antibodies, driven by the production of IL-4 by helper T cells (T$_H$2). The IgE binds to mast cells via FcεRI. Once enough IgE antibody is present on mast cells, exposure to the same pollen induces mast-cell activation and an acute allergic reaction, here allergic rhinitis or hay fever. Allergic reactions require presensitization to the antigen or allergen, and several exposures may be needed before the allergic reaction is initiated.

Within 12 hours of contact with antigen, a late-phase reaction occurs (Fig. 33.3). Arachidonic acid metabolism in the mast cell generates prostaglandins and leukotrienes, which further increase blood flow and vascular permeability (Fig. 33.4). Cytokines such as interleukin-3 (IL-3), IL-4, IL-5, and tumor necrosis factor-α (TNF-α) are produced by both activated mast cells and helper T cells, and these further prolong the allergic reaction. The mediators and cytokines released by mast cells and helper T cells cause an influx of monocytes, more T cells, and eosinophils into the site of allergen entry. The late-phase reaction is dominated by this cellular infiltrate. The cells of the infiltrate, particularly the eosinophils, make a variety of products that are thought to be responsible for much of the tissue damage and overproduction of mucus that is associated with chronic allergic reactions. Mast cells are also reactivated and thus the inflammatory reaction is perpetuated. Cytokine-producing NK T cells have also been implicated in allergic asthma.

Fig. 33.2 Cross-linking of IgE antibody on mast-cell surfaces leads to rapid release of inflammatory mediators by the mast cells. Mast cells are large cells found in connective tissue that can be distinguished by secretory granules containing many inflammatory mediators. They bind stably to monomeric IgE antibodies through the very high-affinity Fcε receptor (FcεRI). Antigen cross-linking of the bound IgE antibody molecules triggers rapid degranulation, releasing inflammatory mediators into the surrounding tissue. These mediators trigger local inflammation, which recruits cells and proteins required for host defense to sites of infection. It is also the basis of the acute allergic reaction causing asthma, hay fever, and the life-threatening response known as systemic anaphylaxis (see Case 32). Photographs courtesy of A.M. Dvorak.

Fig. 33.3 The acute response in allergic asthma leads to T$_H$2-mediated chronic inflammation of the airways. In sensitized individuals, cross-linking of specific IgE on the surface of mast cells by inhaled allergen triggers them to secrete inflammatory mediators, causing bronchial smooth muscle contraction and an influx of inflammatory cells, including eosinophils, and T$_H$2 lymphocytes. Activated mast cells and T$_H$2 cells secrete cytokines that also augment eosinophil activation and deregulation, which causes further tissue injury and influx of inflammatory cells. The end result is chronic inflammation, which may then cause irreversible damage to the airways.

Approximately 15% of the population suffer from IgE-mediated allergic diseases. Many common allergies are caused by inhaled particles containing foreign proteins (or allergens) and result in allergic rhinitis (e.g. hay fever), asthma, and allergic conjunctivitis (Fig. 33.5). In asthma, the allergic inflammatory response increases the hypersensitivity of the airway not only to reexposure to allergen but also to nonspecific agents. So exercise, pollutants, and cold air can cause asthma in people without allergies (this is known as intrinsic asthma; allergic asthma is known as extrinsic asthma).

Class of product	Examples	Biological effects
Enzyme	Tryptase, chymase, cathepsin G, carboxypeptidase	Remodel connective tissue matrix
Toxic mediator	Histamine, heparin	Toxic to parasites Increase vascular permeability Cause smooth muscle contraction
Cytokine	IL-4, IL-13	Stimulate and amplify T$_H$2 cell response
	IL-3, IL-5, GM-CSF	Promote eosinophil production and activation
	TNF-α (some stored preformed in granules)	Promotes inflammation, stimulates cytokine production by many cell types, activates endothelium
Chemokine	CCL3 (MIP-1α)	Attracts monocytes, macrophages, and neutrophils
Lipid mediator	Leukotrienes C4, D4, E4	Cause smooth-muscle contraction Increase vascular permeability Stimulate mucus secretion
	Platelet-activating factor	Attracts leukocytes Amplifies production of lipid mediators Activates neutrophils, eosinophils, and platelets

Fig. 33.4 Molecules released by mast cells on activation. Mast cells produce a wide variety of biologically active proteins and other chemical mediators. The enzymes and toxic mediators listed in the first two rows are released from the preformed granules. The cytokines, chemokines, and lipid mediators are synthesized after activation. GM-CSF, granulocyte/macrophage colony-stimulating factor.

Fig. 33.5 IgE-mediated reactions to extrinsic antigens. All IgE-mediated responses involve mast-cell degranulation, but the symptoms experienced by the patient can be very different depending on whether the allergen is injected, inhaled, or eaten, and depending on the dose of the allergen.

IgE-mediated allergic reactions			
Syndrome	Common allergens	Route of entry	Response
Systemic anaphylaxis	Drugs Serum Venoms Peanuts	Intravenous (either directly or after rapid absorption)	Edema Increased vascular permeability Tracheal occlusion Circulatory collapse Death
Acute urticaria (wheal-and-flare)	Insect bites Allergy testing	Subcutaneous	Local increase in blood flow and vascular permeability
Allergic rhinitis (hay fever)	Pollens (ragweed, timothy, birch) Dust-mite feces	Inhaled	Edema of nasal mucosa Irritation of nasal mucosa
Asthma	Danders (cat) Pollens Dust-mite feces	Inhaled	Bronchial constriction Increased mucus production Airway inflammation
Food allergy	Tree nuts Peanuts Shellfish Milk Eggs Fish Wheat	Oral	Vomiting Diarrhea Pruritis (itching) Urticaria (hives) Anaphylaxis (rarely)

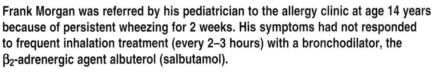

The case of Frank Morgan: a 14-year-old boy with chronic asthma and rhinitis.

14-year-old boy with persistent wheezing.

Frank Morgan was referred by his pediatrician to the allergy clinic at age 14 years because of persistent wheezing for 2 weeks. His symptoms had not responded to frequent inhalation treatment (every 2–3 hours) with a bronchodilator, the β_2-adrenergic agent albuterol (salbutamol).

This was not the first time that Frank had had respiratory problems. His first attack of wheezing occurred when he was 3 years old, after a visit to his grandparents who had recently acquired a dog. He had similar attacks of varying severity on subsequent visits to his grandparents. Beginning at age 4 years, he had attacks of coughing and wheezing every spring (April and May) and towards the end of the summer (second half of August and September). A sweat test at age 5 years to rule out cystic fibrosis, a possible cause of chronic respiratory problems, was within the normal range.

As Frank got older, gym classes, basketball, and soccer games, and just going out of doors during the cold winter months could bring on coughing and sometimes wheezing. He had been able to avoid wheezing induced by exercise by inhaling albuterol 15–30 minutes before taking exercise. Frank had frequently suffered from a night cough and his colds had often been complicated by wheezing.

Frank's chest symptoms had been treated as needed with bronchodilators such as oral theophylline and/or oral or inhaled albuterol. During the past 10 years Frank had been admitted three times to hospital for treatment of his asthma with inhaled bronchodilators and intravenous steroids. He had also made many emergency visits with severe asthma attacks. He had maxillary sinusitis at least three times, ascertained by the presence of fluid in the cavity of his maxillary sinuses, which shows up on the radiographs, each episode associated with exacerbation of his asthma and a green nasal discharge.

Since he was 4 years old Frank had also suffered from intermittent sneezing, nasal itching, and nasal congestion (rhinitis), which always worsened upon exposure to cats and dogs and in the spring and late summer. His face had also swelled up on exposure to a cat or a dog. The nasal symptoms had been treated as needed with oral antihistamines with moderate success. Frank had eczema as a baby but this cleared up by age 5 years.

Family history revealed that Frank's only sibling (a 10-year-old sister), his mother and his maternal grandfather all have asthma. Frank's mother, father, and paternal grandfather suffer from allergic rhinitis.

When he arrived at the allergy clinic, Frank was thin and unable to breathe easily. He had no fever. The nasal mucosa was severely congested, and wheezing could be heard over all the lung fields. Lung function tests were consistent with obstructive lung disease with a reduced peak flow rate (PFR) of 180 liter min^{-1} (normal >350–400 liter min^{-1}), and expiratory volume in the first second of expiration (FEV$_1$) was reduced to 50% of that predicted for his sex, age, and height. A chest radiograph showed hyperinflation of the lungs and increased markings around the airways (Fig. 33.6).

A complete blood count was normal except for a high number of circulating eosinophils (1200 µl^{-1}; normal range <400 µl^{-1}). Serum IgE was high at 1750 IU ml^{-1} (normal <200 units ml^{-1}). Radioallergosorbent assays (RAST) for antigen-specific IgE revealed IgE antibodies against dog and cat dander, dust mites, and tree, grass, and ragweed pollens in Frank's serum. Levels of immunoglobulins IgG, IgA, and IgM were normal. Histological examination of Frank's nasal fluid showed the presence of eosinophils.

Frank was given immediate inhalation treatment of 0.25 ml of the bronchodilator Ventolin (albuterol sulfate) at the clinic, after which he felt better, his PFR rose to 400 liter min^{-1} and his FEV$_1$ rose to 65% of predicted. He was sent home on a 1-week course of the corticosteroid prednisone, taken orally in two doses daily, totalling 1 mg kg^{-1} body weight per day. He was told to inhale albuterol three times a day with extra doses if needed but not more than every 4 hours. He was also put on inhaled Intal (disodium cromoglycate), a drug that decreases mast-cell granule release and inhibits allergen-induced airway reactivity. To relieve his nasal congestion, Frank was given disodium cromoglycate and the steroid beclomethasone (Beconase, Vancenase) to inhale through the nose, and was advised to use an oral antihistamine as needed. He was asked to return to the clinic 2 weeks later for follow-up, and for immediate hypersensitivity skin tests to try to detect which antigens he was allergic to (Fig. 33.7).

On the next visit Frank had no symptoms except for a continually stuffy nose. His PFR and FEV$_1$ were normal. He was maintained on inhaled disodium cromoglycate and beclomethasone, and advised to inhale albuterol as needed. Skin tests for type I hypersensitivity were positive for multiple tree and grass pollens, dust mites, and dog and cat dander. He was advised to avoid contact with cats and dogs. To reduce his exposure to dust mites the pillows and mattresses in his room were covered with zippered plastic covers. Rugs, stuffed toys, and books were removed from his bedroom. He was also started on immunotherapy with injections of grass, tree, and ragweed pollens, and house dust mite antigens, to try to reduce his sensitivity to these antigens.

A year and a half later Frank's asthma continues to be stable with occasional use of albuterol during infections of the upper respiratory tract and in the spring. His rhinitis and nasal congestion now require much less medication.

History of chronic asthma and rhinitis.

Fig. 33.6 Chest radiographs of a patient with asthma. Top: anteroposterior (A-P) view. Bottom: lateral view. The volume occupied by the lungs spans eight to nine rib spaces instead of the usual seven in the A-P view and indicates hyperinflation. The lateral view shows an increased A-P dimension, also reflecting hyperinflation. The bronchial markings are accentuated and can be seen to extend beyond one-third of the lung fields. This indicates inflammation of the airways.

Allergic asthma.

Fig. 33.7 An intradermal skin test. The photograph was taken 20 min after intradermal injections had been made with ragweed antigen (top), saline (middle) and histamine (bottom). A central wheal (raised central swelling), reflecting increased vascular permeability, surrounded by a flare (red area), reflecting increased blood flow, is observed at the sites where the ragweed antigen and the histamine-positive control were introduced. The small wheal at the site of saline injection is due to the volume of fluid injected into the dermis.

Like Frank, millions of adults and children suffer from allergic asthma. About 70% of patients with asthma have a family history of allergy. This genetic predisposition to the development of allergic diseases is called atopy. Wheezing and coughing are the main symptoms of asthma, and both are due to the forced expiration of air through airways that have become temporarily narrowed by constriction of smooth muscle as a result of the allergic reaction. As a consequence of the narrowed airways, air gets trapped in the lung and the lung volume is increased during an attack of asthma.

Once asthma is established, an asthma attack can be triggered not only by the allergen but by viral infection, cold air, exercise, or pollutants such as sulfur dioxide. This is due to a general hyperirritability or hyperresponsiveness of the airway, which leads it to constrict in response to nonspecific shocks and irritants, reducing the air flow. The degree of hyperresponsiveness can be measured by determining the threshold dose of inhaled metacholine (a cholinergic agent) or histamine that results in a 20% reduction in airway flow. Airway irritability correlates positively with serum IgE levels.

It has recently been proposed that NK T cells are crucial to the development of allergen-induced airway hyperresponsiveness and constitute the majority of CD4 T cells producing T_H2-type cytokines in the airways of patients with severe allergic asthma. NK T cells are a subpopulation of thymus-derived T cells that express NK receptors such as NK1.1 and Ly-49 as well as the T-cell receptor:CD3 complex. In humans, most NK T cells express an invariant antigen receptor with a variable region composed of $V_\alpha24$-$J_\alpha15$ paired with $V_\beta11$, and these are called iNK T cells. Unlike conventional T cells, iNK T cells can recognize glycolipid antigens bound and presented by the major histocompatibility complex (MHC) class Ib molecule CD1d, but the natural ligands presented by CD1d *in vivo* remain unclear. On activation, NK T cells rapidly produce large amounts of the T_H1-type cytokine IFN-γ, the T_H2-type cytokines IL-4 and IL-13, TNF-α, and IL-2. Mice deficient in iNK T cells and sensitized to an allergen by intraperitoneal injection do not develop airway inflammation and hyperresponsiveness after challenge with the same allergen in aerosol form. Infusion of NK T cells, however, restores airway hyperresponsiveness. The failure of the iNK T-deficient mice to develop airway hyperresponsiveness is probably due to their lack of the T_H2-type cytokines, such as IL-4 and IL-13, that would normally be secreted by iNK T cells into the lungs and bronchoalveolar lavage fluid in response to inhaled allergen.

Although asthma is a reversible disease, a severe attack can be fatal. The mortality from asthma has been rising alarmingly in recent years. Three classes of drugs are commonly used to treat it. Disodium cromoglycate reduces airway irritability by inhibiting the release of chemical mediators (e.g. histamine) from mast cells and therefore inhibits both the immediate and late phases of the allergic reaction. The precise mechanism of action of this drug is unknown. β_2 agonists (e.g. albuterol) bind to the β_2-adrenergic receptor, which is expressed on the surface of bronchial smooth-muscle cells. β_2 agonists relax smooth muscle, thus rapidly relieving airway constriction, and are helpful in treating the immediate phase of the allergic reaction in the lungs. Anti-inflammatory corticosteroids (e.g. oral prednisone and inhaled beclomethasone) inhibit the cells involved in airway inflammation and are most useful in the late phase of the allergic reaction. The treatment of allergic asthma also includes minimizing exposure to allergens and trying to desensitize the patient by immunotherapy.

Questions.

Fig. 33.8 Eosinophilia. The arrows indicate eosinophils.

1 Explain the basis of Frank's chest tightness and radiograph findings.

2 Explain the failure of Frank's asthma to improve despite the frequent use of bronchodilators, and his response to steroid therapy.

3 Many members of Frank's family are predisposed to the development of IgE-mediated allergic responses, a condition known as atopy. What is the basis for this familial predisposition?

4 Eosinophilia (Fig. 33.8) is often detected in the blood and in the nasal and bronchial secretions of patients with allergic rhinitis and asthma. What is the basis for this finding?

5 What is the basis of the wheal-and-flare reaction that appeared 20 minutes after Frank had had a skin test for hypersensitivity to ragweed pollen?

6 Frank called 24 hours after his skin test to report that redness and swelling had recurred at several of the skin test sites. Explain this observation.

7 Frank asks to be skin tested for sensitivity to rabbits because he wishes to have a rabbit as a pet. What do you do?

8 How would the immunotherapy that Frank received help to alleviate his allergies?

9 Although atopic children are repeatedly immunized with protein antigens such as tetanus toxoid, they almost never develop allergic reactions to these antigens. Explain.

CASE 34 | Atopic Dermatitis

Skin as a target organ for allergy.

As we saw in Cases 32 and 33, allergic or hypersensitivity reactions to otherwise innocuous antigens occur in certain individuals. The site of such reactions and the symptoms they produce vary depending on the type of allergen and the route by which it enters the body. When inhaled or ingested allergens enter the bloodstream they can be carried to the skin, where they induce a chronic inflammation—known as atopic dermatitis—in sensitized (atopic) individuals who possess T cells and IgE specific for an allergen. Allergens that act by direct skin contact can also incite inflammation. The immunologic response of the skin to antigen is complex, resulting in inflammation and localized tissue destruction. This chronic inflammation is called eczema and is the most prominent symptom of atopic dermatitis.

Topics bearing on this case:

IgE and allergic reactions

Inflammatory responses

Cytokines and chemokines

T_H2 cell response

Migration and homing of lymphocytes

Anti-inflammatory effects of corticosteroids

Bacterial superantigens

The immunologic mechanisms underlying atopic dermatitis have much in common with other IgE-mediated allergies (type I hypersensitivity reactions) such as allergic asthma. The T cells involved in atopic dermatitis are principally T_H2 cells, which in this case home to the skin. These and other inflammatory cells are recruited to the skin by the release of cytokines and chemokines from the affected tissue. Cytokines induced by the allergen, or as a result of scratching and damaging the skin, stimulate the keratinocytes, the epidermal skin cells, to produce chemokines and cytokines. Interleukin (IL)-1 and tumor necrosis factor (TNF)-α secreted by keratinocytes in turn stimulate dermal fibroblasts to secrete chemokines that are involved in recruiting T cells and inflammatory cells to the site of injury.

Chemokines are small polypeptides that are synthesized by many cells, including keratinocytes, fibroblasts, T cells, eosinophils, and macrophages (Fig. 34.1). They act through receptors that are members of the G-protein-coupled seven-span receptor family. In atopic dermatitis, activated keratinocytes and dermal fibroblasts secrete the chemokines CCL5 (RANTES), CCL11 (eotaxin), and CCL13 (MCP-4), which attract lymphocytes, eosinophils, and macrophages into the skin. The chemokines and receptors known to be involved in atopic dermatitis are listed in Fig. 34.2.

A chronic inflammatory reaction is sustained by the lymphocytes, eosinophils, and other inflammatory cells that are attracted out of the blood vessels at the site of inflammation. The first step in the process of lymphocyte homing to skin is the reversible binding (rolling) of lymphocytes to the vascular endothelium through interactions between cutaneous lymphocyte antigen (CLA) on the lymphocyte with E-selectin on the endothelial cells. The rolling cells are brought into contact with chemokines retained on heparan

Fig. 34.1 Properties of selected chemokines. Chemokines fall into three related but distinct structural subclasses: CC chemokines have two adjacent cysteines at a particular point in the amino acid sequence; CXC chemokines have the same two cysteine residues separated by another amino acid; and C chemokines (not shown) have only one cysteine residue at this site.

Chemokine class	Chemokine	Produced by	Receptors	Major effects
CXC	CXCL8 (IL-8)	Monocytes Macrophages Fibroblasts Keratinocytes Endothelial cells	CXCR1 CXCR2	Mobilizes, activates, and degranulates neutrophils Angiogenesis
CC	CCL3 (MIP-1α)	Monocytes T cells Mast cells Fibroblasts	CCR1, 3, 5	Competes with HIV-1 Anti-viral defense Promotes T_H1 immunity
	CCL4 (MIP-1β)	Monocytes Macrophages Neutrophils Endothelium	CCR1, 3, 5	Competes with HIV-1
	CCL2 (MCP-1)	Monocytes Macrophages Fibroblasts Keratinocytes	CCR2B	Activates macrophages Basophil histamine release Promotes T_H2 immunity
	CCL5 (RANTES)	T cells Endothelium Platelets	CCR1, 3, 5	Degranulates basophils Activates T cells Chronic inflammation
	CCL11 (Eotaxin)	Endothelium Monocytes Epithelium T cells	CCR3	Role in allergy

Target cell		Receptor	Chemokine
T cells	T$_H$1	CXCR3	CXCL10 (IP-10), CXCL9 (MIG)
	T$_H$2	CCR3 CCR4	CCL11, CCL5, CCL7, CCL13 (MCP-3, MCP-4), CCL5, CCL3, CCL2, CCL17 (TARC), CCL22 (MDC)
	Both	CCR10	CCL27 (CTACK), CCL28
Eosinophils		CCR1	CCL7
		CCR2	CCL13
		CCR3	CCL11, CCL5, CCL7, CCL13

Fig. 34.2 Chemokines that act on T cells and eosinophils.

sulfate proteoglycans on the endothelial cell surface. Chemokine signaling activates the lymphocyte integrin lymphocyte function-associated antigen-1 (LFA-1), leading to firm adherence to intercellular cell adhesion molecule-1 (ICAM-1) on the endothelium followed by extravasation, or departure from the blood vessel, into the skin (see Fig. 3.1). Eosinophils migrate into tissues in a similar way, via an interaction between the integrin VLA-4 on the eosinophil and vascular cell adhesion molecule-1 (VCAM-1) on the vascular endothelium. Once the lymphocytes and other cells have crossed the endothelium into the dermis, their migration to the focus of inflammation is directed by a gradient of chemokine molecules bound to the extracellular matrix.

If the T cells are activated by specific antigen in the tissue that they enter, they produce chemokines such as CCL5, as well as cytokines such as TNF-α, which activates endothelial cells to express E-selectin, VCAM-1, and ICAM-1. The chemokines act on other T cells to upregulate their adhesion molecules, thus increasing the recruitment of T cells into the affected tissue. At the same time, monocytes and polymorphonuclear leukocytes are recruited to these sites by adhesion to E-selectin. The TNF-α and interferon (IFN)-γ released by the activated T cells also act synergistically to change the shape of endothelial cells, resulting in increased blood flow, increased vascular permeability, and increased immigration of leukocytes, fluid, and protein into the site of inflammation. Thus, a few allergen-specific T cells encountering antigen in a tissue can initiate and amplify a potent local inflammatory response that recruits both antigen-specific and accessory cells to the site.

The case of Tom Joad: the itch that rashes.

Tom was admitted to hospital when he was 2 years old because of his worsening eczema. In the week before admission he had developed many open skin lesions (erosions), increased itching (pruritis), redness (erythema), and swelling (edema) of the skin. The lesions oozed a clear fluid, which formed crusts around them.

Tom had suffered from eczema since he was 2 months old, when he developed a scaly red rash over his cheeks and over his knees and elbows (Fig. 34.3). He was breast-fed until 3 months old, when he was given a cow's milk-based formula.

2-year-old boy with severe eczema and family history of allergy.

Fig. 34.3 An infant with severe eczema due to atopic dermatitis. Note the reddened and broken skin especially in places, such as over the knees, where it is subjected to continual stretching and stress. Photograph courtesy of S. Gellis.

Reduce exposure to allergens.

After 24 hours on the formula he started to vomit and to scratch his skin. A casein hydrolyzate formula was substituted for milk and he tolerated this well, but as new foods were added to his diet, the eczema worsened. At 9 months old he developed a wheeze and was treated with bronchodilators. At 2 years old he had hives after eating peanut butter.

Tom's mother suffers from hay fever and his father has atopic dermatitis. The family lives in an old house with 20-year-old carpeting. Tom slept on a 10-year-old mattress surrounded by lots of stuffed animals. There were no pets and his parents did not smoke.

During physical examination Tom was evidently uncomfortable and scratched continuously at his skin. His temperature was 37.9°C, pulse 96, respiratory rate 24, blood pressure 98/58 mmHg, weight 12 kg (10th percentile), and height 90 cm (25th percentile). His skin was very red, with large scales, and with scratched and infected lesions on his face, trunk, and extremities. Pustules were present on his arms and legs, and there were thick scales in his scalp. Thickened plaques of skin (lichenification) with a deep criss-cross pattern were seen around the creases on the insides of his elbows and knees, and on the backs of his hands and feet.

A skin culture was positive for *Staphylococcus aureus* and *Streptococcus pyogenes* Group A. Tom was treated with intravenous oxacillin (an antibiotic), antihistamines, topical steroids, and skin emollients such as coal tar. The infection resolved and his skin healed. Laboratory studies during hospitalization showed a white blood cell count of 9600 μl^{-1}, with 41% polymorphonuclear leukocytes, 26% lymphocytes, and 25% eosinophils (normal 0–5%), 13.1 g dl^{-1} hemoglobin, and a hematocrit of 37.2%. The absolute eosinophil count was elevated at 2400 μl^{-1} (normal 0–500 μl^{-1}). Serum IgE was also much elevated at 32,400 IU ml^{-1} (normal 0–200 IU ml^{-1}).

After discharge from hospital, Tom attended the allergy clinic, where he was tested for sensitivity to a range of allergens. He showed a positive type I allergic skin response (see Fig. 33.7) to numerous inhaled allergens including dust mites, mold spores, animal dander, and a variety of pollen. Tom also tested positive in a type I skin response to milk, cod, wheat, egg white, peanut, and tree nuts (cashew, almond, pecan, walnut, Brazil nut, and hazel nut) but had no reaction to rice or soybean. To determine whether any of these foods could be causing his atopic dermatitis, double-blind placebo-controlled food challenges were carried out. He developed hives and wheezing after eating 1 g of egg white, and hives and eczema after drinking 2 g of powdered milk, but had no reaction to wheat and cod. Four hours after the milk challenge his serum tryptase level was raised, indicating IgE-induced mast-cell degranulation.

His parents were advised to cover his mattress and pillows with a plastic covering and to remove the carpet and stuffed animals from his bedroom to decrease exposure to dust and mite allergens. Tom had a history of allergic reaction to peanuts, so continued avoidance of peanuts and tree nuts was recommended; he was therefore placed on a diet that excluded milk, eggs, peanuts, and tree nuts.

Tom's eczema improved significantly in response to these environmental and dietary control measures, together with the use of emollients and low-potency topical steroids on his skin. The family reduced the risk of skin irritation by avoiding all perfumed soaps and lotions, double-rinsing Tom's laundry, and dressing him in cotton clothes. He did well on this regimen until he was 12 years old when he awoke one day with itchy vesicles on his lower left leg and ankle. The lesions progressed to become painful punched-out erosions, and were diagnosed as herpes simplex infection. He was given the antiviral drug acyclovir and the lesions resolved. Tom is now 15 years old.

Atopic dermatitis.

Atopic dermatitis (AD) is a chronic itchy inflammatory skin disorder that affects about 10% of children to some extent, and can have a significant impact on the quality of life for the patient and his or her family. Many children with AD develop other symptoms of atopy such as food allergies, asthma, and allergic rhinitis (hay fever).

Atopic dermatitis is characterized by a number of immunologic abnormalities including increased IgE levels and impaired cell-mediated immunity. The importance of IgE-mediated hypersensitivity in the condition is suggested by the following clinical findings: most patients with AD have significantly elevated IgE levels; allergens exacerbate AD; and there is usually a personal or family history of atopic disease. Defective cell-mediated immunity is present in up to 80% of patients with AD, as manifested by increased susceptibility to severe viral skin infections, increased susceptibility to chronic fungal infections, and decreased responsiveness to delayed-type hypersensitivity skin testing.

There is evidence that atopic dermatitis is associated with a defective skin barrier. In mice, application of the protein ovalbumin to the surface of skin that has been tape-stripped to remove the fur results in an allergic inflammation of the skin that has features of atopic dermatitis. A mutation that weakens the integrity of the skin has been found in some people with AD. Two separate studies have found that about 20% of patients with AD have mutations in the gene *FLG*, a member of the epidermal differentiation complex (EDC) located on chromosome 1. *FLG* codes for the filament-aggregating protein filaggrin, which is an epidermal barrier protein essential to the integrity and impermeability of the epidermis. Two loss-of-function mutations in *FLG* have been described (R501X and 2282del4) in patients with AD and in people with the related disease ichthyosis vulgaris (in which the skin is finely scaly), which often coexists with atopic dermatitis. Mutations in other skin-specific genes may result in alterations in the permeability and integrity of the skin that are critical in the development of allergic skin inflammation in AD.

The skin lesions in AD contain a mononuclear cell infiltrate that is predominantly in the dermis (Fig. 34.4) and is composed of activated memory CD4 T cells and macrophages. T cells infiltrating the skin lesions express high levels of CLA, which functions as a skin-homing receptor for T lymphocytes by binding to E-selectin. Increased numbers of T_H2 CD4 cells are found in AD, and might be responsible for the raised IgE levels and the weak cell-mediated immunity. Cells infiltrating acute AD skin lesions contain increased levels of the mRNAs for IL-4, IL-5, and IL-13, which are characteristic of a T_H2 response. Consistent with this, allergen-specific T-cell clones from affected skin produce IL-4 but little or no IFN-γ (the latter is characteristic of a T_H1 response). Activated T_H2 cells found in acute skin lesions secrete the cytokines IL-3, IL-4, IL-5, IL-6, IL-10, IL-13, and granulocyte/macrophage colony-stimulating factor (GM-CSF).

IL-4, IL-13, and IL-5 in particular have key roles in the condition. IL-4 and IL-13 cause skin-resident cells such as keratinocytes to secrete chemokines such as CCL11 and CCL17 (TARC), the ligands for the receptors CCR3 and CCR4, which are expressed on eosinophils and skin-homing T cells, respectively.

Fig. 34.4 A section through an acute skin lesion from a patient with atopic dermatitis. The section is stained with hematoxylin and eosin. The dermis has been infiltrated by mononuclear cells (arrowed), which are mostly T cells.

This causes an influx of eosinophils and a further influx of T cells into the skin. IL-4 and IL-13 also increase IgE synthesis by promoting B-cell isotype switching to IgE, and IL-4 stimulates the preferential differentiation of T_H2 cells from naive CD4 T cells on encounter with antigen. IL-5 promotes the differentiation and survival of eosinophils, which secrete a range of inflammatory mediators.

Langerhans cells and macrophages that infiltrate the AD skin lesions have IgE bound to their surface through CD23, a low-affinity receptor for IgE. Langerhans cells and monocytes can act as antigen-presenting cells, which can process and present the allergen to naive T cells and activate them. The presence of bound IgE on the surface of these cells allows them to concentrate the antigen and renders them more efficient at antigen presentation. IgE is also bound to mast cells in the tissues through the high-affinity IgE receptor FcεRI. Signaling through this receptor after the cross-linking of IgE by antigen leads to the production and secretion of IL-4 and IL-5 by the mast cells, thus further biasing the T-cell response to a T_H2 phenotype. In chronic AD, the dermis is also infiltrated by dendritic cells that bear IgE bound to the high-affinity IgE receptor they express on their surface. These cells secrete large amounts of IL-12 upon IgE cross-linking and therefore may be critical in inducing infiltrating T cells to secrete IFN-γ, a hallmark of the chronic lesions in AD.

Treatment of AD is aimed at softening the underlying dry skin and reducing inflammation. AD skin has increased water loss through the epidermis, which reduces its barrier function and facilitates sensitization through the skin. Emollients are the first line of topical therapy for AD because they improve the skin's barrier function. Avoidance of irritants such as soaps and synthetic fabrics, which disrupt this barrier, is crucial in controlling AD. Acute flare-ups of AD require treatment with topical corticosteroid ointments or creams. Antihistamines can be helpful, especially at night, to control the itching.

Questions.

1 Topical steroids were effective in reducing the eczema associated with AD. Why?

2 What other immunomodulatory agents might be effective in AD?

3 Why did Tom develop an extensive herpesvirus infection?

4 Atopic dermatitis is described as the 'itch that rashes.' If patients are prevented from scratching, no rash occurs. What is the relationship of scratching to the rash?

5 Skin infections with staphylococci and other bacteria exacerbate AD. Can you suggest a possible explanation for this?

6 How do environmental allergens such as dust mites, molds, and animal danders trigger atopic dermatitis?

CASE 35 | Drug-Induced Serum Sickness

An adverse immune reaction to an antibiotic.

The intravenous administration of a large dose of antigen can evoke in some individuals a type III hypersensitivity reaction or immune-complex disease (Fig. 35.1). Antigen administration produces a rapid IgG response and the formation of antigen:antibody complexes (immune complexes) that can activate complement.

Owing to the large amount of antigen present and the rapid IgG response, small immune complexes begin to be formed in conditions of antigen excess (Fig. 35.2). Unlike the large immune complexes that are formed in conditions of antibody excess, which are rapidly ingested by phagocytic cells and cleared from the system, the smaller immune complexes are taken up by endothelial cells in various parts of the body and become deposited in tissues. Local activation of the complement system by these immune complexes provokes localized inflammatory responses.

The experimental model for immune-complex disease is the Arthus reaction, in which the subcutaneous injection of large doses of antigen evokes a brisk IgG response. The activation of complement by the IgG:antigen complexes generates the complement component C3a, a potent stimulator of histamine release from mast cells, and C5a, one of the most active chemokines produced by the body. The local endothelial cells are activated by the inter-actions in blood vessels between the immune complexes, complement and circulating leukocytes and platelets. They upregulate their expression of adhesion molecules such as selectins and integrins, which facilitates the emigration of white blood cells from the bloodstream and the initiation of a local inflammatory reaction (Fig. 35.3). Platelets also accumulate at the site, causing blood clotting; the small blood vessels become plugged with clots and burst, producing hemorrhage in the skin (Fig. 35.4).

Topics bearing on this case:
Hypersensitivity reactions
Immune-complex formation
Properties of IgG antibodies
Activation of complement by antigen:antibody complexes
Inflammatory reactions

Type III immune-mediated tissue damage		
Immune reactant	IgG antibody	
Antigen	Soluble antigen	
Effector mechanism	Complement FcR+ cells	
	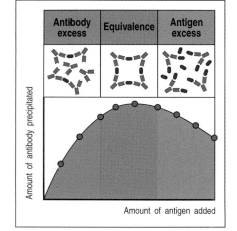	
Example of hyper-sensitivity reaction	Serum sickness, systemic lupus erythematosus	

Fig. 35.1 Type III hypersensitivity reactions. These can be caused by large intravenous doses of soluble antigens (serum sickness) or by an autoimmune reaction against some types of self antigen. The IgG antibodies produced form small immune complexes with the antigen in excess. The tissue damage involved is caused by complement activation and the subsequent inflammatory responses, which are triggered by immune complexes deposited in tissues.

When an antigen is injected intravenously, the immune complexes formed can be deposited at a wide range of sites. When deposited in synovial tissue, the resulting inflammation of the joints produces arthritis; in the kidney glomeruli they cause glomerulonephritis; and in the endothelium of the blood vessels of the skin and other organs they provoke vasculitis (Fig. 35.5).

In the early years of the twentieth century, the most common cause of immune-complex disease was the administration of horse serum, which was used as a source of antibodies to treat infectious diseases, and so this type of hypersensitivity reaction to large doses of intravenous antigen is still known as serum sickness.

This case describes a 12-year-old boy who received massive intravenous injections of penicillin and ampicillin (one of its analogues) to treat pneumonia. He developed a serum sickness reaction to the antibiotics.

The case of Gregory Barnes: serum sickness precipitated by penicillin.

When Gregory was brought to the Children's Hospital Emergency Room, his parents told the physicians that for 2 days he had had high fever (>39.5°C), a cough, and shortness of breath. Before then he had enjoyed excellent health. On physical examination he was pale, looked dehydrated and was breathing rapidly with flaring nostrils. His respiratory rate was 62 min⁻¹ (normal 20 min⁻¹), pulse was 120 beats min⁻¹ (normal 60–80 beats min⁻¹), blood pressure was 90/60 mmHg (normal). When his chest was examined with a stethoscope the emergency room doctors heard crackles (bubbly sounds) over the lower left lobe of his lungs. A chest radiograph revealed an opaque area over the entire lower lobe of the left lung. A diagnosis of lobar pneumonia was made.

A white blood count revealed 19,000 cells μl^{-1} (normal 4000–7000 cells μl^{-1}) with an increase in the percentage of neutrophils to 87% of total white blood cells (normal 60%) and the abnormal presence of immature forms of neutrophils. A Gram stain of

Fig. 35.2 Antibody can precipitate soluble antigen in the form of immune complexes. *In vitro*, the precipitation of immune complexes formed by antibody cross-linking the antigen molecules can be measured and used to define zones of antibody excess, equivalence, and antigen excess. In the zone of antigen excess, some immune complexes are too small to precipitate. When this happens *in vivo* such soluble immune complexes can produce pathological damage to blood vessels.

| Locally injected antigen in immune individual with IgG antibody | Local immune-complex formation | Activation of complement releases inflammatory mediators C5a, C3a, and C4a. C5a also induces mast-cell degranulation | Local inflammation, increased fluid and protein release, phagocytosis, and blood-vessel occlusion |

1–2 hours

Fig. 35.3 The deposition of immune complexes in local tissues causes a local inflammatory response known as an Arthus reaction. In individuals who have already made IgG antibodies to an allergen, the same allergen injected into the skin forms immune complexes with IgG antibody that has diffused out of the capillaries. Since the dose of antigen is low, the immune complexes are only formed close to the site of injection, where they activate complement, releasing inflammatory mediators such as C5a, which in turn can activate mast cells to release inflammatory mediators. As a result inflammatory cells invade the site and blood vessel permeability and blood flow are increased. Platelets also accumulate at the site, ultimately leading to occlusion of the small blood vessels, hemorrhage, and the appearance of purpura.

Gregory's sputum revealed Gram-positive cocci. Sputum and blood cultures grew *Streptococcus pneumoniae* (the pneumococcus).

Gregory was admitted to the hospital and treated with intravenous ampicillin at a dose of 1 g every 6 hours. Gregory gave no history of allergy to penicillin so ampicillin was used to cover both Gram-positive and Gram-negative bacteria. On the fourth day of treatment, he felt remarkably better, his respiratory rate had decreased to 40 min^{-1}, and his temperature to 37.5°C. His white cell count had decreased to 9000 μl^{-1}. Because the *S. pneumoniae* grown from his sputum and blood was sensitive to penicillin, the ampicillin was replaced by penicillin. On his ninth day in hospital, Gregory had no fever, his white cell count was 7000 μl^{-1} and his chest radiograph had improved. Plans for discharge from hospital were made for the following day.

The next morning, Gregory had puffy eyes, and welts resembling large hives on his abdomen. He was given the anti-histamine Benadryl (diphenhydramine hydrochloride) orally and penicillin was discontinued. Two hours later he developed a tight feeling in the throat, a swollen face, and widespread urticaria (hives). With a stethoscope, wheezing could be heard all over his lungs. The wheezing responded to inhalation of the β$_2$-adrenergic agent, albuterol. That evening Gregory developed a fever (a temperature of 39°C), swollen and painful ankles, and his urticarial rash became more generalized. He appeared once again acutely ill.

The rash spread over his trunk, back, neck, and face, and in places had become confluent (Fig. 35.6). Gregory also had reddened eyes owing to inflamed conjunctivae, and had swelling around the mouth. The anterior cervical, axillary and inguinal lymph nodes on both sides were enlarged, measuring 2 × 1 cm in diameter.

Allergy developing; discontinue penicillin immediately.

Fig. 35.4 Hemorrhaging of the skin in the course of a serum sickness reaction.

Route	Resulting disease	Site of immune complex deposition
Intravenous (high dose)	Vasculitis	Blood vessel walls
	Nephritis	Renal glomeruli
	Arthritis	Joint spaces
Subcutaneous	Arthus reaction	Perivascular area
Inhaled	Farmer's lung	Alveolar/capillary interface

Fig. 35.5 The dose and route of antigen delivery determines the pathology observed in type III allergic reactions.

Serum complement components low. Serum sickness.

Fig. 35.6 Urticarial rash as a consequence of a serum sickness reaction.

The spleen was also enlarged, with its tip palpable 3 cm below the rib margin. Ankles and knee joints were swollen and tender to palpation, and were too painful to move very far. The child was alert and his neurologic examination was normal.

Laboratory analysis of a blood sample revealed a raised white blood cell count (19,800 μl^{-1}) in which the predominant cells were lymphocytes (72% compared with the normal 30%). Plasma cells were detected in a blood smear, although plasma cells are normally not present in blood. The erythrocyte sedimentation rate, an indicator of the presence of acute-phase reactants in the blood, was elevated at 30 mm h^{-1} (normal <20 mm h^{-1}). His total serum complement level and his serum C1q and C3 levels were decreased.

A presumptive diagnosis of serum sickness was made and Gregory was given Benadryl and Naprosyn (naproxen), a nonsteroidal anti-inflammatory agent. On the following day, the rash and joint swellings were worse and the child complained of abdominal pain. There were also purpuric lesions, caused by hemorrhaging of small blood vessels under the skin, on his feet and around his ankles. There was no blood in his stool.

Later in the day, Gregory became agitated, and had periods of disorientation when his speech was unintelligible and he could not recognize his parents. A CT scan of his brain proved negative, as did an examination of his cerebrospinal fluid for the presence of inflammatory cells, increased protein concentration, and decreased sugar concentration, all of which are indicators of infection and inflammation. However, his electroencephalogram was abnormal, with a pattern that suggested diminished circulation in the posterior part of the brain.

His white blood count rose to 23,700 cells ml^{-1} and his erythrocyte sedimentation rate to 54 mm h^{-1}. Red cells and protein were now present in the urine. A skin biopsy from a purpuric area on his foot showed moderate edema (swelling) around the capillaries and in the dermis, as well as perivascular infiltrates of lymphocytes in the deeper dermis. Immunofluorescence microscopy of the biopsy tissue with the appropriate antibodies revealed deposition of IgG and C3 in the perivascular areas.

Gregory was started on the anti-inflammatory corticosteroid prednisone and all his symptoms improved progressively; the joint swelling and splenomegaly resolved over the next few days. He was soon able to walk and was discharged 7 days after the onset of his serum sickness on a slowly decreasing course of prednisone and Benadryl. On follow-up examination 2 weeks later, Gregory had no IgE antibodies to penicillin or ampicillin, as detected by both immediate hypersensitivity skin tests and by an *in vitro* radioallergosorbent test (RAST). His parents were instructed that Gregory should never be given any penicillin, penicillin derivatives, or cephalosporins.

Serum sickness.

The classic symptoms of serum sickness that Gregory showed were first described in great detail by Clemens von Pirquet and Bela Schick in a famous monograph entitled *Die Serumkrankheit* (Serum Sickness), published in 1905. Schick subsequently translated this monograph into English and it was reissued by Williams and Wilkins in 1951. It is fascinating to read this short work in the light of current knowledge. In the 1890s it had become common practice to treat diphtheria with horse serum containing antibodies taken from horses that had been immunized with diphtheria toxin. Immune horse serum was also used to treat scarlet fever, which was then a life-threatening illness. Von Pirquet and Schick made systematic observations on tens of children who developed the symptoms and signs of serum sickness at the St Anna's Children's Hospital in Vienna and described the classic symptoms of the disease. They correctly surmised that serum sickness was due to an immunologic reaction to horse serum proteins in their patients (Fig. 35.7).

Experimental models of serum sickness were developed in the 1950s by Hawn, Janeway, and Dixon, who injected rabbits with large amounts of bovine serum albumin or bovine gamma globulin. They noted that the rabbits developed glomerulonephritis just at the time when antibody to the foreign protein first appeared in the rabbit serum, accompanied by a profound and transitory fall in the serum complement level. By this time immunochemistry had advanced to the point where it was possible to show that the disease was caused by the formation and deposition of small immune complexes.

Although horse serum is no longer used in therapy, other foreign proteins are still administered to patients. Antitoxins to snake venom are produced in various animal species and mouse monoclonal antibodies are used in clinical practice. However, the commonest causes of serum sickness today are antibiotics, particularly penicillin and its derivatives, which act as haptens. These drugs bind to host proteins that serve as carriers and thus can elicit a rapid and strong IgG antibody response.

Serum sickness, although very unpleasant, is a self-limited disease that terminates as the immune response of the host moves into the zone of antibody excess. It can prove fatal if it provokes kidney shutdown or bleeding in a critical area such as the brain. Its course can be ameliorated by anti-inflammatory drugs such as prednisone and anti-histamines. It is also unlike the other types of hypersensitivity in that a reaction can appear on first encounter with the antigen, if it is long-lived and given in a sufficiently large dose. This seems to have been the case for Gregory.

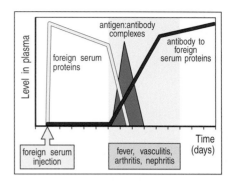

Fig. 35.7 Serum sickness is the classic example of a transient syndrome mediated by immune complexes. An injection of large amounts of foreign proteins, in this case derived from horse serum, leads to an antibody response. These antibodies form immune complexes with the circulating foreign proteins. These complexes activate complement and phagocytes, inducing fever, and are deposited in small blood vessels, inducing the symptoms of vasculitis, glomerulonephritis, and arthritis. All these effects are transient and resolve when the foreign protein is cleared from the system.

Questions.

1   Hives (urticaria) and edema about the mouth and eyelids were the first symptoms of serum sickness developed by Gregory. What caused these early symptoms?

2   At one point Gregory became confused and disoriented and did not recognize his parents. His cerebrospinal fluid was normal and a CT scan of his brain was normal. However, an electroencephalogram displayed an abnormal pattern of brain waves. What produced these clinical and laboratory findings?

3   What other manifestations of vasculitis were noted in Gregory?

4   Gregory had enlarged lymph nodes everywhere and his spleen was also enlarged. If you had a biopsy of a node what would you expect to see?

5   Gregory had a brisk 'acute-phase response'. What is it and what causes it?

6   Penicillin can cause more than one type of hypersensitivity reaction. What laboratory test gave the best evidence that Gregory was suffering from a disease caused by immune complexes?

7   When Gregory returned for a follow-up clinic visit, a skin test for immediate hypersensitivity was performed by intradermal injection of penicillin. He did not respond. Does this mean that an incorrect diagnosis was made and that he did not have serum sickness due to penicillin?

CASE 36 Celiac Disease

Chronic tissue damage in the gut caused by hypersensitivity to a common food antigen.

Celiac disease (also known as celiac sprue or gluten-sensitive enteropathy) is an immune-mediated inflammatory disease of the gastrointestinal tract caused by a permanent sensitivity to wheat, barley and rye. The gluten in wheat and related proteins in barley and rye are degraded into antigenic peptides that trigger an immunologic process that leads to diarrhea, malabsorption and, ultimately, nutritional deficiencies and failure to thrive. There is a description of celiac disease dating back to the 1st century AD, but it was not until the 20th century that the trigger for the disease was identified. Willem Karel Dicke, a Dutch pediatrician, noted that children with celiac disease had a clear improvement in their symptoms during the Second World War when cereals were scarce. After the war and reinstatement of cereals in their diet, children with celiac disease relapsed, and wheat was recognized as the major culprit.

Topics bearing on this case:
Mucosal immunity
Allergy and hypersensitivity

Celiac disease is perhaps the most common genetic disorder, with a prevalence of 0.5–1.0% in many populations. The principal genetic determinants of the disease have been identified as the highly variable HLA class II DQA and DQB genes located on chromosome 6 in the major histocompatibility complex. Celiac disease is singular among immunologic diseases in that the exogenous antigen, the autoantigen, and the genetic susceptibility have all been clearly described and therefore it provides a useful model for studying the genesis of immunologically mediated diseases.

Demeter O'Reilly: a little girl who lost weight and became weak.

Demeter O'Reilly was a healthy baby who grew normally until 12 months old, when her parents noticed that she was less energetic than previously and seemed less interested in eating. She was observed closely by her pediatrician for the next 3 months during which time she developed a distended abdomen, passed large, foul-smelling stools, and lost nearly 2 kg (4 lb) in weight. Her parents also noticed that Demeter was becoming progressively weaker and more lethargic, to the point where she no longer wanted to walk.

At birth, Demeter weighed 3.8 kg, was 52 cm long, and had a head circumference of 36 cm (all within the normal range). She grew normally until nearly 15 months of age and reached all the developmental milestones appropriately.

Both of her parents were of Irish American ancestry and were healthy, as was her 3-year-old sister. There was no known history of metabolic, neurologic, or gastrointestinal disease in children in the family. There was no history of autoimmune diseases, including type 1 diabetes mellitus.

Given the progression of her symptoms and the weight loss, Demeter was referred to the Children's Hospital Emergency Room at 15 months. When she arrived she appeared pale and chronically ill. She weighed 9 kg (5th percentile), her height was 77.5 cm (50th percentile) and her head circumference was 46 cm (50th percentile). The examination made Demeter fretful but she could be consoled. She was mildly dehydrated, and had a protuberant abdomen and significant muscle wasting in her buttocks, legs, and arms (Fig. 36.1). No enlargement of liver, spleen or lymph nodes was evident.

Laboratory tests revealed anemia, with a hemoglobin level of 10 g dl^{-1} (normal 12–14 g dl^{-1}). Her white cell count and platelets were normal. She had a mild elevation of liver enzymes (transaminitis) with aspartate aminotransferase at 48 U l^{-1} (normal range 2–40 U l^{-1}) and alanine aminotransferase at 46 U l^{-1} (normal range 3–30 U l^{-1}). Her albumin was normal at 3.7 g dl^{-1}. No evidence of metabolic disease was found, and so serologic samples for testing for celiac disease were sent to the laboratory. Demeter tested positive for anti-endomysial IgA antibodies (autoantibodies against the connective tissue sheathing muscle—the endomysium) at a titer of 1:2560 and for anti-tissue transglutaminase IgA antibodies at a level greater than 118 units (U) ml^{-1} (>25 U ml^{-1} is considered positive). She was positive for anti-gliadin IgA antibodies at greater than 104 U ml^{-1} (>23 U ml^{-1} is considered positive) and anti-gliadin IgG antibodies at greater than 77 U ml^{-1} (>28 U ml^{-1} is considered positive).

Given Demeter's symptoms and the positive serologic results, the decision was made to carry out a biopsy of her gastrointestinal tract. An upper gastrointestinal endoscopy revealed edema and flattening of the mucosal folds in the duodenal

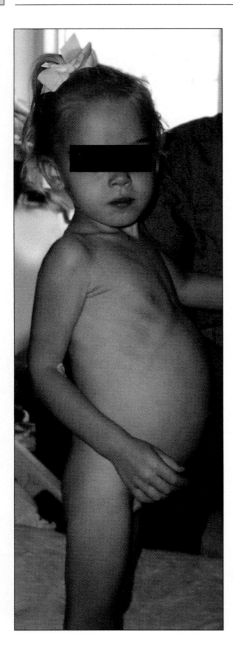

Fig. 36.1 A child with celiac disease. Note the protuberant abdomen, muscle wasting and malnourished appearance. Photograph courtesy of Dascha Weir.

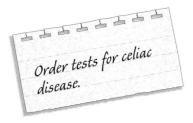

Order tests for celiac disease.

| Celiac disease | Control |

Fig. 36.2 Comparison of mucosal tissue taken by internal biopsy from a patient with celiac disease and from a healthy intestine. The celiac disease tissue is on the left, the healthy tissue on the right. Note the flattened mucosal surface, crypt hyperplasia, and extensive inflammatory infiltrate in the celiac disease biopsy. Photograph courtesy of Donald A. Antonioli.

bulb. Multiple biopsies were taken from the stomach and duodenum. Review of the pathology of the tissues showed subtotal to total villous atrophy and increased intraepithelial lymphocytes in the second portion of the duodenum and in the duodenal bulb (Fig. 36.2). The clinical picture, serology and biopsy results confirmed the diagnosis of celiac disease.

Demeter's family received extensive nutritional counseling and she was started on a gluten-free diet. At the 6-month follow up, she was doing extremely well, with a good appetite and an excellent energy level. She had normal bowel movements once a day and had no abdominal pain or vomiting. Her weight was now in the 75th percentile and her height was at the 50th percentile for her age. She appeared a robust and healthy toddler with a normal abdomen and muscle bulk. Serologic tests revealed no anti-endomysial and anti-tissue transglutaminase antibodies. Her family had availed themselves of the many sources of advice and support available to families with celiac disease, and felt comfortable with her gluten-free diet.

Tests for celiac disease positive. Do duodenal biopsy.

Celiac disease (gluten-sensitive enteropathy).

The prevalence of celiac disease varies by population, and it has been estimated that 0.5–1.0% of the general population in Europe and the United States are affected. However, variability in presentation and delay in onset of symptoms probably lead to the disease being under-diagnosed. Ninety to 95% of patients with celiac disease express HLA class II DQ2, and the remainder express DQ8. The disease clusters in families, and first-degree relatives of affected patients have a prevalence of 5–10%. In addition, patients with type I diabetes mellitus, Down syndrome (trisomy 21), Turner syndrome (45,XO), Williams syndrome (congenital facial and cardiac abnormalities with hypercalcemia), autoimmune thyroiditis, and selective IgA deficiency are at higher risk than the rest of the population of developing celiac disease.

Children with celiac disease classically come to the physician's attention between 6 and 24 months of age with abdominal distension, diarrhea, malabsorption, and weight loss. Rarely, children in celiac crisis are seen, with severe diarrhea, dehydration, and marked disturbance in electrolyte balance; they require immediate treatment with intravenous corticosteroid. Older children may have more subtle symptoms but will also have diarrhea, nausea, vomiting, and abdominal pain. Interestingly, celiac disease with constipation is sometimes seen. Atypical features of the disease include hypoplasia of the dental enamel of the permanent teeth, osteopenia/osteoporosis, short stature, and delayed puberty. Dermatitis herpetiformis, an intensely itchy rash consisting of small solid papules and water-filled vesicles, is a dermatologic manifestation of celiac disease; small-bowel biopsies in patients with dermatitis herpetiformis reveal lesions that are consistent with gluten-sensitive enteropathy. Both the rash and the gastrointestinal lesions remit if gluten is removed from the diet.

Adults can also develop celiac disease. An iron-deficiency anemia that does not respond to oral therapy is the most common presentation in adults; gastrointestinal symptoms such as diarrhea and flatulence sometimes predominate. It is also important to consider celiac disease in patients with unexplained bone fractures, transaminitis, or neurologic symptoms including peripheral neuropathy or ataxia, as well as in women with infertility.

The development of highly sensitive and specific serologic tests for celiac disease has greatly aided its diagnosis. Serologic tests are also important to monitor compliance with a gluten-free diet, as no anti-gluten antibodies can be detected when patients avoid gluten completely. Many different serologic tests are in use; for initial screening, the current guidelines recommend testing for IgA antibody against human recombinant tissue transglutaminase (TTG). Testing for anti-endomysial IgA antibody is frequently used, but measurement of this test is observer-dependent and is therefore subject to error. Positive titers of anti-TTG and anti-endomysial antibodies are almost 100% successful in predicting celiac disease lesions in symptomatic patients. Anti-gliadin antibodies are less predictive, because elevated levels of anti-gliadin IgG antibodies are found in patients with inflammatory bowel disease and in healthy people. IgA deficiency occurs in 2% of children with celiac disease and can lead to misinterpretation of low antibody levels, so it is important to measure IgA levels. Patients with known IgA deficiency should be screened for anti-TTG IgG antibodies. Regardless of the serologic findings, all suspected cases should be confirmed by a biopsy of the small intestine. Characteristic pathology includes partial to total villous atrophy and lymphocyte infiltration with crypt hyperplasia in the small intestine.

The treatment for celiac disease is lifelong adherence to a gluten-free diet. This is challenging to comply with and is also more expensive than a normal diet. It is essential to give patients and their families extensive nutritional counseling as well as support in their attempts to maintain the diet.

Untreated celiac disease leads to significant illness and an increased risk of mortality. Children with untreated celiac disease are most likely to have problems with growth and decreased bone mineralization, which will be corrected on a gluten-free diet. Women with untreated celiac disease are more likely to have spontaneous abortions, babies of low birth weight and a shorter duration of breastfeeding. Lastly, there is an increase in mortality due to cancer in adults with celiac disease, primarily enteropathy-associated T-cell lymphoma. Recent results suggest that strict compliance with a gluten-free diet may reduce the risk of cancer compared with those patients who do not comply.

The pathogenesis of celiac disease is now quite well understood. The exogenous antigen is gluten, a mixture of gliadin and glutenin polypeptides present in wheat. Ingested proteins are usually degraded into small residues by gastric, pancreatic, and intestinal enzymes during transit through the gastrointestinal tract. In most cases, these residues are not appropriate stimulators of the immune system because they are not big enough to bind major histocompatibility complex (MHC) molecules. However, gluten contains a proline-rich peptide of 33 amino acids (the 33-mer) that survives transit through

Transamidation of gliadin peptide and activation of gliadin-specific T cell. Activation of plasma cells and other lymphocytes.

Cytokines released by activated lymphocytes lead to damage to intestinal epithelium

Fig. 36.3 The pathogenesis of celiac disease. Inflammation of the small intestine is caused by CD4 T cells responding to peptides derived from gluten that are deamidated by tissue transglutaminase and presented by HLA-DQ8 or HLA-DQ2 molecules. APC, antigen-presenting cell.

the gastrointestinal tract and arrives intact in the small bowel. This peptide passes through the gastrointestinal lining into the subepithelial space. Such trafficking is usually prevented by the competent tight intercellular junctions of the gastrointestinal epithelium. A number of factors may damage the epithelial barrier, including infection, gluten-induced upregulation of zonulin (an intestinal peptide involved in the regulation of tight junctions), or mechanical stress.

In the subepithelial space the peptides are modified by enzymes, including TTG, that increase their antigenicity, and they are picked up and presented by antigen-presenting cells to CD4 T cells, thereby initiating an inflammatory response that is self-perpetuating in the presence of gluten and damages the gastrointestinal tract. The enzyme TTG deamidates the peptide making it more antigenic, and has also been identified as an autoantigen in the disease (Fig. 36.3). The key genetic determinants of CD have been identified as HLA DQ2 or DQ8. The interplay of these three factors underpins the immune-mediated damage that is seen in patients with celiac disease (Fig. 36.4).

After presentation of the immunogenic peptides to CD4 T cells, an inflammatory response is initiated that is characterized by a CD4 T_H1 response, with interferon-γ (IFN-γ) as the predominant cytokine. We know relatively little about the effector mechanisms that give rise to the celiac lesion. It is likely that some of the intestinal damage is mediated via a bystander effect of IFN-γ, but there may be other mechanisms involved. Examination of gastrointestinal lesions in patients shows infiltration of the intestinal mucosa by both CD4 lamina propria lymphocytes and intraepithelial CD8 T cells. As described above, there is significant evidence for the role of CD4 cells, but the role of the CD8 cells remains elusive.

The anti-TTG antibodies produced in celiac disease are of clinical importance in diagnosis and in tracking disease activity, but the mechanism of their production is not yet fully understood. It has been proposed that hapten–carrier-like complexes form between TTG and complexes of gliadin, which activate helper T cells that then provide help to TTG-specific B cells. When gluten is removed from the diet, antibody levels decrease and the inflammation and injury resolve.

Fig. 36.4 Immunofluoresence staining of the mucosa of the small intestine from an untreated patient with celiac disease. Tissue transglutaminase (TTG, pink), HLA-D (green), and T cells (CD3, purple) have been specifically stained for. Note the close spatial relationship between TTG, the antigen-presenting cells expressing HLA-DQ, and T cells just beneath the intestinal epithelium.

The innate immune system may also be stimulated by gluten. There is evidence that gluten-derived peptides induce the release of IL-15 by intestinal epithelial cells. IL-15 activates antigen-presenting dendritic cells in the lamina propria and also leads to the upregulation of expression of the cell-surface protein MIC-A by epithelial cells. Potentially cytotoxic CD8 intraepithelial lymphocytes in the mucosal epithelium can be activated via their NKG2D receptors, which recognize MIC-A, and they then kill MIC-A-expressing epithelial cells. These innate immune responses may create some intestinal damage and may also induce the co-stimulatory molecules necessary for initiating an antigen-specific CD4 T-cell response to other parts of the α-gliadin molecule. The ability of gluten to stimulate both innate and adaptive immune responses may explain its unique ability to induce celiac disease.

Questions.

1 Given the high prevalence of celiac disease and the difficulty in maintaining a gluten-free diet, are there any new therapies on the horizon for this condition?

2 What are the current recommendations regarding screening of first-degree relatives and high-risk groups for celiac disease?

3 How is a food determined to be gluten free?

4 Can patients with celiac disease eat oats?

CASE 37 Contact Sensitivity to Poison Ivy

A delayed hypersensitivity reaction to a hapten.

Allergic or hypersensitivity reactions can be elicited by antigens that are not associated with infectious agents, for example pollen, dust, food, and chemicals in the environment. They do not usually occur on the first encounter with the antigen, but a second or subsequent exposure of a sensitized individual causes an allergic reaction whose symptoms will depend on the type of antigen, the route by which it enters the body, and the cells involved in the immune response. These unwanted responses can cause distressing symptoms, tissue damage, and even death. These are the same reactions that would be provoked by a pathogenic antigen that behaved in the same way. When they are not helping to clear up an infection, however, these damaging side-effects are clearly unwanted.

Topics bearing on this case:

Allergic reactions

T-cell priming (sensitization)

Preferential activation of T_H1 cells

Inflammatory response

Apoptosis

Anti-inflammatory effects of corticosteroids

Type IV immune-mediated tissue damage		
Immune reactant	T cells	
Antigen	Soluble antigen	Cell-associated antigen
Effector mechanism	Macrophage activation	Cytotoxicity
	T$_H$1	CTL
	cytotoxins	
Example of hyper-sensitivity reaction	Contact dermatitis, graft rejection	
	Rheumatoid arthritis	Diabetes mellitus

Fig. 37.1 There are four types of immune-mediated tissue damage. Types I–III are antibody-mediated and are distinguished by the different types of antigens recognized and the different classes of antibody involved. We have seen examples of type I responses in Case 32 and Case 33 and of type III in Case 35. Type IV reactions are T cell-mediated and can be subdivided into two classes. In the first class, tissue damage is caused by T$_H$1 cells, which activate macrophages, leading to an inflammatory response. On encounter with antigen, effector T$_H$1 cells secrete cytokines, such as interferon-γ, that activate macrophages, and to a lesser extent mast cells, to release cytokines and inflammatory mediators that cause the symptoms. In the second class of type IV reactions, damage is caused directly by cytotoxic T cells that attack tissue cells presenting the sensitizing antigen on their surface. The delay in the appearance of a type IV hypersensitivity reaction is due to the time it takes to recruit antigen-specific T cells and other cells to the site of antigen localization and to develop the inflammatory response. This is in contrast to the IgE-mediated type I allergic reactions, where the mast-cell degranulation that causes many of the symptoms is achieved by native antigen directly cross-linking the preformed IgE molecules on the surface of the mast cells. These reactions occur within 20–30 minutes of contact with antigen. Because a delayed hypersensitivity response involves antigen processing and presentation to achieve T-cell activation, quite large amounts of antigen need to be present at the site of contact. The amount of antigen required is two or three orders of magnitude greater than that required to initiate an immediate allergic reaction.

There are four main types of hypersensitivity reactions, which are distinguished by the type of immune cells and antibodies involved, and the pathologies produced. The one discussed here is an example of a type IV (delayed hypersensitivity) reaction (Fig. 37.1). Many allergic reactions occur within minutes or a few hours of encounter with the antigen, but some take a day or two to appear (Fig. 37.2). These are the delayed hypersensitivity reactions (type IV hypersensitivity). Delayed hypersensitivity reactions are mediated by T cells only, either T$_H$1 CD4 T cells or cytotoxic CD8 T cells, or sometimes both. Antibodies are not involved. The reactions can be triggered by foreign proteins or by self proteins that have become modified by attachment of a hapten, such as a small organic molecule or metal ions. A common type of delayed hypersensitivity reaction is contact dermatitis, a skin rash caused by direct contact with the antigen.

Fig. 37.2 The time course of a delayed-type hypersensitivity reaction. The first phase involves uptake, processing, and presentation of the antigen by local antigen-presenting cells. In the second phase, T$_H$1 cells that were primed by a previous exposure to the antigen migrate into the site of injection and become activated. Because these specific cells are rare, and because there is no inflammation to attract cells into the site, it may take several hours for a T cell of the correct specificity to arrive. These cells release mediators that activate local endothelial cells, recruiting an inflammatory cell infiltrate dominated by macrophages and causing the accumulation of fluid and protein. At this point, the lesion becomes apparent.

| Antigen is injected into subcutaneous tissue and processed by local antigen-presenting cells | A T$_H$1 effector cell recognizes antigen and releases cytokines that act on vascular endothelium | Recruitment of phagocytes and plasma to site of antigen injection causes visible lesion |

24–72 hours

Delayed hypersensitivity reactions fall into two classes (see Fig. 37.1). In the first, the damage is due to an inflammatory response and tissue destruction by T_H1 cells and the macrophages they activate. In the second class of delayed hypersensitivity reactions, tissue damage is caused mainly by the direct action of antigen-specific cytotoxic CD8 T cells on target cells displaying the foreign antigen. Some antigens may cause a combination of both types of reaction.

This case describes the most frequently encountered delayed hypersensitivity reaction in the United States—contact dermatitis due to the woodland plant poison ivy.

The case of Paul Stein: a sudden appearance of a severe rash.

Fig. 37.3 Blistering skin lesions on hand of patient with poison ivy contact dermatitis.

Paul Stein was 7 years old and had enjoyed perfect health until 2 days after he returned from a hike with his summer camp group, when itchy red skin eruptions appeared all along his right arm. Within a day or two, the rash had spread to his trunk, face, and genitals. His mother gave him the antihistamine Benadryl (diphenhydramine hydrochloride) orally to suppress the itching, but this only gave partial relief. The rash did not improve and a week after it first appeared he attended the dermatology clinic at the Children's Hospital.

Physical examination revealed large patches of raised, red, elongated blisters, oozing a little clear fluid, on his body and extremities (Fig. 37.3). Paul also had swollen eyelids and a swollen penis. There was no history of fever, fatigue, or any other symptom. A contact sensitivity reaction to poison ivy was diagnosed.

He was given a corticosteroid-containing cream to apply to the skin lesions three times a day, and Benadryl (25 mg) to take orally three times a day. He was asked to shampoo his hair, wash his body thoroughly with soap and water, and cut his nails short.

Two days later, his parents reported that, although no new eruptions had appeared, the old lesions were not significantly better. Paul was then given the corticosteroid prednisone orally, starting at a dose of 2 mg kg^{-1} body weight per day, which was gradually decreased over a period of 2 weeks. The topical steroid cream was discontinued.

Within a week, the rash had almost disappeared. Upon stopping the prednisone there was a mild flare-up of some lesions and this was controlled by application of topical steroid for a few days. Paul was shown how to identify poison ivy in order to avoid further contact with it, and told to wear long pants and shirts with long sleeves on any future hikes in the woods.

Contact sensitivity to poison ivy.

The reaction to poison ivy is the most commonly seen delayed hypersensitivity reaction in those parts of the United States where the plant grows wild. The absence of fever or general malaise accompanying the rash, and Paul's otherwise excellent health except for the skin lesions, point to a contact sensitivity reaction rather than to a viral or bacterial infection or some other underlying

Pentadecacatechol

$(CH_2)_7CH = CHCH_2CH = CH(CH_2)_2CH3$

Fig. 37.4 The chemical formula of pentadecacatechol, the causative agent of contact sensitivity to poison ivy.

long-term illness. The appearance of the rash just 2 days after Paul returned from a hiking trip where he could easily have been in contact with poison ivy virtually clinches the diagnosis.

Contact dermatitis due to poison ivy is caused by a T-cell response to a chemical in the leaf called pentadecacatechol (Fig. 37.4). On contact with the skin, this small, highly reactive, lipid-like molecule penetrates the outer layers, and binds covalently and nonspecifically to proteins on the surfaces of skin cells, in which form it functions as a hapten. Most people are susceptible and sensitivity, once acquired, is life-long.

On the very first contact with the plant, the individual becomes sensitized. The haptenated self proteins are ingested by specialized phagocytic cells in the skin (Langerhans' cells) into intracellular vesicles, where they are cleaved into peptides. Some of these peptides will have hapten attached. The peptides bind to MHC class II molecules in the vesicles and are presented as MHC: peptide complexes on the Langerhans' cell surface. If these Langerhans' cells reach a lymph node, they become antigen-presenting cells that can activate naive hapten-specific T cells to become recirculating hapten-specific effector T_H1 cells. The activated effector T_H1 cells can then react with haptenated peptides presented by Langerhans' cells and skin macrophages at the site of contact with the plant to initiate a local delayed hypersensitivity reaction in the skin. As the haptenated peptides can persist in the skin for days, it is possible to get a hypersensitivity reaction to poison ivy even after touching it just once.

The appearance of Paul's rash 2 days after his suspected exposure to poison ivy is typical of a delayed hypersensitivity reaction. The haptenated self peptides presented on skin macrophages and Langerhans' cells at the site of contact with poison ivy are initially recognized by the small number of activated hapten-specific T_H1 cells within the pool of recirculating T cells. On encounter with the haptenated peptides, these T_H1 cells release chemokines (cell-attractant molecules), cytokines, and cytotoxins that initiate an inflammatory reaction and also kill cells directly (Fig. 37.5).

One of the cytokines produced by the T_H1 cells is interferon-γ (IFN-γ), whose main effect in this context is to activate macrophages. The subsequent macrophage activity causes many of the symptoms of the delayed hypersensitivity reaction. Macrophages activated by IFN-γ release cytokines and inflammatory mediators such as interleukins, prostaglandins, nitric oxide (NO), and leukotrienes. The combined effects of T-cell and macrophage activity cause a local inflammatory response and tissue damage at the site of the contact with poison ivy.

Antigen is processed by tissue macrophages and stimulates T_H1 cells

T_H1

chemokines   cytokines   cytotoxins

IFN-γ	TNF-α/LT
Induces expression of vascular adhesion molecules. Activates macrophages increasing release of inflammatory mediators	Local tissue destruction. Increased expression of adhesion molecules on local blood vessels
Chemokines	**IL-3/GM-CSF**
Macrophage recruitment to site of antigen	Monocyte production by bone marrow stem cells

Fig. 37.5 The delayed-type (type IV) hypersensitivity response is directed by cytokines released by T_H1 cells stimulated by antigen. Antigen in the local tissues is processed by antigen-presenting cells and presented on MHC class II molecules. Antigen-specific T_H1 cells can recognize the antigen locally at the site of injection, and release chemokines and cytokines that recruit macrophages to the site of antigen deposition. Antigen presentation by the newly recruited macrophages then amplifies the response. T cells may also affect local blood vessels through the release of TNF-α and LT (lymphotoxin) and stimulate the production of macrophages through the release of IL-3 and GM-CSF (granulocyte–macrophage colony-stimulating factor). Finally, T_H1 cells activate macrophages through the release of IFN-γ, and kill macrophages and other sensitive cells through the release of LT or by the expression of Fas ligand.

The red, raised, blistering skin lesions of poison ivy dermatitis are due to the infiltration of large numbers of blood cells into the tissue at the site of contact, combined with the localized death of tissue cells and the destruction of the extracellular matrix that holds the layers of skin together. One of the first actions of effector T_H1 cells on contact with their antigen is to release the cytokines TNF-α and LT (lymphotoxin), and chemokines such as CCL5 (formerly known as RANTES). TNF-α in particular increases the expression of adhesion molecules on the endothelium lining near post-capillary venules and increases vascular permeability so that macrophages and other leukocytes adhere to the sides of the blood vessel. This aids their migration from the bloodstream into the tissues in response to the secreted chemokines.

Once at the contact site, macrophages are activated and themselves release cytokines and other inflammatory mediators, which attract more monocytes, T cells, and other leukocytes to the site, thus helping to amplify and maintain the inflammatory reaction. The blood vessels also dilate, which causes the redness associated with the rash. The inflammatory mediators also act on mast cells to cause degranulation and the release of histamine, which is the main cause of the itching that accompanies the reaction.

Tissue destruction, which is a feature of delayed hypersensitivity reactions, is caused both by cytokines and by direct cell–cell interactions. The TNF-α and LT released by T_H1 cells and macrophages act at the same TNF receptors, which are expressed on virtually all types of cell, including skin cells. Stimulation of these receptors induces a 'suicide' pathway in the cells—apoptosis—which causes their death. Activated CD4 T cells also express the Fas ligand (FasL) in their plasma membrane, which interacts with the ubiquitously expressed cell-surface molecule, Fas, to cause the death of the target cell by apoptosis. Other mediators released by activated T cells, such as the enzyme stromelysin, degrade the proteins of the extracellular matrix, which maintain the integrity of the skin.

Lipid-like haptens such as pentadecacatechol can also cause the priming and activation of cytotoxic CD8 T cells, as small fat-soluble molecules can enter the cytosol of skin cells directly by diffusing through the plasma membrane. Once inside, pentadecacatechol binds to intracellular proteins. Peptides generated from the haptenated proteins in the cytosol are delivered to the cell surface associated with MHC class I molecules. These target cells are recognized and attacked by antigen-specific CD8 cytotoxic T cells, which have become primed and activated on a previous encounter with the antigen. The outcome of all these reactions are the raised, red, weeping blisters characteristic of sensitivity to poison ivy.

Corticosteroids are the standard treatment for hapten-mediated contact sensitivity as they inhibit the inflammatory response by inhibiting the production of many of the cytokines and chemokines. Corticosteroids are lipid-like molecules that can diffuse freely across plasma membranes. Once inside the cell, they bind to receptor proteins in the cytoplasm. The receptor:steroid complex enters the nucleus, where it controls the expression of a number of genes. Of relevance here, it induces the production of an inhibitor of the transcription factors required to switch on transcription of the cytokine and chemokine genes. In mild cases of poison ivy dermatitis, topical steroids applied locally are sufficient. In more severe cases such as Paul's, oral steroids are needed to achieve a concentration necessary to inhibit the inflammatory response.

Paul was given antihistamines to block the histamine receptors and counteract the action of the histamine released from mast cells. Antihistamines also counteract itching caused by substances other than histamine (e.g., prostaglandins released from macrophages).

Questions.

1 Paul had lesions not only on the exposed skin but also on areas that would be covered, like the trunk and penis and in areas that were not in obvious contact with poison ivy leaves. How do you explain that?

2 How do you explain the recurrence of the lesions after discontinuation of the corticosteroids?

3 Paul must take great care to avoid poison ivy all his life, as subsequent reactions to it could be even more severe. Why would this be?

4 How would you confirm that Paul's contact dermatitis was caused by poison ivy rather than by another chemical such as the one found in the leaves of poison sumac (Toxicodendron vernix), another plant that gives rise to contact dermatitis?

5 One of Paul's friends, Brian, has X-linked agammaglobulinemia. What is the likelihood that this boy will develop poison ivy sensitivity?

6 Delayed hypersensitivity reactions are a rapid, inexpensive, and easy measure of T-cell function. What antigens might you use to test people for T-cell function in this way?

7 What are some common causes of contact sensitivity?

CASE 38 Autoimmune Polyendocrinopathy-Candidiasis-Ectodermal Dystrophy

Role of the thymus in the negative selection of T lymphocytes.

Millions of different T-cell antigen receptors are generated during the development of thymocytes in the thymus gland. This stochastic process inevitably leads to the formation of some T-cell antigen receptors that can bind to self antigens. When a T-cell receptor of a developing thymocyte is ligated by antigen on stromal cells in the thymus, the thymocyte dies by apoptosis. This response of immature T lymphocytes to stimulation by antigen is the basis of negative selection (Fig. 38.1). Elimination of these T cells in the thymus aborts their subsequent potentially harmful activation.

Topics bearing on this case:

Autoimmune disease

Lymphocyte development

Negative selection of T lymphocytes

Fig. 38.1 T cells specific for self antigens are deleted in the thymus. In mice transgenic for a T-cell receptor that recognizes a known peptide antigen complexed with self MHC, all the T cells have the same specificity. In the absence of the peptide, most thymocytes mature and emigrate to the periphery. This can be seen in the bottom left panel, where a normal thymus is stained with antibody to identify the medulla (in green), and by the TUNEL technique to identify apoptotic cells (in red). If the mice are injected with the peptide that is recognized by the transgenic T-cell receptor, then massive cell death occurs in the thymus, as shown by the increased numbers of apoptotic cells in the right-hand bottom panel. Photographs courtesy of A. Wack and D. Kioussis.

Experiments in mice have shown that self-MHC:self-peptide complexes encountered in the thymus purge the T-cell repertoire of immature T cells bearing self-reactive receptors. The bone marrow derived dendritic cells and macrophages of the thymus are thought to have the most significant role in negative selection (Fig. 38.2), but thymic epithelial cells may also be involved.

Most of what we know about negative selection has come from studying transgenic mice. How the myriad of self antigens, many of them organ specific, get to the thymus gland to be 'seen' by developing thymocytes has never been adequately explained. Recently, the genetic defect in the human autosomal recessive disease APECED (autoimmune polyendocrinopathy-candidiasis-ectodermal dystrophy) has been identified. The gene that is mutated in this disease is called *AIRE* (for *autoimmune regulator*) and its discovery has given new insight into the process of negative selection of thymocytes.

Fig. 38.2 Bone marrow derived cells mediate negative selection in the thymus. When MHC$^{a \times b}$ F$_1$ bone marrow is injected into an irradiated MHC^a mouse, the T cells mature on thymic epithelium expressing only MHC^a molecules. Nevertheless, the chimeric mice are tolerant to skin grafts expressing MHC^b molecules (provided that these grafts do not present skin-specific peptides that differ between strains a and b). This implies that the T cells whose receptors recognize self antigens presented by MHC^b have been eliminated in the thymus. As the transplanted MHC$^{a \times b}$ F$_1$ bone marrow cells are the only source of MHC^b molecules in the thymus, bone marrow derived cells must be able to induce negative selection.

The case of Robert Jordan: a young man with a lifelong history of autoimmune diseases of various endocrine glands.

When Robert was 18 months old his mother took him to the pediatrician, complaining that his skin was very dry and that his movements seemed sluggish compared with those of a normal active infant. Robert's pediatrician made the diagnosis of hypothyroidism. He prescribed Synthroid (thyroid hormone) and in a few weeks the child was much improved. But by the time Robert was almost 6 years old his mother felt that he was not growing at a normal rate. The pediatrician referred Robert to Dr Hemingway, an endocrinologist at the Children's Hospital. On examination he found that Robert's height and weight were below the third centile for his age. An X-ray of Robert's wrist revealed a bone age of 3 years 6 months when he was, in fact, 5 years and 10 months old.

A blood test showed that Robert's level of thyroid-stimulating hormone (TSH) was > 60 ng dl^{-1}. This is several times the upper limit of normal TSH levels and indicated that Robert was receiving inadequate thyroid hormone replacement. The dose of Synthroid was increased from 0.05 to 0.075 mg per day. Robert resumed normal growth and his bone age subsequently caught up with his actual age.

As Dr Hemingway pondered the cause of Robert's hypothyroidism, he learned from Mrs Jordan that she had another child, a daughter 2 years older than Robert. This girl had hypoparathyroidism, serum antibodies against islet cells of the pancreas (but no clinically apparent diabetes), and Graves' disease (autoimmune hyperthyroidism). This led Dr Hemingway to suspect that Robert had the inherited disease APECED, an autosomal recessively inherited abnormality. He tested Robert's serum calcium levels to make sure that he had not developed hypoparathyroidism like his sister; they were normal. Neither did Robert's serum contain antibodies against any other endocrine glands.

Dr Hemingway had noticed that Robert's fingernails were thickened, with longitudinal notches and ridging (Fig. 38.3). He referred Robert to a dermatologist, who agreed that this abnormality was consistent with autoimmune polyglandular syndrome. Nail scrapings cultured for *Candida albicans* were negative. The dermatologist also noticed two patches of hair loss at the top and back of Robert's scalp. He could easily pull out the hair, whose roots looked atrophied under the microscope, and made the diagnosis of alopecia areata (patchy hair loss).

Robert continued to grow satisfactorily and at 8 years old his TSH level was 2.5 ng dl^{-1}, a normal level which indicated adequate thyroid hormone replacement. But he continued to lose hair in patches and lost his eyebrows. He developed fissures at the angle of his mouth as a result of infection with *C. albicans*. He was taunted at school about his bizarre appearance and his schoolwork deteriorated. With his parents' divorce pending, Robert became depressed and took an overdose of Synthroid in a suicide attempt. He received intensive psychotherapy for his depression.

As Robert approached puberty, his scrotum became darkly pigmented, as did the areolae around his nipples. Dr Hemingway suspected that Robert was developing adrenal insufficiency (Addison's disease). A blood test revealed a level of adrenocorticotropic hormone (ACTH) three times the upper limit of normal. The doctor prescribed the steroid prednisone at 5 mg per day and Fluorinef (which conserves sodium and potassium excretion) at 0.1 mg a day.

18-month-old boy with lethargic movements and dry skin.

Age 6, growth less than normal. Increase thyroid hormone.

Older sibling with various autoimmune conditions. Inherited autoimmune disease? APECED?

APECED confirmed in the patient and sister by genetic testing.

Age 16, Addison's disease and idiopathic thrombocytopenic purpura have developed.

Fig. 38.3 Abnormal growth (dysplasia) of fingernails in APECED. Photograph courtesy of Mark S. Anderson.

When he was 18 years old Robert noticed that he had started to bruise easily and that his gums bled after brushing his teeth. Dr Hemingway sent him to a hematologist who found a low platelet count of 34,000 per cubic millimeter. The hematologist decided that Robert had developed yet another autoimmune disease, idiopathic thrombocytopenic purpura (ITP).

Autoimmune polyendocrinopathy-candidiasis-ectodermal dystrophy (APECED).

APECED is also known as autoimmune polyglandular syndrome (APS) type I. Although there are several polyglandular diseases of unknown origin, APECED is the only one known to have a pattern of autosomal recessive inheritance. Affected individuals such as Robert Jordan develop a wide range of autoantibodies not only against organ-specific antigens of the endocrine glands but also against antigens in the liver and skin, and, as we have seen in this case, against blood cells such as platelets. In addition to these autoimmune diseases, affected individuals have abnormalities of various ectodermal elements such as fingernails, teeth, and skin. They also have an increased susceptibility to infection with the yeast *Candida albicans*.

Fig. 38.4 Ovarian follicles stained with serum from a patient with APECED (AIRE$^{-/-}$). Serum was reacted with a microscopic section of ovary and then with fluorescent antibody against human IgG. Photograph courtesy of Mark S. Anderson.

APECED has a high incidence among Finns, Sardinians, and Iranian Jews, as high as one in 9000 births in some populations. This made it possible to map the gene responsible to chromosome 21 in affected Finnish families. The gene was cloned and named *AIRE* for *autoimmune regulator*. It was the first identified gene outside the major histocompatibility complex to be associated with autoimmune disease, and seems to encode a transcriptional regulator. Robert and his sister were found to have a deletion of 13 base pairs in exon 8 of the gene. Their parents were both heterozygous for this mutation.

When the equivalent gene *aire* was 'knocked out' in mice by homologous recombination, they developed autoimmune diseases just like the human patients; the extent and severity of these autoimmune diseases progressed as the mice aged. The protein AIRE was found normally to be expressed predominantly in the epithelial cells of the medulla of the thymus and only weakly in peripheral lymphoid tissue. AIRE was also found to act as a transcriptional activator in the thymus, where it induces the expression of 200–1200 genes, including genes that encode organ-specific antigens of the salivary glands and the zona pellucida of ova (Fig. 38.4), among others. In normal circumstances, the expression of these self antigens in the thymus permits negative selection of those developing T cells that react to them. When AIRE is lacking, these antigens are also not present. The potentially self-reactive T cells are not removed from the repertoire in the thymus and leave to cause autoimmune reactions in peripheral organs.

It is not yet clear how a deficiency of AIRE causes the ectodermal abnormalities and susceptibility to candidiasis (*C. albicans* infection).

Questions.

1 It seems paradoxical that Robert developed hypothyroidism while his sister had hyperthyroidism (Graves' disease). How can this be explained?

2 What caused the deep pigmentation of Robert's scrotum and the areolae around his nipples that led to the suspicion that he had developed an autoantibody against his adrenal cortical cells (Addison's disease)?

3 What abnormality might be found in the lymph nodes of the mice that lacked the aire gene?

4 When lymphocytes from aire-deficient mice were transferred into Rag-deficient mice, which have no mature lymphocytes of their own, autoimmune disease developed in the recipient. This did not occur when lymphocytes from normal mice of the same strain were transferred into Rag-deficient mice. What can be deduced from this experiment?

CASE 39   Autoimmune Hemolytic Anemia

Autoimmune disease triggered by infection.

There are various ways in which an infection could induce autoimmunity: disruption of a tissue barrier might expose a normally sequestered autoantigen; the infecting microorganism might act as an adjuvant; microbial antigen might bind to self proteins and act as haptens; and the microorganism might share cross-reactive antigens with the host (molecular mimicry). The major histocompatibility antigens carried by an individual may also confer susceptibility to, or protection against, certain autoimmune diseases that can be incited by infection (Fig. 39.1).

Topics bearing on this case:
Coombs test
Humoral autoimmunity
Mechanisms for breaking tolerance

Fig. 39.1 Association of infection with autoimmune diseases. Several autoimmune diseases occur after specific infections and are presumably triggered by the infection. The case of post-streptococcal disease is best known. Most of these post-infection autoimmune diseases also show susceptibility linked to the MHC.

Associations of infection with immune-mediated tissue damage		
Infection	**HLA association**	**Consequence**
Group A *Streptococcus*	?	Rheumatic fever (carditis, polyarthritis)
Chlamydia trachomatis	HLA-B27	Reiter's syndrome (arthritis)
Shigella flexneri, Salmonella typhimurium, Salmonella enteritidis, Yersinia enterocolitica, Campylobacter jejuni	HLA-B27	Reactive arthritis
Borrelia burgdorferi	HLA-DR2, -DR4	Chronic arthritis in Lyme disease

There are only a few conditions in which the infection has been identified as the direct cause of the autoimmune disease. We have seen how some infections with *Streptococcus pyogenes* may lead to rheumatic fever (see Case 29). Some strains of *S. pyogenes* lead to a type III autoimmune disease (caused by immune-complex formation; see also Case 43) of the renal glomeruli called post-streptococcal acute glomerulonephritis. These streptococcal strains are said to be nephritogenic (causing nephritis). About 3–4% of children infected with a nephritogenic strain of streptococcus will develop acute glomerulonephritis within a week or two of the onset of the streptococcal infection. What predisposes this subset of children to develop the complication is unknown.

A direct association of infection with autoimmune disease also occurs in patients with pneumonia caused by *Mycoplasma pneumoniae*, the case we shall discuss here. About 30% of patients with this infection develop a transient increase in serum antibody to a red blood cell antigen, and a small proportion of these patients develop hemolytic anemia, a type II autoimmune disease (see Fig. 40.1). In this case, the decrease in the number of red blood cells (anemia) results from their immunological destruction (hemolysis) as a result of the binding of an IgM autoantibody to a carbohydrate antigen on the red cell surface. When the infection subsides, either spontaneously or after treatment, the autoimmune disorder disappears.

The IgM autoantibody agglutinates red blood cells at temperatures below 37°C. Thus the antibodies are called cold hemagglutinins or cold agglutinins, and the hemolytic disorder is known as cold agglutinin disease. Low titers of cold agglutinins (detectable at up to 1 in 30 dilution of serum) occur in healthy people. In about one-third of patients with mycoplasma infection there is a transient increase in the titer, but without symptoms; the rise usually goes unnoticed unless the patient happens to have a blood test, when clumping of the erythrocytes will be apparent.

The autoantibodies in the transient cold agglutinin syndrome resemble those in the chronic autoimmune disorder known as chronic cold agglutinin disease, which is of unknown etiology. This runs a protracted course in which the autoantibodies are characteristically oligoclonal, and lymphoproliferative disorders commonly develop. Thus chronic cold agglutinin disease behaves as a variant of an IgM gammopathy called Waldenström's macroglobulinemia. A gammopathy is an abnormal monoclonal or oligoclonal spike in the immunoglobulin electrophoretic pattern; see also Case 19.

The case of Gwendolen Fairfax: the sudden onset of fever, cough, and anemia.

Gwendolen Fairfax was a healthy, unmarried 34-year-old who worked as a manager in a bank. She had never had any illness other than minor colds and the usual childhood infections. After developing a feverish cough her symptoms got progressively worse over the next few days and she decided to seek the advice of her physician, Dr Wilde.

34-year-old female with respiratory infection; anemia.

Dr Wilde noted that Gwendolen was extremely pale; this was particularly noticeable in the palms of her hands, which were completely white. (Look at your palms; they have a healthy reddish pink color, unless your hemoglobin level is <10 g dl^{-1}.) Her temperature was raised, at 38.5°C, and her respiratory rate was 30 per minute (normal 20). On listening to her chest, the physician heard scattered rhonchi (harsh breath sounds) at the bases of both lungs.

The physician ordered blood counts to be performed immediately. Gwendolen's hematocrit and hemoglobin level were low—hematocrit 26% (normal 38–46%) and hemoglobin 9.5 g dl^{-1} (normal 13–15 g dl^{-1}). Her white blood cell count was elevated at 11,300 μl^{-1}, with 70% neutrophils, 24% lymphocytes, 6% monocytes, and 1% eosinophils, revealing an increased absolute neutrophil count, and the platelet count was 180,000 μl^{-1}. The chest radiograph showed patchy infiltrates in both lower lungs (Fig. 39.2). He suspected a diagnosis of mycoplasma pneumonia (also known as atypical pneumonia because there was no complete opacity over a whole lobe of the lung such as typically occurs in pneumococal and other bacterial pneumonias). He advised Gwendolen to go to Reading Hospital.

Pneumonia with marked anemia; hemolytic anemia following mycoplasma infection?

In the hospital, a sputum sample proved negative for bacteria but contained abundant neutrophils. A blood sample was tested for the presence of antibodies against the red blood cells. In the presence of sodium citrate as an anticoagulant, the sample was chilled to 4°C; the red cells became clumped and the blood looked as though it had clotted. When warmed to 37°C, however, the red cells dispersed. Thus, the red cells were reversibly agglutinated in the cold. A blood sample was spun in a warmed centrifuge and the red cells and plasma separated. Gwendolen's red cells were tested using direct and indirect Coombs tests for the presence of anti-red cell IgG antibody (see Fig. 45.2); these tests proved negative. Her red cells, however, were strongly agglutinated by an antibody to the complement component C3.

When Gwendolen's plasma, or dilutions of the plasma, were incubated in the refrigerator with red cells from type O normal blood, the red cells were strongly agglutinated; when the same test was performed with red cells obtained from an umbilical cord in the delivery room of the hospital, agglutination was weaker, or negligible.

Gwendolen was started on erythromycin by intravenous administration. In 3 days her symptoms had abated; she had no fever and her cough had greatly improved. Her hemoglobin had risen to 12 g dl^{-1}. She was discharged from the hospital, and advised to continue erythromycin orally and to avoid exposure to cold weather. When she returned to Dr Wilde's office 2 weeks later for a check-up, her fever had not returned, her cough had gone, and her hemoglobin was now a normal 14.5 g dl^{-1}.

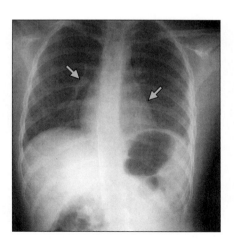

Fig. 39.2 Chest radiograph of a patient with mycoplasma pneumonia. Patchy opacities (arrowed) are evident in the lungs. Courtesy of T. Griscom.

Fig. 39.3 Autoantibodies specific for the surface antigens of red blood cells can cause hemolytic anemia. Antibody-coated red blood cells (RBCs) can be destroyed in several ways. Cells coated with IgG autoantibodies are cleared predominantly by uptake by Fc receptor-bearing macrophages in the fixed mononuclear phagocytic system (left panel). Cells coated with IgM autoantibodies and complement are cleared by macrophages bearing complement receptors CR1 and CR3 in the fixed mononuclear phagocytic system (not shown). The binding of certain rare autoantibodies that fix complement extremely efficiently causes the formation of the membrane-attack complement complex on the red cells, leading to intravascular hemolysis (right panel).

Autoimmune hemolytic anemia.

Gwendolen's autoimmune hemolytic anemia was transient, persisting only as long as the infection persisted. In contrast, autoantibodies to red blood cells in chronic autoimmune disorders of unknown origin tend to be persistent. Transient autoantibodies most frequently develop after nonbacterial infection. Autoimmune hemolytic anemia can also be induced by certain chemicals and drugs.

The autoantibodies in autoimmune hemolytic anemias are of the IgM or IgG class. They bring about hemolysis of the red blood cells either by complement fixation or adherence of the red cells to the Fcγ receptors on cells of the fixed mononuclear phagocytic system—principally in the spleen, but also in the liver and other organs (Fig. 39.3).

Anti-erythrocyte antibodies can differ in their physical properties. Some bind to the red blood cells at 37°C (warm autoantibodies); some at lower temperatures (cold hemagglutinins or hemolysins, like those in Gwendolen's case). In 1904, Donath and Landsteiner (who later got the Nobel Prize for the discovery of the ABO blood groups) reported antibodies in the blood of patients with syphilis that bound to human red blood cells at cold temperatures. These antibodies differ from cold agglutinins in that they cause the lysis of erythrocytes by complement when the blood is warmed to 37°C. These antibodies are still called Donath–Landsteiner or DL antibodies.

A transient increase in cold agglutinins is most frequently encountered during mycoplasma infection, infectious mononucleosis (the result of infection with the Epstein–Barr virus; see Case 27), and, rarely, during other viral infections. In mycoplasma pneumonia the cold autoantibodies react with a red cell surface antigen called the I antigen. This is a branched carbohydrate structure that occurs on glycoproteins and glycolipids (Fig. 39.4). Cord blood cells express only small amounts of the I antigen and that is why Gwendolen's plasma did not agglutinate them, or agglutinated only weakly. Cord erythrocytes express a related antigen, i (see Fig. 39.4); the I antigen is not fully expressed on red blood cells until 6 months of age.

It has not been possible to find with certainty a cross-reacting I antigen on *Mycoplasma pneumoniae*, which would provide a simple explanation for this autoimmune disease. Exactly how *M. pneumoniae* triggers the transient production of an autoantibody directed to red cells is not known, although there is a strong biochemical similarity between the attachment site (ligand) for the mycoplasma on the host cell surface and the red cell antigen to which the autoantibody binds (Fig. 39.5).

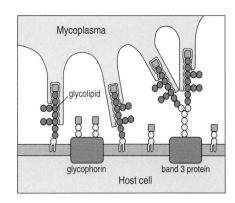

Fig. 39.4 Structure of the I and i antigen regions with which cold agglutinins react. Gal, galactose; GlcNAc, N-acetylglycosamine.

Questions.

1   *The hemolytic anemia in Gwendolen Fairfax's case was not severe and it subsided rapidly. In similar cases the hemolysis can be extensive and the dramatic fall in hemoglobin can be life threatening. What emergency measure might be undertaken to stop severe hemolysis?*

2   *Autoimmune hemolytic anemia is caused, in general, by autoantibodies of the IgG and IgM classes. Could antibodies of other classes such as IgA, IgD, and IgE cause autoimmune hemolytic anemia?*

3   *In the direct and indirect Coombs tests (see Case 45), agglutination of the antibody-coated red blood cells is brought about by an antibody to human IgG that cross-links the IgG on the red cell surface and causes the cells to agglutinate. In the so-called 'non-gamma' Coombs test an antibody to C3 or C4 is used to agglutinate red blood cells coated with an IgM antibody that has fixed complement. Why is it important to have an antibody to the C3d or C4d portions of C3 or C4 to carry out the test?*

4   *Most normal individuals have a low titer (up to 1:30) of IgM anti-I cold agglutinins in their blood. As mentioned above, the I antigen is little expressed until 6 months of age. We must develop peripheral B-cell tolerance to this antigen in infancy. How does this happen? How might the mycoplasma infection break this tolerance?*

Fig. 39.5 *Mycoplasma pneumoniae* **interacts with long carbohydrate chains of I antigen type on the human red cell membrane.** These chains include the carbohydrates of band 3 protein and certain glycolipids, but not glycolipids and glycoproteins with short carbohydrate chains. Sequences of sugars that act as I antigen are shown as orange circles; the orange squares depict sialic acid.

CASE 40 | Myasthenia Gravis

The immune response turns against the host.

The specific adaptive immune response can, in rare instances, be mounted against self antigens and cause autoimmune disease. Injury to body tissues can result from antibodies directed against cell-surface or extracellular-matrix molecules, from antibodies bound to molecules circulating in the plasma that deposit as immune complexes, or from clones of T cells that are reactive with self antigens. A special class of autoimmune diseases are caused by autoantibodies to cell-surface receptors (Fig. 40.1). Graves' disease and myasthenia gravis are two well-studied examples. Graves' disease is caused by autoantibodies to the receptor on thyroid cells for thyroid-stimulating hormone (TSH), secreted by the pituitary gland. In this disease, the binding of the autoantibody to the TSH receptor acts like TSH itself, and stimulates the thyroid gland to produce thyroid hormones. In myasthenia gravis, the opposite effect is observed: antibodies to the acetylcholine receptor at the neuro-muscular junction impede the binding of acetylcholine and stimulate internalization of the receptor, thereby blocking the transmission of nerve impulses by acetylcholine (Fig. 40.2).

Topics bearing on this case:
Humoral autoimmunity
Transfer of maternal antibodies
Mechanisms for breaking tolerance

Fig. 40.1 Autoimmune diseases caused by antibody to surface or matrix antigens. These are known as Type II autoimmune diseases. Damage by IgE-mediated responses (Type I) does not occur in autoimmune disease. In most Type II diseases, autoantibodies bind to the cell surface or extracellular matrix and target them for destruction by phagocytes (often with the help of complement) and/or natural killer cells. A special class of autoimmune diseases is caused by autoantibodies that bind cellular receptors and either stimulate or block their normal function. Immune-complex disease (Type III) is discussed in Case 43. T cell-mediated disease (Type IV) is discussed in Cases 37 and 46.

Some common Type II autoimmune diseases caused by antibody to surface or matrix antigens		
Syndrome	Autoantigen	Consequence
Autoimmune hemolytic anemia (see Case 39)	Rh blood group antigens, I antigen	Destruction of red blood cells by complement and phagocytes, anemia
Autoimmune thrombocytopenic purpura	Platelet integrin GpIIb:IIIa	Abnormal bleeding
Goodpasture's syndrome	Noncollagenous domain of basement membrane collagen type IV	Glomerulonephritis Pulmonary hemorrhage
Pemphigus vulgaris (see Case 41)	Epidermal cadherin	Blistering of skin
Acute rheumatic fever (see Case 29)	Streptococcal cell wall antigens, antibodies cross-react with cardiac muscle	Arthritis, myocarditis, late scarring of heart valves
Graves' disease	Thyroid-stimulating hormone receptor	Hyperthyroidism
Myasthenia gravis	Acetylcholine receptor	Progressive weakness
Insulin-resistant diabetes	Insulin receptor (antagonist)	Hyperglycemia, ketoacidosis
Hypoglycemia	Insulin receptor (agonist)	Hypoglycemia

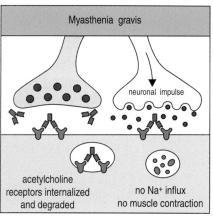

Myasthenia gravis means severe (gravis) muscle (my) weakness (asthenia). It was first realized that this disease was an autoimmune disease when an immunologist immunized rabbits with purified acetylcholine receptors in order to obtain antibodies to this receptor. He noticed that the rabbits developed floppy ears, like the droopy eyelids (ptosis) that are the most characteristic symptom of myasthenia gravis in humans. It was then shown that patients with this disease indeed have antibodies to the acetycholine receptor. It has been known for a long time that pregnant women with myasthenia gravis transfer the disease to their newborn infants. As IgG is the only maternal serum protein that crosses the placenta from mother to fetus, the occurrence of neonatal myasthenia gravis is clear evidence that myasthenia gravis is caused by an IgG antibody.

Fig. 40.2 Autoantibodies to the acetylcholine receptor weaken the reception of the signal from nerve ends that cause the muscle cell to contract. At the neuromuscular junction, acetylcholine is released from stimulated neurons and binds to acetylcholine receptors, triggering muscle contraction. The acetylcholine is destroyed rapidly by the enzyme acetylcholinesterase after release. In myasthenia gravis, autoantibodies to the acetylcholine receptor induce its endocytosis and degradation, and prevent muscles from responding to neuronal impulses.

The case of Mr Weld: from floppy ears to droopy eyelids.

Mr Weld is a 71-year-old retired engineer. He developed double vision (diplopia), which gradually worsened over the course of 4 months and caused him to seek medical attention. He had been in good health and active his entire life.

On examination, the doctor noticed that Mr Weld had ptosis of both eyelids so that they covered the upper third of the irises of his eyes. When the doctor asked Mr Weld to look to the right and then to the left, he noticed limitations in the ocular movements of both eyes, as shown in Fig. 40.3.

The remainder of the neurological examination was normal. No other muscle weakness was found during the examination.

A radiological examination of the chest was performed, and it was normal. There was no evidence in the radiograph of enlargement of the thymus gland. A blood sample was taken from Mr Weld and his serum was tested for antibodies to the acetylcholine receptor. The serum contained 6.8 units of antibody to the acetylcholine receptor (normal <0.5 units).

Mr Weld was told to take pyridostigmine, an inhibitor of cholinesterase.

His double vision steadily improved but he developed diarrhea from the pyridostigmine, and this limited the amount he could take.

Ptosis of both eyelids. Limitation in ocular movements of both eyes.

Prescribe Pyridostigmine.

Three years later

Mr Weld developed a severe respiratory infection. Soon after, his ptosis became so severe that he had to lift his eyelids by taping them with adhesive tape.

His diplopia recurred and his speech became indistinct. He developed difficulty with chewing and swallowing food. He could only tolerate a diet of soft food and it would take him several hours to finish a meal.

The neurologist noted that he now had weakness of his facial muscles and his tongue, and the abnormality in ocular movements again became apparent.

Because of the diarrhea he was only able to tolerate a quarter of the prescribed dose of pyridostigmine. He also developed difficulty in breathing. His vital capacity (the amount of air he could exhale in one deep breath) was low, at 3.5 liters.

Mr Weld was admitted to the hospital and treated with azathioprine. Thereafter he showed steady improvement.

His ptosis and diplopia improved remarkably and he was able to eat normally. His vital capacity returned to normal and was measured to be 5.1 liters.

Symptoms recur during severe respiratory infection, difficulty chewing and swallowing vital capacity 3.5 l.

Fig. 40.3 Diagram of ocular movement limitation.

Myasthenia gravis.

Mr Weld illustrates the features of a common type of myasthenia gravis, which is observed in older people and is called the oculobulbar form because it principally involves the muscles of the eye. In younger people, the disease is much more severe and the antibody titers to the acetylcholine receptor are higher. In very severe cases, impaired breathing may even cause death, and difficulty in swallowing can cause aspiration of food particles into the lung. Chest radiographs of younger people with myasthenia gravis frequently reveal enlargement of the thymus gland, and sometimes tumors of the thymus (thymoma). Removal of the thymus gland (thymectomy) in younger patients may lead to rapid improvement in their symptoms. The relationship of the thymus gland to myasthenia gravis is a mystery, and it is equally unclear why thymectomy improves the symptoms of this disease. Abnormalities of the thymus are not observed in older patients and thymectomy does not help them.

Questions.

1   Newborn infants of mothers with myasthenia gravis exhibit symptoms of myasthenia gravis at birth. How long would the disease be likely to last in these infants?

2   Pyridostigmine is an ideal drug for the treatment of myasthenia gravis. It inhibits the enzyme cholinesterase, which normally cleaves and inactivates acetylcholine. In this way, pyridostigmine prolongs the biological half-life of acetylcholine. Unfortunately, it also causes diarrhea by raising the amount of acetylcholine in the intestine. Acetylcholine binds to the muscarinic receptors in the intestine and increases intestinal motility. Since he could not tolerate full therapeutic doses of pyridostigmine and was getting worse, Mr Weld was given azathioprine (Fig. 40.4) and showed marked improvement. What did the azathioprine do? What would concern you about prolonged use of this drug?

3   Mr Weld had a severe relapse in his disease following a respiratory infection. Many autoimmune diseases appear to be triggered by infection, and relapses in autoimmune diseases frequently follow an infection. Can you explain how this might happen?

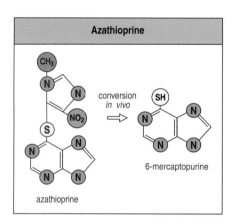

Azathioprine

conversion in vivo

6-mercaptopurine

azathioprine

Fig. 40.4 The structure of azathioprine and its active product.

CASE 41 Pemphigus Vulgaris

Autoimmune attack on the integrity of the skin.

Autoimmune disease results when the adaptive immune response is directed against self antigens. The particular pathology of the disease depends on the nature of the self antigen and the type of immune response that is mounted against it. In the type II autoimmune diseases (see Fig. 40.1) the pathology is caused by autoantibodies that interact with self antigens in the extracellular matrix or on cell surfaces. In myasthenia gravis, for example, autoantibodies against the acetylcholine receptor on skeletal muscle block its function at the neuromuscular junction and cause paralysis (see Case 40), whereas in Graves' disease an autoantibody against the receptor for thyroid-stimulating hormone acts as an agonist, stimulating receptor activity and causing hyper-thyroidism. In the autoimmune disease pemphigus vulgaris, the self-reactive agent is an autoantibody against a structural protein of the epidermal cells of the skin. Its actions result in the skin cells coming apart from each other; the affected skin blisters and is destroyed.

Pemphigus is derived from the Latin word for blister, and vulgaris means com-mon or ordinary. Patients with this disease have autoantibodies against desmoglein-3, which is a protein component of the desmosome—one of the intercellular junctions that link skin cells and other epithelial cells tightly to each other. Desmogleins are members of the cadherin family of cell adhesion molecules, proteins that effect intercellular adhesion in a calcium-dependent manner. Disruption of desmoglein causes blisters to form in the skin and on the mucous membranes; extensive sloughing of the skin may ensue (Fig. 41.1).

Topics bearing on this case:
Peptide binding by MHC class II antigens
Humoral autoimmunity
Epitope spreading
HLA associations with disease

Fig. 41.1 **Lesions of pemphigus vulgaris that have coalesced to produce a large plaque.** The plaque has a moist surface that is oozing and has crusts of dried serum from blister fluid. Photograph courtesy of Razzaque Ahmed.

Evidence that pemphigus vulgaris is caused by an autoantibody comes from the observation that infants born to mothers with the disease have a transient period of skin blistering during the neonatal period. There is also a good animal model in which human IgG from patients with pemphigus injected into neonatal mice reproduces pemphigus.

The case of Arthur Sammler, Esq.: a hoarse lawyer with a sore mouth and blistering skin.

Arthur Sammler is a lawyer of Ashkenazi Jewish descent. He was 55 years of age when an irritation in his throat became persistent enough for him to consult a physician. The physician suggested that he gargle with warm saline, which seemed to help. A month later, however, after a long day in court he became very hoarse and the hoarseness persisted for a week. A week later he noticed a 'sore' on his right cheek. He told the physician he was under considerable stress over a custody case, and the physician diagnosed a herpes simplex infection and advised him to take oral acyclovir. The lesion did not improve.

At a more thorough examination 2 weeks later, numerous erosions and ulcers in the mucosa of his mouth and gums were also revealed. These were thought to result from a yeast infection with *Candida albicans* and he was treated with a Mycostatin (nystatin) mouthwash.

However, new lesions soon appeared on the palate, uvula, and tongue, and Mr Sammler returned to the physician. This time, a biopsy of the lesions was taken by an oral surgeon. While awaiting the results of the biopsy Mr Sammler developed erosions of the skin on his scalp and neck.

Microscopic examination of the biopsy revealed disruption of the epidermal layer of the skin characteristic of pemphigus vulgaris (Fig. 41.2). Mr Sammler was sent to a dermatologist, who took another biopsy for immunofluorescence studies; blood samples were taken for analysis. The immunofluorescence studies revealed deposits of IgG in an intercellular pattern within the epidermis (Fig. 41.3). His serum contained an antibody against desmoglein-3 (known commonly as 'pemphigus antibody') in a titer of 1:640.

Mr Sammler was treated with large doses of prednisone, 120 mg per day. After 3 weeks of this therapy, during which time no improvement occurred, new lesions appeared on his back and scalp. The dose of prednisone was increased to 180 mg per day. Two weeks later he developed a persistent cough, fever and chest pain. A chest radiograph during an emergency room visit suggested that he had developed interstitial pneumonia due to *Pneumocystis carinii*. He was admitted to hospital and treated for this infection.

It was found that his pemphigus antibody titer had risen to 1:1280. He was started on cyclophosphamide at 150 mg per day, and was told to decrease his prednisone dose by 10 mg every 10 days and to rinse his mouth daily with hydrogen peroxide and elixir of Decadron (a corticosteroid). New lesions kept appearing, so the cyclophosphamide was increased to 200 mg per day.

Over the next 8 weeks the pemphigus antibody titer decreased to 1:320; no new lesions appeared, and the older ones healed. The cyclophosphamide dose was decreased over an 8-month period and he remained symptom-free after

Persistent skin sore and ulceration of mouth; resistant to usual therapies; do biopsy.

Diagnosis of pemphigus vulgaris; start immunosuppressive therapy.

Condition not responding to steroids; try cyclophosphamide.

cyclophosphamide was stopped. In the year since then, he has noticed the occasional new lesion after damage to his skin and these new lesions are promptly injected with a corticosteroid. His pemphigus antibody titer has remained at 1:80 during this asymptomatic period.

Responding to cyclophosphamide; decrease dose gradually.

Pemphigus vulgaris.

The case of Arthur Sammler illustrates how an autoimmune disease that was frequently fatal in the past can now be controlled by immunosuppressive drugs. The epidermis, with its cells held tightly together, forms an important barrier to the entry of pathogens into the body. Extensive blistering and sloughing of skin destroy this barrier and, in the past, often resulted in blood-stream infection (septicemia) with *Staphylococcus aureus*, a common contaminant of human skin.

The mechanism of autoimmune tissue destruction seems to be somewhat unusual in pemphigus vulgaris. In some autoimmune diseases, the autoantibodies cause tissue destruction by stimulating the complement system, but the autoantibody in pemphigus vulgaris is of the IgG4 subclass and does not fix complement. Some other mechanism must therefore be invoked to explain the disruption of desmoglein-3 adhesion. It is thought that the binding of the autoantibody to its antigen in some way causes upregulation of serine proteinase activity on the surface of the epidermal cells and that this results in the proteolytic digestion of desmoglein-3. The particular enzyme that digests the desmoglein-3 is as yet unknown.

Long-term study of the antibody response in some patients with pemphigus vulgaris has shown that the response evolves to involve epitopes other than the one against which antibodies are initially directed. This is known as epitope spreading and has been found to correlate with the clinical progression of disease. Antibodies present before the onset of disease are specific for epitopes in the extracellular part of the molecule nearest to the cell membrane. These antibodies do not bind to cell-surface desmoglein, nor do they transfer disease when injected into mice. With disease onset, antibodies specific for the domains furthest away from the cell surface become detectable (Fig. 41.4). These IgG antibodies bind to cell-surface desmoglein, and when injected into mice induce disease. The process by which the immune response initially targets epitopes in one part of an antigenic molecule and then progresses to other, non-cross-reactive, epitopes of the same molecule is called intramolecular epitope spreading. In pemphigus, it is only with epitope spreading that disease-causing antibodies are made.

Pemphigus vulgaris is encountered most frequently in Ashkenazi Jews and has a strong association with an HLA haplotype found mainly in that ethnic group. This HLA haplotype includes HLA-DR4 and HLA-DQ3, which are present in virtually all Ashkenazi Jews with pemphigus vulgaris. Patients are usually heterozygous for this haplotype and only in rare cases are they homozygous. Unaffected relatives of pemphigus patients who bear the same HLA haplotype frequently also have antibodies against desmoglein-3. However, these antibodies are of the IgG1 subclass and do not cause disease. In other ethnic groups, pemphigus vulgaris is usually associated with HLA-DR14 and HLA-DQ5.

Fig. 41.2 Histopathology of an early skin lesion of pemphigus vulgaris. Stained with hematoxylin and eosin. An epidermal, suprabasal vesicle (blister) is clearly seen. Photograph courtesy of Razzaque Ahmed.

Fig. 41.3 Direct immunofluorescence study of tissue around a pemphigus vulgaris lesion. The normal-appearing tissue around the lesion contains deposits of IgG, other immunoglobulins, and complement components in the intercellular spaces of the entire stratum malpighii of the epidermis. Photograph courtesy of Razzaque Ahmed.

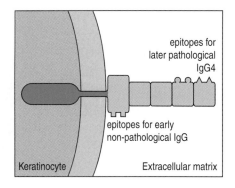

Fig. 41.4 The progressive involvement of desmoglein epitopes in pemphigus vulgaris. This skin-blistering disease is caused by antibodies that bind to desmoglein-3, an adhesion molecule in the cell junctions that hold keratinocytes together. When the autoimmune B-cell response begins, harmless antibodies are made against the extracellular region of desmoglein-3 nearest to the cell surface. In time the response spreads and antibodies are made against the region furthest away from the cell surface. These antibodies cause disease and are of the IgG4 subclass.

A molecular explanation has been discovered for the association of pemphigus vulgaris with HLA-DR4. Each major histocompatibility complex (MHC) class II variant can bind a subset of all possible peptides; the actual peptides that can be bound by any given MHC molecule are determined by the particular pattern of amino acid binding sites in the peptide-binding groove of the MHC molecule (Fig. 41.5). The β chain of the MHC class II molecule DR is designated DRB1; 22 known subtypes of this chain have been identified by DNA typing among those bearing DR4 (DRB1*04). Only one of these variants—DRB1*0402—is associated with pemphigus vulgaris. It differs from the other 21 variants in that the amino acid residue at position 71, which lies in one of the binding pockets in the peptide-binding groove of the MHC molecule, is negatively charged (glutamic acid), whereas all the other variants have positively charged amino acid residues (lysine or arginine) at this position. Only DRB1*0402 will bind the antigenic peptide derived from desmoglein-3 and present it to T lymphocytes.

Questions.

1   The intravenous administration of gamma globulin frequently improves autoimmune disease. It has been observed that the administration of gamma globulin to patients with pemphigus vulgaris significantly lowers the titer of IgG4 anti-desmoglein-3 and improves the clinical course of the disease. How do you explain this effect?

2   Relatives of patients with pemphigus vulgaris who share the DRB1*0402 MHC class II type have IgG1 antibodies against

Fig. 41.5 Peptides bind to MHC class II molecules by interactions along the length of the binding groove. A peptide (yellow; shown as the peptide backbone only, with the amino terminus to the left and the carboxy terminus to the right), is bound by an MHC class II molecule through a series of hydrogen bonds (dotted blue lines) that are distributed along the length of the peptide. The hydrogen bonds towards the amino terminus of the peptide are made with the backbone of the class II polypeptide chain, whereas throughout the peptide's length bonds are made with residues that are highly conserved in MHC class II molecules. The side chains of these residues are shown in gray on the ribbon diagram of the MHC class II binding groove. Most DR4 alleles have a positively charged arginine (shown here in purple) or lysine at position 71 of the β chain. The DRB1*0402 allele that is linked to susceptibility to pemphigus has a negatively charged glutamic acid residue instead.

desmoglein-3. These antibodies fix complement but do not cause disease. Can you speculate why this antibody does not cause disease, whereas the IgG4 antibody causes disease?

3   What might explain the difference between unaffected people who make only the IgG1 antibody against desmoglein-3 and the patients who make IgG4 antibody? Can you design an experiment to test your hypothesis?

4   We have seen cyclophosphamide used to eradicate bone marrow cells in Cases 3 and 18. Why is it useful in the treatment of pemphigus vulgaris?

5   Why did Mr Sammler develop pneumonia caused by Pneumocystis carinii during his initial treatment for pemphigus vulgaris?

CASE 42 | Rheumatoid Arthritis

A common, severe inflammatory disease of unknown cause.

By inciting a severe inflammatory response, the adaptive immune system can cause extensive tissue damage and destruction, even when the utility of this response to the host is not apparent and may even be detrimental. A subset of autoimmune diseases causes rheumatic disease—inflammation and damage to joints (Fig. 42.1). One of the commonest of these inflammatory diseases is rheumatoid arthritis, a chronic debilitating disease, characterized chiefly by inflammation of the synovium (the thin lining of a joint). As the disease progresses, the inflamed synovium invades and damages the cartilage, followed by erosion of the bone, leading to the destruction of the joint (Fig. 42.2). Patients with rheumatoid arthritis suffer chronic pain, loss of function, and disability.

Rheumatic diseases caused by autoimmunity
Systemic lupus erythematosus (SLE)
Rheumatoid arthritis
Juvenile arthritis
Sjögren's syndrome
Scleroderma (progressive systemic sclerosis)
Polymyositis–dermatomyositis
Behçet's disease
Ankylosing spondylitis
Reiter's syndrome
Psoriatic arthritis

Fig. 42.1 Rheumatic diseases are autoimmune in nature.

Rheumatoid arthritis was originally considered an autoimmune disease that was mainly driven by B cells producing anti-IgG autoantibodies called rheumatoid factor. However, rheumatoid factor occurs in some healthy individuals, and is absent from some patients with rheumatoid arthritis, indicating that it cannot be primarily responsible for the pathology. The discovery that rheumatoid arthritis has an association with particular alleles of the major histocompatibility complex (MHC) suggested that T cells were involved in the pathogenesis of this disease. It is now thought likely that autoreactive CD4 T cells initially become activated by dendritic cells or by inflammatory cytokines produced by macrophages. Once activated, autoreactive CD4 T cells can engage with antigens in the joint, maintaining and exacerbating the inflammatory reaction, and can also provide help to B cells to differentiate into plasma cells producing arthritogenic antibodies.

The case of Ariadne Underwood: swollen joints that are stiff and painful in the morning portend a common diagnosis.

Fig. 42.2 X-ray of the right hand of a patient with rheumatoid arthritis. It shows extensive destruction and dislocation of the metacarpophalangeal joints.

Ariadne Underwood teaches a class of sixth-graders in school. She was in good health until the age of 30 years, when she noticed stiffness in her shoulders, hands, knees, and hips every morning when she got up. The stiffness lasted for about an hour. After her wrists and fingers began to swell she went to her local physician, who suspected the onset of rheumatoid arthritis and advised her to take four aspirin tablets a day. He took a blood sample, which proved positive for rheumatoid factor.

As her joints grew more stiff and painful she was referred to Dr Shaw, a rheumatologist at the Brigham and Women's Hospital, who ascertained that she had no family history of rheumatoid disease. On physical examination he felt spongy swelling indicating thickening of the synovium (the lining of the joint) in both elbows, and decreased flexion of her wrists. Her blood pressure, pulse, and rate of respiration were all normal, and she had no fever. Dr Shaw prescribed the non-steroidal anti-inflammatory drug Motrin, 600 mg, four times a day. When Ariadne returned to see Dr Shaw 3 months later she felt much better, but her synovia had thickened further, and Dr Shaw added the antimalarial agent hydroxychloroquine (Plaquenil) to her daily medication.

Blood tests during a return visit to Dr Shaw when Ariadne was 43 years old revealed that her rheumatoid factor titer was 1:2560 (normal 0). Her hemoglobin was 11.4 g dl^{-1} and her hematocrit was 33.4% (she was slightly anemic). Her white blood cell count was 4720 μl^{-1} (slightly low). Her sedimentation rate was 87 mm h^{-1} (very high). A urinalysis was normal. X-rays of her hands and feet revealed erosion in a number of joints. Two years later Ariadne's arthritic symptoms flared up and Dr Shaw decided to start her on methotrexate, 2.5 mg, four times a week. Her symptoms improved a little, but by the time she was 55 years old her feet had become so deformed by the arthritis that she needed surgery to enable her to walk normally.

Because her symptoms continued to worsen, Dr Shaw decided to give Ariadne infusions of a monoclonal antibody against tumor necrosis factor-α (TNF-α) called infliximab (Remicade), which she received every 8 weeks. She felt remarkably better, her joint swellings disappeared and she had no further morning stiffness. She has two grown-up children, neither of whom suffers from arthritis at present.

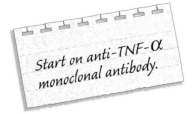

Start on anti-TNF-α monoclonal antibody.

Criteria for the diagnosis of rheumatoid arthritis
Morning stiffness lasting at least 1 hour before maximal improvement
Arthritis of three or more joints simultaneously with swelling and/or fluid in the joints
Arthritis in hand joints with swelling in the wrists, or metacarpophalangeal joints or proximal interphalangeal joints
Symmetrical arthritis of the same joint areas
Rheumatoid nodules
Serum rheumatoid factor
Typical radiographic changes

Fig. 42.3 The American Rheumatism Association criteria for the diagnosis of rheumatoid arthritis.

Rheumatoid arthritis.

As there is no single diagnostic test for rheumatoid arthritis, the American Rheumatism Association has set out criteria for establishing a diagnosis (Fig. 42.3). Ariadne Underwood met all the criteria with the exception of rheumatoid nodules, which she did not have.

There is a strong association between the MHC alleles HLA-DRB1*0404 and DRB1*0401 and the occurrence of rheumatoid arthritis; it is also three times as common in women as in men. Autoantigens such as type II collagen, proteoglycans, aggrecan, cartilage link protein, and heat-shock proteins have been proposed as potential antigens, because of their ability to induce arthritis in mice. Their pathogenic role in humans, however, remains to be ascertained.

In rheumatoid joints, the synovial membrane is thickened as a result of hyperplasia, and there is increased growth of blood vessels. The synovium is infiltrated by inflammatory cells; prominent among these are activated CD4 T cells. These T cells stimulate the activation of macrophages and synovial fibroblasts to produce pro-inflammatory cytokines such as TNF-α, interleukin-1 (IL-1), IL-6, IL-17, and IFN-γ, the chemokines CXCL8 (IL-8) and CCL2 (MCP-1), and matrix metalloproteinases (MMPs), which are responsible for the tissue destruction. B cells in the joints become activated by the binding of their CD40 and B7.1 (CD80) by CD40 ligand and CD28, respectively, on the T cells. The activated B cells secrete immunoglobulins, including rheumatoid factor (which activates complement), into the synovial fluid. Osteoclasts and chondrocytes become activated and secrete enzymes that destroy cartilage and bone.

TNF-α is the dominant cytokine found in rheumatoid joints. Inhibition of TNF-α has proved beneficial in subduing arthritic symptoms both in experimental animal models of arthritis and in human patients. Although the cause of rheumatoid arthritis remains elusive, understanding the inflammatory mechanism has led to useful medical interventions. In patients with rheumatoid arthritis, inhibition of TNF-α has brought about major relief of symptoms (Fig. 42.4). Three protein drugs to do this are in widespread clinical use. Infliximab (Remicade) is a chimeric (mouse–human) IgG1 antibody against TNF-α. Adalimumab (HUMIRA) is a human anti-TNF monoclonal antibody with a completely human sequence. Etanercept (Enbrel) is a fusion protein of an IgG1 Fc region and a TNF-α receptor. They all act as a trap or sump for released TNF-α.

Fig. 42.4 The effects of treatment of rheumatoid arthritis with anti-TNF-α. For each parameter measured (pain, swollen joints, and level of C-reactive protein), the values for patients given a placebo treatment are depicted by the blue curve and the values for patients treated with anti-TNF-α antibody are depicted by the red curve.

Several animal models have shed light on the pathogenesis of rheumatoid arthritis. In one that was discovered relatively recently, mice bearing a transgenic T-cell receptor from the K/BxN strain spontaneously develop arthritis after making antibody against a ubiquitously expressed antigen, glucose-6-phosphate isomerase. The arthritis can be transferred by this antibody. To generate arthritis in recipient animals, an intact alternative pathway of complement activation is needed, specifically the generation of C5a via this pathway, as well as intact FcγRIII receptors. The arthritis cannot be induced in mice lacking mast cells. Thus, it seems in this important model that a specific adaptive immune response to a widely distributed antigen is amplifying the innate immune response to the extent of extreme hyperreactivity, and a disease that resembles human rheumatoid arthritis ensues.

Which cells trigger rheumatoid arthritis in humans is not yet known. However, mouse models of rheumatoid arthritis demonstrate that both T cells and B cells are needed to initiate the disease, as mice lacking either CD3+ T cells or B cells are resistant to developing it.

Questions.

1 The presence of rheumatoid factor in the serum of a patient is not diagnostic of rheumatoid arthritis. Why is this?

2 Why is the integrity of the FcγRIII important in the appearance of arthritis in K/BxN mice?

3 How do TNF-α and IL-1 enhance the infiltration of leukocytes into the joint space?

4 What are the risks of infliximab therapy?

CASE 43 | Systemic Lupus Erythematosus

A disease caused by immune complexes.

Immune complexes are produced whenever there is an antibody response to a soluble antigen. As the immune response progresses, larger immune complexes are formed that trigger the activation of complement, activated components of which bind to the complexes. These are then efficiently cleared by binding to complement receptor 1 (CR1) on erythrocytes, which convey the immune complexes to the liver and spleen. There the complexes are removed from the red cell surface by Kupffer cells and other phagocytes (Fig. 43.1) and ingested via a variety of complement and Fc receptors on the cells lining the sinusoids of the hepatic and splenic circulation. When antigen is released repeatedly so as to sustain the formation of small immune complexes, these complexes tend to be trapped in the small blood vessels of the renal glomerulus and synovial tissue of the joints.

Topics bearing on this case:

Clearance of immune complexes by complement

Immune-complex disease

Bound C3b binds to the receptor CR1 on erythrocyte surfaces

CR1

C3b

C4b2a3b

In the spleen and liver, phagocytic cells remove the immune complexes

FcR

Fig. 43.1 Immune complexes are cleared from the circulation by binding to complement and Fc receptors. Immune complexes activate C3 in the serum, and bind activated complement components C3b, C4b, and C2a. C3b binds to complement receptors on erythrocytes, which transport the immune complexes to the spleen and liver, where complement receptors and Fc receptors on phagocytic cells bind to complement components and to the Fc portion of antibodies and are thereby stimulated to engulf the complexes and degrade them.

The most common immune-complex diseases are listed in Fig. 43.2. In subacute bacterial endocarditis, bacteria reside for a protracted period on the valves of the heart. The antibody response to the prolonged presence of the bacteria becomes intense and immune complexes are formed with the bacterial antigens. The immunoglobulin in the immune complexes provokes the formation of anti-IgG antibodies or rheumatoid factor. These complexes are trapped in the renal glomeruli and cause glomerulonephritis. In a similar fashion, hepatitis virus in the liver can become a chronic infection that provokes a marked antibody response and the formation of rheumatoid factor. These immune complexes with hepatitis virus precipitate in the cold and are hence called cryoglobulins (see Case 28, Mixed Essential Cryoglobulinemia). They can also be entrapped in the renal glomeruli as well as in small blood vessels of the skin, nerves, and other tissues, where they cause inflammation of the blood vessels (vasculitis). The most prevalent immune-complex disease is systemic lupus erythematosus (SLE), which is characterized by the formation of antibody to DNA. Every day, many millions of nuclei are extruded from erythroblasts in the bone marrow as they turn into mature red blood cells (erythrocytes). This event, among others, provides a rich source of DNA in those individuals prone to making an immune response to DNA and thereby to developing SLE.

Fig. 43.2 Three autoimmune diseases that result in damage by immune complexes.

Immune-complex disease		
Syndrome	Autoantigen	Consequence
Subacute bacterial endocarditis	Bacterial antigen	Glomerulonephritis
Mixed essential cryoglobulinemia (see Case 28)	Rheumatoid factor IgG complexes (with or without hepatitis C antigens)	Systemic vasculitis
Systemic lupus erythematosus	DNA, histones, ribosomes, snRNP, scRNP	Glomerulonephritis, vasculitis, arthritis

Sixteen-year-old girl, butterfly rash and symmetrical morning stiffness.

The case of Nicole Chawner: too much sun at the beach.

Nicole Chawner was 16 years of age and had enjoyed good health all her life until August 1994. A few days after excessive exposure to the sun on the beach, Nicole developed a red rash on her cheeks. Her parents took her to the family doctor who recognized that the butterfly pattern of the rash on her cheeks and over the bridge of her nose was typical of systemic lupus erythematosus (SLE) (Fig. 43.3).

He referred Nicole to the Children's Hospital, where she was asked about any other problems she might have noticed, and said that when she woke up in the morning her finger joints and hips were stiff, though they got better as the day wore on. She felt equally stiff, she said, in both hands and both hips.

A blood sample was taken from Nicole to ascertain whether she had anti-nuclear antibodies (ANA). This was positive at a titer of 1:1280. In view of this result, further tests were performed for antibodies characteristically found in SLE. An elevated level of antibodies to double-stranded DNA was also found. Her serum C3 level was 73 mg dl^{-1} (normal 100–200 mg dl^{-1}). Her platelet count was normal at 225,000 μl^{-1} and her direct and indirect Coombs tests were negative, as was a test for anti-phospholipid antibody. A urine sample was taken and found to be normal.

Nicole was advised to take an antimalarial agent, Plaquenil (hydroxychloroquine sulfate) and to avoid direct sunlight. (Antimalarials have a beneficial effect on SLE for unknown reasons.) She did well for a while but, after a month, the pain in her fingers and hips in the morning got worse. She developed fever of 39°C every evening accompanied by shaking chills. Enlarged lymph nodes were felt behind her ears and in the back of her neck. She lost 4.6 kg of body weight over the course of the next 2 months.

When she returned to the Children's Hospital for a check-up, it was noted that her butterfly rash had disappeared. She had diffuse swelling of the proximal joints in her fingers and toes. Blood was drawn at this time and the level of anti-DNA antibodies was found to have increased. The serum C3 level was 46 mg dl^{-1}. Nicole was advised to take 10 mg of prednisone twice a day, as well as the nonsteroidal anti-inflammatory drug Naprosyn (naproxen), 250 mg twice a day. This quickly controlled her symptoms and she remained well. At her next visit to the Children's Hospital her serum C3 level was 120 mg dl^{-1}.

Fig. 43.3 The so-called butterfly rash typical of systemic lupus erythematosus. Photograph courtesy of M. Walport.

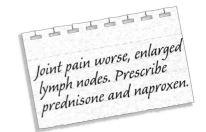

Joint pain worse, enlarged lymph nodes. Prescribe prednisone and naproxen.

Systemic lupus erythematosus (SLE).

Systemic lupus erythematosus (SLE) is currently the most prevalent immune-complex disease in developed countries. For reasons that are not known it affects ten times as many females as males. Patients with SLE usually have antibodies to many autoantigens. The commonest autoantibody, which is found in the serum of 60% of all SLE patients, is to double-stranded DNA. Other antibodies very commonly found are to small ribonucleoproteins. Autoantibodies to blood cells, such as platelets and red blood cells, as well as to the phospholipid complex that is formed by the activation of the proteins of the clotting system (antiphospholipid antibodies), are common but less frequent. Most patients have a range of these autoantibodies.

The immune complexes in SLE are small and tend to be trapped or formed inside tissues, primarily in the kidney and, to a smaller extent, in the synovial tissues of joints. For this reason, glomerulonephritis and arthritis are two of the most frequently encountered symptoms of SLE. The immune complexes encountered in patients with SLE fix complement efficiently. Tissue injury in the kidney or joints is mediated by activation of the complement system.

The word 'lupus' is Latin for wolf and this word is applied to a common symptom of SLE, the butterfly rash on the face. In the 19th century, the severe scarring rash on the face was named lupus because it was said to resemble the bite of a wolf. At that time, it was not possible to distinguish lupus erythematosus from lupus vulgaris, a scarring rash caused by tuberculosis. For unknown reasons, the rash is evoked by exposure to the sun (ultraviolet light). There is a seasonal variation to the onset of SLE, which is greatest in the Northern Hemisphere between March and September when the greatest amount of ultraviolet light penetrates the atmosphere.

Questions.

1 Why do you think Nicole's serum C3 was measured, both on her first visit to the hospital and after therapy?

2 What are the direct and indirect Coombs tests and what did they tell us in this case?

3 Why was Nicole told to avoid direct exposure to sunlight?

4 Repeated analysis of Nicole's urine was negative. What does this mean?

5 Nicole had a serum IgG level of 1820 mg dl^{-1}. This is a substantially elevated level of IgG; this is commonly found in patients with SLE. How could you explain this? And what would you expect to find if we had a biopsy of Nicole's swollen lymph nodes?

6 The antigen in the immune complexes formed in SLE is often a complex antigen, such as part of a nucleosome or a ribonucleoprotein particle, which contains several different molecules. Patients often produce autoantibodies against each of these different components. What is the reason for the production of this variety of autoantibodies, and what type of failure in tolerance could be responsible for autoantibody production?

CASE 44 | Multiple Sclerosis

Autoimmune attack on the integrity of the central nervous system.

As we saw in Case 40, autoimmune diseases can be caused by antibodies against self antigens. This can be proved from observations that newborn infants of affected mothers acquire the disease from maternal immunoglobulins that cross the placenta, and from the fact that the disease can be transmitted to experimental animals by immunoglobulin from patients.

Autoimmune disease can also be caused by activated effector T cells specific for self peptides (see Case 42). When T cells recognize self-peptide:MHC complexes and become activated, they can cause local inflammation by activating macrophages, for example, with consequent tissue damage. An example of T-cell mediated autoimmune disease is the neurologic disease multiple sclerosis (MS). However, unlike in autoimmune diseases caused by autoantibodies, it has been difficult to prove the involvement of T cells in MS, because T cells do not cross the placenta into the fetus and T-cell transfer is not possible in humans.

Seventy years ago, an experimental model of MS was established in mice, in which the injection of myelin in adjuvant caused the development of neurologic symptoms similar to those of MS. This disease is called experimental autoimmune encephalomyelitis (EAE). The antigens in myelin that can induce EAE are myelin basic protein (MBP), proteolipid protein (PLP), and myelin oligodendrocyte glycoprotein (MOG). The disease can be transferred to syngeneic animals by cloned antigen-specific T-cell lines derived from animals with EAE (Fig. 44.1). When the recipient animals are immunized with MBP, for example, they develop active disease. T cells specific for MBP, PLP, and MOG have been found in the blood and cerebrospinal fluid (CSF) of patients with MS.

Topics bearing on this case:

Inflammatory reactions

Interactions of co-stimulatory molecules with their receptors

The development of tolerance to self antigens

Activation of self-reactive T cells

Immunologically privileged sites

Experimental autoimmune encephalomyelitis

Induction of oral tolerance

Fig. 44.1 T cells specific for myelin basic protein mediate inflammation of the brain in experimental autoimmune encephalomyelitis (EAE). This disease is produced in experimental animals by injecting them with isolated spinal cord homogenized in complete Freund's adjuvant. EAE is due to an inflammatory reaction in the brain that causes a progressive paralysis affecting first the tail and hind limbs before progressing to forelimb paralysis and eventual death. One of the autoantigens identified in the spinal cord homogenate is myelin basic protein (MBP). Immunization with MBP alone in complete Freund's adjuvant can also cause these disease symptoms. Inflammation of the brain and paralysis are mediated by T$_H$1 cells specific for MBP. Cloned MBP-specific T$_H$1 cells can transfer symptoms of EAE to naive recipients provided that the recipients carry the correct MHC allele. In this system it has therefore proved possible to identify the peptide:MHC complex recognized by the T$_H$1 clones that transfer disease.

29-year-old female with sudden loss of vision in one eye.

Family history of MS; corticosteroid therapy; order MRI brain scan.

Muscle weakness developing on left side; repeat MRI scan; MS diagnosed.

Relapse; give weekly IFN-β.

The case of Vivie Warren: an oboist who has difficulty reading a musical score.

Mrs Vivie Warren is a 29-year-old professional oboe player. She was in good health until one morning she noticed a loss of vision in her left eye. Her physician referred her to a neurologist, who found that her eye movement was normal and not accompanied by any pain. The visual acuity in Vivie's left eye was 20/100 and in her right eye 20/200. Her retina was normal, and a detailed neurologic examination also proved normal. The neurologist diagnosed optic neuritis (inflammation of the optic nerve). Her family history was informative, however, in that her mother had severe MS and was permanently disabled, and a magnetic resonance imaging (MRI) brain scan was ordered.

Vivie was given a 5-day course of intravenous corticosteroids and her vision returned to normal over the next 3 weeks. The MRI scan revealed multiple lesions in the white matter of the brain under the cortex and around the ventricles (Fig. 44.2). Intravenous injection of gadolinium, a contrast agent that leaks from blood vessels in recently inflamed tissue, showed that some of the brain lesions were probably of recent origin (Fig. 44.3). The neurologist told Vivie that she had a high probability of developing MS and advised her to return for frequent neurologic examination.

Vivie remained well for a further 3 years and then developed weakness of the muscles on the left side of her face that were innervated by the seventh cranial nerve. A repeat MRI scan with gadolinium enhancement showed new lesions in the left middle cerebellar peduncle and in the pons. CSF was obtained by lumbar puncture. It contained 28 mg dl^{-1} of protein (normal) and 8 lymphocytes ml^{-1} (normal 0–3 ml^{-1}). At this point a firm diagnosis of MS was made. Despite the normal level of protein in the CSF, the IgG content was raised. Upon electrophoresis, discrete bands of IgG were observed, indicating clonal expansion of restricted B-cell populations in the central nervous system (CNS). Another 5-day course of corticosteroids was administered intravenously and Vivie's symptoms improved. Weekly intramuscular injections of interferon (IFN)-β were started to prevent progression of the disease.

Vivie did well for 3 more years, after which she developed a weakness in her left leg and left hand. Her speech became slurred. She developed nystagmus (rapid uncontrolled horizontal jerking eye movements when attempting to fix the gaze on something) and ataxia (wide-based staggering gait). Vivie was given another course of corticosteroids, after which her symptoms improved, but 8 months later they recurred. The injections of IFN-β were stopped and she was put on high doses of cyclophosphamide and corticosteroids at monthly intervals. After 3 months of this therapy, the cyclophosphamide and corticosteroid injections were gradually reduced to every 12 weeks. Her neurological examination became normal and no new lesions were observed on gadolinium-enhanced MRI.

Multiple sclerosis.

Multiple sclerosis (MS) was first described by the great French neurologist Jean-Martin Charcot in the 1860s. It was noted at autopsy that patients who died of this disease had multiple hard (sclerotic) plaques scattered throughout the white matter of the CNS. The disease is 10 times more frequent in women than in men and is associated with HLA-DR2. Those affected have a variety of nervous symptoms, such as urinary incontinence, blindness, ataxia, muscle weakness, and paralysis of limbs. The plaques characteristic of the disease show dissolution of myelin along with infiltrates of lymphocytes and macrophages, particularly along blood vessels. The inflammatory exudate causes increased vascular permeability.

The CNS is a relatively immunologically privileged site from which antigens do not normally reach the lymphoid tissues, and so there is no negative selection of T cells with the potential to react against CNS antigens. In MS, an unknown injurious event is presumed to provoke the release of CNS antigens and their presentation to lymphocytes in the peripheral lymphoid organs. This results in the expansion of clones of autoreactive T cells and their differentiation into T_H1 cells, which home to the CNS and initiate inflammation. These T_H1 cells can be readily identified in the CNS of patients with MS.

Lymphocytes and other blood cells do not normally cross the blood–brain barrier. If tissue becomes inflamed, however, activated CD4 T cells autoreactive for a brain antigen and expressing $\alpha_4{:}\beta_1$ integrin, which binds vascular cell adhesion molecules (VCAM) on the surface of activated venule endothelium, can migrate out of the blood into the brain. There they re-encounter their specific autoantigen presented by MHC class II molecules on microglial

Fig. 44.2 Three-dimensional magnetic resonance images of the brain at three different time points in the course of MS. Left, early; center, during acute exacerbation; right, after therapy. The technique used causes fluid to appear white. The lateral ventricles in the middle of the brain scan and the sulci of the cerebral cortex around the edge appear white as a result of normal cerebrospinal fluid. The white spots, which are due to edema fluid and decreased myelin, are MS lesions. Photograph courtesy of David Hafler.

Relapse; aggressive immunosuppressive therapy started.

Fig. 44.3 Three computer-generated images of MRI scans of the brain shown in Fig. 44.2 at the same time points. Several levels of MRI scans have been integrated by computer. The lateral and middle ventricles appear in purple. Old MS lesions appear yellow-green. A new, gadolinium-enhanced, lesion appears orange. On the right, this lesion has diminished in size after immunosuppressive therapy. Photograph courtesy of David Hafler.

| Unknown trigger sets up initial focus of inflammation in brain and blood–brain barrier becomes locally permeable to leukocytes and blood proteins | T cells specific for CNS antigen and activated in peripheral lymphoid tissues re-encounter antigen presented on microglia or dendritic cells in brain | Inflammatory reaction in brain due to mast-cell activation, complement activation, antibodies and cytokines | Demyelination of neurons |

Fig. 44.4 The pathogenesis of multiple sclerosis. At sites of inflammation, activated T cells autoreactive for brain antigens can cross the blood–brain barrier and enter the brain, where they re-encounter their antigens on microglial cells and secrete cytokines such as IFN-γ. The production of T-cell and macrophage cytokines exacerbates the inflammation and induces a further influx of blood cells (including macrophages, dendritic cells, and B cells), and blood proteins (such as complement), into the affected site. Mast cells also become activated. The individual roles of these components in demyelination and loss of neuronal function is still not well understood.

cells and produce pro-inflammatory cytokines such as IFN-γ (Fig. 44.4). Microglia are phagocytic macrophage-like cells of the innate immune system resident in the CNS and, like macrophages, can act as antigen-presenting cells. Inflammation causes increased vascular permeability and the site becomes heavily infiltrated by activated macrophages and T_H1 cells, which produce pro-inflammatory cytokines that exacerbate the inflammation, resulting in the further recruitment of T cells, B cells, macrophages, and dendritic cells to the site of the lesion. Autoreactive B cells produce autoantibodies against myelin antigens with help from T cells. Activated mast cells release histamine, contributing to the inflammation. In some way not yet fully understood, these combined activities lead to demyelination and interference with neuronal function.

In mice, feeding with MBP before immunization with MBP in adjuvant protects against the development of EAE. MBP-specific T cells can be identified in the brain of protected mice, but they secrete transforming growth factor (TGF)-β rather than IFN-γ and fail to initiate an inflammatory reaction. Furthermore, adoptive transfer of T cells from mice fed with MBP can confer protection against EAE. Therapeutic attempts to treat patients with MS by oral MBP have not been successful, however. This suggests that therapeutic interventions that induce oral tolerance may not be effective in already established disease.

Questions.

1   *Oligoclonal immunoglobulins were found in Mrs Warren's central nervous system. How do you explain this?*

2   *Mrs Warren was treated with corticosteroids, cyclophosphamide, and IFN-β. What was the aim of this therapy?*

3   *An attempt has been made to treat MS patients with IFN-γ. Can you predict what the outcome was and why?*

4   *What is the rationale behind feeding MBP to mice to prevent EAE?*

5   *Can you predict whether EAE can be induced in CD28 knockout mice?*

CASE 45 | Hemolytic Disease of the Newborn

The adaptive immune response distinguishes different individuals within a species.

Adaptive immune responses evolved to protect vertebrate species against the world of microorganisms. However, anything discerned as 'nonself' may become a target of such responses, which can also be directed at molecular differences between individuals within a species. An antibody directed against an antigenic determinant present in some members of a species but not others is said to show polymorphic variation, and antibodies directed against such determinants are called alloantibodies. Perhaps the best known alloantibodies are those that are used to determine our blood groups. Individuals who have red cells of type A have alloantibodies that react with the red blood cells of individuals who are type B and vice versa. Individuals who have red blood cells of type AB have no alloantibodies and those with type O red blood cells have antibodies to both A and B red blood cells (Fig. 45.1). These alloantibodies arise from the fact that the capsules of Gram-negative bacteria, which inhabit our gut, bear antigens that stimulate antibodies that cross-react with the carbohydrate antigens of the ABO blood groups.

A serious problem is posed frequently by alloantibodies induced by a fetus in the pregnant mother. Alloimmunization most often results from Rhesus (Rh) incompatibility between mother and fetus. Approximately 13% of women are Rh-negative; that is, their red blood cells do not bear the Rh antigen. A woman who is Rh-negative has an 87% chance of marrying an Rh-positive man and their chances of having an Rh-positive baby are very high. Not infrequently, during delivery of the newborn, some blood escapes from the fetal circulation into the maternal circulation and the mother develops alloantibodies to the Rh antigen as a result. During a subsequent pregnancy with an Rh-positive fetus, the maternal IgG alloantibodies cross the placenta and cause hemolysis of the fetal red cells. As we shall see, the consequences of this can be very grave and result in fetal death or severe damage to the newborn infant.

Topics bearing on this case:
Antibodies can be produced against almost any substance
ABO blood groups
Rhesus blood group antigens
Hemagglutination assays
Anti-immunoglobulin antibodies are a useful tool for detecting bound antibodies
Antibody can suppress naive B-cell activation

Fig. 45.1 Hemagglutination is used to type blood groups and match compatible donors and recipients for blood transfusion. Common gut bacteria bear antigens that are similar or identical to blood group antigens, and these stimulate the formation of antibodies to these antigens in individuals who do not bear the corresponding antigen on their own red blood cells (left column); thus, type O individuals, who lack A and B, have both anti-A and anti-B antibodies, whereas type AB individuals have neither. The pattern of agglutination of the red blood cells of a transfusion donor or recipient with anti-A and anti-B antibodies reveals the individual's ABO blood group. Before transfusion, the serum of the recipient is also tested for antibodies that agglutinate the red blood cells of the donor, and vice versa, a procedure called a cross-match, which may detect potentially harmful antibodies to other blood groups that are not part of the ABO system.

Serum from individuals of type	Red blood cells from individuals of type O — Express the carbohydrate structures R–GlcNAc–Gal Fuc	A — R–GlcNAc–Gal–GalNAc Fuc	B — R–GlcNAc–Gal Fuc	AB — R–GlcNAc–Gal–GalNAc Fuc + R–GlcNAc–Gal–Gal Fuc
O — Anti-A and anti-B antibodies	no agglutination	agglutination	agglutination	agglutination
A — Anti-B antibodies	no agglutination	no agglutination	agglutination	agglutination
B — Anti-A antibodies	no agglutination	agglutination	no agglutination	agglutination
AB — No antibodies to A or B	no agglutination	no agglutination	no agglutination	no agglutination

The Rh antigenic determinants are spaced very far apart on the red cell surface. As a consequence, IgG antibodies to the Rh antigen do not fix complement and therefore do not hemolyze red blood cells *in vitro*. For reasons that are less well understood, IgG antibodies to the Rh antigen do not agglutinate Rh-positive red blood cells. Because of this it was very difficult to detect Rh antibodies until Professor Robin Coombs at the University of Cambridge devised a solution to the problem by developing antibodies against human immunoglobulin. He showed that Rh-positive red blood cells coated with IgG anti-Rh antibodies could be taken from a fetus and agglutinated by antibodies to IgG. Furthermore, he showed that when the serum of an alloimmunized woman was incubated with Rh-positive red blood cells, these red blood cells could then be agglutinated by antibody to IgG (Fig. 45.2). The former is called the direct Coombs test and the latter the indirect Coombs test. This application of immunology to a vexing clinical problem led ultimately to a treatment and prevention of the problem.

The case of Cynthia Waymarsh: a fetus in immunological distress.

Mrs Waymarsh was 31 years old when she became pregnant for the third time. She was known to have blood group A, Rh-negative red cells. Her husband was also type A but Rh-positive. Their first-born child, a male, was healthy. During her second pregnancy Mrs Waymarsh was noted to have an indirect Coombs titer at a 1:16 dilution of her serum. The fetus was followed closely and the delivery of a healthy baby girl was induced at 36 weeks of gestation.

It was 5 years later when Mrs Waymarsh became pregnant again. At 14 weeks gestation her indirect Coombs titer was 1:8 and at 18 weeks 1:16. Amniotic fluid was obtained at 22, 24, 27, and 29 weeks of gestation and was found to have increasing amounts of bilirubin (a pigment derived from the breakdown of heme, indicating that the fetus' red blood cells were being hemolyzed). At 29 weeks of gestation a blood sample was obtained from the umbilical vein and found to have a hematocrit of 6.2% (normal 45%). (The hematocrit is the proportion of blood that is composed of red cells, and because the volume of white cells is comparatively negligible, this is simply ascertained by centrifuging whole unclotted blood in a tube.) Upon finding that the fetus was profoundly anemic, 85 ml of type O, Rh-negative packed red blood cells were transfused into the umbilical vein. At 30.5 weeks of gestation another sample of blood from the umbilical vein was obtained; the hematocrit was 16.3%. The fetus was transfused with 75 ml of type O, Rh-negative packed red blood cells.

The fetus was examined at weekly intervals for the appearance of hydrops (see below) and none was observed. At 33.5 weeks of gestation the hematocrit of a blood sample from the umbilical vein was 21%, so 80 ml of O Rh-negative packed red blood cells were again transfused into the umbilical vein. At 34.5 weeks of gestation it was determined that the fetus was sufficiently mature to sustain extrauterine life without difficulty; labor was induced and a normal female infant was born. The hematocrit in the umbilical vein blood was 29%. The baby did well and no further therapeutic measures were undertaken.

30.5 weeks: hematocrit up to 16.3%. Further transfusion requested.

33.5 weeks: hematocrit up to 21%. Further transfusion requested.

Hemolytic disease of the newborn.

Although hemolytic disease of the newborn is most commonly the result of alloimmunization with Rh antigen, other red blood cell alloantigens, such as Lewis, Kell, Duffy, Kidd, Lutheran, and still others, may cause alloimmunization. In any case, the maternal IgG antibodies cross the placenta in increasing amounts during the second trimester of pregnancy, and hemolyze the fetal red blood cells. The resulting anemia may become so severe that, if untreated, the fetus goes into heart failure and develops massive edema; this is called hydrops fetalis and results in fetal death. The risk of fetal development of hydrops rises from 10% when the indirect Coombs titer of the mother is 1:16 to 75% when the maternal titer is 1:128. If the anemia is not so severe as to cause hydrops, the affected infant at birth is still massively hemolyzing red blood cells. Now the newborn must dispose of the heme breakdown pigments rapidly because an excessive accumulation of bilirubin results in the deposition of this pigment in the brain, and severe neurological impairments. In response to the

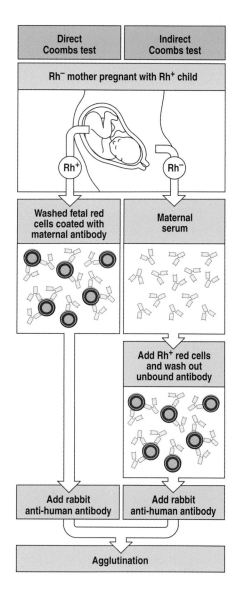

Fig. 45.2 The Coombs direct and indirect anti-globulin tests for antibody to red blood cell antigens. An Rh-negative (Rh⁻) mother of an Rh-positive (Rh⁺) fetus can become immunized to fetal red blood cells that enter the maternal circulation at the time of delivery. In a subsequent pregnancy with an Rh⁺ fetus, IgG anti-Rh antibodies can cross the placenta and damage the fetal red blood cells. In contrast to anti-Rh antibodies, anti-ABO antibodies are of the IgM isotype and cannot cross the placenta, and so do not cause harm. Anti-Rh antibodies do not agglutinate red blood cells, but their presence can be shown by washing the fetal red blood cells and then adding antibody to human immunoglobulin, which agglutinates the antibody-coated cells. The washing removes unrelated immunoglobulins that would otherwise react with the anti-human immunoglobulin antibody. Anti-Rh antibodies can be detected in the mother's serum in an indirect Coombs test; the serum is incubated with Rh⁺ red blood cells, and once the antibody binds, the red cells are treated as in the direct Coombs test.

profound anemia, the number of red blood cell precursors (erythroblasts) in the spleen, liver, and bone marrow expands rapidly; for this reason, hemolytic disease of the newborn has also been called erythroblastosis fetalis.

As we have seen in this case, the extent of hemolysis can be determined easily by obtaining amniotic fluid into which the fetus commences to urinate by 20 weeks of gestation. The quantity of bilirubin excreted into the amniotic fluid correlates with the amount of hemolysis in the fetus. Secondly, the fetus can be followed by ultrasonography for the development of hydrops. Thirdly, the degree of anemia can be ascertained directly, but with some difficulty, by obtaining a sample of blood from the umbilical vein.

It has become possible in the past few decades to eliminate to a very great extent hemolytic disease of the newborn. All Rh-negative women are given 300 µg of purified immunoglobulin specific for the rhesus antigen (Rhogam) at 28 weeks of gestation and again within 72 hours of delivery. The amount of Rhogam in one vial (300 µg) is sufficient to neutralize 30 ml of fetal blood in the maternal circulation. This procedure fails to prevent alloimmunization in only 0.1% of Rh-negative women, such as Mrs Waymarsh. It must be presumed in such failures that the fetus bled more than 30 ml of blood into its mother.

Questions.

1 It was stated that the Rh antigens are so sparsely scattered on the red cell surface that IgG molecules bound to the Rh antigens are too separated to fix C1q. Therefore complement-mediated hemolysis cannot be invoked to explain hemolytic disease of the newborn. By what mechanism are the red cells destroyed?

2 When an Rh-negative woman is ABO compatible with her husband, as Mr and Mrs Waymarsh are, the risk of Rh alloimmunization is 16%. When they are ABO incompatible the risk falls to 7%. How do you explain this difference?

3 Why were Rh-negative red blood cells used for the intrauterine transfusion?

4 Do you have concerns about administering Rhogam to women at 28 weeks of gestation?

5 The serum of an Rh-negative woman who is pregnant gives a negative indirect Coombs test but her serum agglutinates Rh-positive cells suspended in saline. What is your interpretation of this phenomenon and what do you do about it?

CASE 46 | **A Kidney Graft for Complications of Autoimmune Insulin-Dependent Diabetes Mellitus**

Mechanisms of graft rejection and HLA associations in autoimmunity.

The transplantation of solid organs is becoming increasingly common as surgeons have become more skilled in transplantation techniques and immunosuppressive drugs have become more effective. Only the supply of donor organs seems to limit this growing surgical practice. The first successful solid organ transplant was a kidney graft from one identical twin brother to the other at the Peter Bent Brigham Hospital in Boston in the mid-1950s. Since then, solid organ transplantation has expanded to successful transplants of liver, heart, lungs, and pancreas.

Ideally, solid organ transplantation should be performed between histoidentical donors and recipients but this, for all practical purposes, is not possible. Perhaps only 2 in 10,000 unrelated individuals are histocompatible at the major histocompatibility locus (MHC). Among sibling donor recipient pairs, 1 in 4 should be histoidentical at the MHC but there are further problems with compatibility of ABO blood groups and minor histocompatibility loci. It is now clear that, in solid organ transplantation, absolute matching at the MHC loci offers only a small incremental advantage for survival of the transplanted organ. This realization has led to the widespread use of unrelated cadaveric organ donors.

The first risk to be considered is the problem of hyperacute rejection. It is important that the recipient matches the donor at the ABO blood group locus lest antibodies in the recipient react to A or B blood group substances in the endothelium of the graft. Such antigen–antibody interactions in the small blood vessels of the graft induce clotting in these small blood vessels; the graft becomes hemorrhagic and quickly dies. It is also very important to ascertain that the recipient has no antibodies to the cell-surface antigens of donor white cells. Such antibodies might be present, for example, if the recipient has had a blood transfusion on an earlier occasion: leukocytes are present in most transfused blood, and prior sensitization to leukocyte antigens of the donor can cause hyperacute rejection of the graft within a few hours. Hyperacute rejection is impossible to treat and when it occurs the graft cannot be saved.

Topics bearing on this case:

Tissue matching for organ transplants

Transplant rejection and immunosuppressive drugs

HLA-disease association

T cell-mediated autoimmune disease

Mechanisms of action of immunosuppressive drugs

Between 7 and 15 days after solid organ transplantation the host T cells will mount an immune response to the histocompatibility antigens of the donor. There is usually a general inflammatory reaction, with massive lymphocyte infiltration culminating in infiltration by CD8 cytotoxic T cells and first-set graft rejection. Unlike hyperacute rejection, first-set graft rejection can be treated satisfactorily, and immunosuppressive drugs such as azathioprine, prednisone, cyclosporin A, and tacrolimus (FK506) allow long-term survival of grafts.

The case of Christopher Goodwood: a saga of organ rejection and acceptance.

Christopher, 12 years old, frequent urination, thirst and weight loss.

Christopher Goodwood was the only child born to his parents. His growth and development were normal and he excelled at school. When he was 12 years old he began to lose weight and developed such an enormous thirst that he was consuming over 3 liters of liquid a day. The thirst and excess fluid consumption was accompanied by frequent urination. He was taken to the family doctor who diagnosed diabetes mellitus. It turned out that several relatives of Christopher's father had also developed juvenile onset diabetes.

Christopher was started on daily insulin injections but his blood sugar was always difficult to control. Despite this handicap, Christopher continued to do well in school and he eventually entered college as an engineering student. Shortly after he graduated he found a job with a civil engineering firm and was married.

Hypertension, proteinuria and elevated serum creatinine.

When he was 35 years old, during a routine annual physical examination, his physician found that his blood pressure was elevated and that his urine tested positive for the presence of protein. A serum creatinine test was obtained and found to be 7.5 mg dl^{-1} (normal <1.0 mg dl^{-1}). On the basis of these results, Christopher's physician suspected that he had developed renal complications of diabetes.

The physician prescribed drugs to control the high blood pressure and recommended that a kidney biopsy be performed. The biopsy, which entails the removal of a small sample of tissue, revealed extensive diabetic glomerulo-sclerosis (scarring of the glomeruli by fibrous tissue) (Fig. 46.1), showing that Christopher was in imminent danger of kidney failure. He was therefore scheduled for hemodialysis (passing blood over a semipermeable membrane to remove waste products such as urea) twice a week. He was also put on a waiting list for a cadaveric kidney transplant.

Christopher was blood group B, Rh-positive. His HLA type was A2,24; B50,51; DR3,4. After a waiting period of 6 months during which he had to continue on hemodialysis, a cadaveric kidney was found from a 20-year-old man who was fatally injured in a motorcycle accident. The donor was blood type B, Rh-positive. Only HLA-A and -B typing were available from the donor. He was HLA-A 2,11; B7,35. Christopher's serum was tested for antibodies against the white blood cells of the potential donor and was found to be nonreactive.

Fig. 46.1 Glomerulus from a diabetic with glomerulosclerosis. Photograph courtesy of R. Colvin and A. Perez-Atayde.

Christopher was admitted to the hospital and, on the following morning, his own kidneys were removed by a surgeon and replaced with the donor kidney. During the procedure Christopher was given 100 mg methylprednisolone and 40 mg of anti-ICAM monoclonal antibody. During the first day after his operation the transplanted kidney put out 200 ml of urine. This is only about one-fifth the normal output for a healthy person but usual for a newly transplanted kidney and showed that the organ was functioning.

Christopher was given 450 mg cyclosporin, 100 mg azathioprine, 160 mg methylprednisolone, and a further 40 mg of anti-ICAM antibodies. The serum creatinine returned to normal during the next 5 days. After the fifth day anti-ICAM antibodies were stopped and Christopher was continued on 100 mg azathioprine, 400 mg cyclosporin, and 20 mg prednisone. Christopher was discharged from the hospital 10 days after the transplant was performed. His blood pressure and serum creatinine were normal. The transplant appeared normal by ultrasonography.

One week after discharge from the hospital Christopher noticed that his urine output had fallen to about half normal. When he went back to the hospital, his creatinine proved to have risen to 2.5 mg dl^{-1} and the graft was found to be enlarged and tender. A renal (kidney) biopsy revealed the presence of many lymphocytes in the transplant (Fig. 46.2) compared with a normal kidney, indicating that the kidney was undergoing rejection.

His immunosuppressive therapy was increased to 500 mg methylprednisolone intravenously, the cyclosporin was continued, and intravenous injections of 5 mg of a monoclonal antibody to CD3 were given for 10 days. His peripheral blood after this regimen contained 15% CD3$^+$ cells. This immunosuppressive regimen successfully suppressed the symptoms and Christopher continued to do well for another 6 weeks.

Six weeks later, however, while he was recuperating at his parents' home, Christopher developed fever and a severe cough that produced yellow, thick sputum. His physician was called and, on examining Christopher, heard crackles and wheezes over both lungs. A sample of Christopher's sputum was tested for *Mycobacterium tuberculosis*, which could not be detected, and for Gram-negative and Gram-positive bacteria, which were present but sparse; this is normal. A silver stain for *Pneumocystis carinii* was also negative. The fungus *Aspergillus fumigatus* was cultured from all his sputum specimens. He was treated with the fungicide amphotericin B, and recovered.

Within a year of the transplant Christopher returned to his civil engineering job. His blood pressure remains normal, his serum creatinine is <1.0 mg dl^{-1} and he continues to take low doses of azathioprine, prednisone, and cyclosporin.

Fig. 46.2 First-set rejection in a kidney graft. Top panel shows lymphocytes around an arteriole in the kidney. The middle panel shows lymphocytes surrounding the renal tubules of a kidney undergoing rejection and the bottom panel shows the staining of lymphocytes with anti-CD3 in the same section.

Insulin-dependent diabetes mellitus and kidney grafting.

Insulin-dependent diabetes mellitus (IDDM), also called type 1 diabetes mellitus, is a T cell-mediated autoimmune disease (Fig. 46.3). Like many other autoimmune diseases, it occurs with a markedly higher frequency in individuals with certain specific MHC class II alleles (Fig. 46.4). The reasons for this association are obscure. Although the MHC alleles seem to predispose to the development of IDDM, it is clear that other factors also have a role. For example, if IDDM develops in one of two identical twins, the chances are only one in three that the other will develop it as well. This suggests that some environmental factor determines whether the disease is actually triggered.

Fig. 46.3 Common T cell-mediated autoimmune diseases.

T cell-mediated autoimmune disease		
Syndrome	Autoantigen	Consequences
Insulin-dependent diabetes mellitus	Pancreatic β-cell antigen	β-cell destruction
Rheumatoid arthritis (see Case 42)	Unknown synovial joint antigen	Joint inflammation and destruction
Experimental autoimmune encephalomyelitis (EAE), multiple sclerosis	Myelin basic protein, proteolipid protein, myelin oligodendrocyte glycoprotein	Brain invasion by CD4 T cells, weakness

IDDM is also known as juvenile onset diabetes because it usually appears in childhood or early adolescence. At the beginning of the disease, the islets of Langerhans in the pancreas become infiltrated with lymphocytes, which selectively destroy the β cells of the islets (Fig. 46.5). The antigen that provokes the autoimmune attack on the β cells is not known and has been difficult to identify. An excellent animal model of IDDM is provided by NOD (non-obese diabetic) mice, in which, as in humans, a specific MHC class II allele is associated with the development of diabetes. A very high proportion of these mice develop diabetes but not all, which, as in humans, indicates an environmental factor.

It has been possible to retard the progression of diabetes in NOD mice by administering immunosuppressive agents or by depleting these mice of T cells. It has also been possible to transfer the disease with T cells. These observations have led to various immunosuppressive therapeutic trials in humans with IDDM but success has not yet been achieved convincingly.

The case of Christopher Goodwood illustrates one of the long-term complications of poorly controlled IDDM (that is, IDDM in which blood sugar is poorly regulated): namely diabetic glomerulosclerosis. The kidneys are damaged irreversibly by this complication of IDDM, renal failure ensues and the only available therapy, other than hemodialysis, is renal transplantation. At best, the kidney recipient may expect perhaps 20 years of useful life from the graft before it too develops glomerulosclerosis. In practice, ill-matched grafts usually do not survive that long, and may last no longer than 5 years.

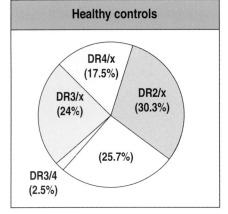

Healthy controls

DR4/x (17.5%)
DR2/x (30.3%)
DR3/x (24%)
(25.7%)
DR3/4 (2.5%)

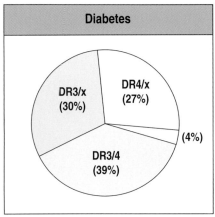

Diabetes

DR3/x (30%)
DR4/x (27%)
(4%)
DR3/4 (39%)

Fig. 46.4 Population studies show association of susceptibility to insulin-dependent diabetes mellitus (IDDM) with HLA genotype. The HLA genotypes (determined by serotyping) of diabetic patients (bottom panel) are not representative of those found in the population (top panel). Almost all diabetic patients express HLA-DR3 and/or HLA-DR4, and HLA-DR3/DR4 heterozygosity is greatly overrepresented among diabetics compared with controls. These alleles are linked tightly to HLA-DQ alleles that confer susceptibility to IDDM. By contrast, HLA-DR2 protects against the development of IDDM and is only found extremely rarely in diabetic patients. The small letter x represents any allele other than DR2, DR3, or DR4.

Fig. 46.5 Pancreatic β cells with lymphocytic infiltration. The left panel is an islet from a 6-week-old non-obese diabetic (NOD) mouse showing the invasions of the islet with lymphocytes. The right panel is the islet of an NOD mouse treated with complete Freund's adjuvant. This causes massive and dramatic lymphocyte invasion of the islet. Photographs courtesy of J. Katz (left) and O. Kanagawa (right).

Questions.

1 Why did Christopher develop an opportunistic infection with Aspergillus fumigatus?

2 How does cyclosporin A work?

3 When the graft was undergoing first-set rejection, it became swollen and tender. How do you explain this? What would second-set rejection be?

4 Why was it considered more important to match the potential donor's blood group with Christopher's blood group and to ascertain that Christopher had no cytotoxic antibodies to the donor than to worry about complete HLA matching?

5 Why did the surgeon choose to give Christopher anti-ICAM monoclonal antibodies and subsequently to treat impending rejection with anti-CD3?

6 Several members of Christopher's father's family had also developed juvenile onset diabetes. Can you suggest why?

7 Christopher is HLA-DR3. This MHC type makes him four times more susceptible to IDDM (as well as to other autoimmune diseases) as the average person. But he is also HLA-DR4, and the combination of HLA-DR3 and HLA-DR4 makes him 50 times more susceptible. What HLA-DR type in place of HLA-DR4 would have lowered his risk of acquiring IDDM?

8 Would you expect to find the mechanism of Christopher's disease susceptibility by looking at the properties of molecules belonging to the HLA-DR3 and HLA-DR4 types?

9 How might specific HLA-D alleles confer susceptibility to IDDM, and conversely how might other alleles lower this risk?

10 How might environmental factors contribute to the development of IDDM?

CASE 47 Graft-Versus-Host Disease

Alien T cells react against their new host.

Bone marrow transplantation has proved to be useful therapy for some forms of leukemia, bone marrow failure (aplastic anemia), and primary immuno-deficiency diseases. More recently, other sources of hematopoietic stem cells, such as peripheral blood stem cells and cord blood, have also been used for these purposes. Bone marrow and most other sources of hematopoietic stem cells contain mature T lymphocytes, which may recognize the tissues of their new host as foreign and cause a severe inflammatory disease in the recipient. This is known as graft-versus-host disease (GVHD) and is characterized by a rash, which often starts on the face (Fig. 47.1), diarrhea, pneumonitis (inflam-mation in the lung), and liver damage.

To achieve successful engraftment of bone marrow and avoid rejection of the transplant by the host, the immune system of the recipient must be destroyed and the recipient rendered immunoincompetent. This is usually accomplished with lethal doses of radiation or injection of radiomimetic drugs such as busulfan. In children who are already immunoincompetent because of a primary immunodeficiency disease, such as severe combined immuno-deficiency (see Case 14), this preparative treatment is not needed.

Topics bearing on this case:
T-cell recognition of nonself MHC molecules
Reactions of T cells against foreign antigens
Transplantation
Minor histocompatibility antigens
Mixed lymphocyte reaction

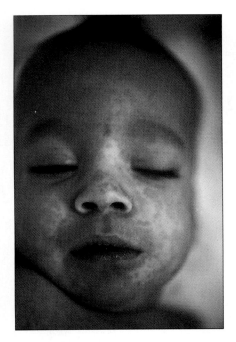

Fig. 47.1 The rash characteristic of GVHD often starts on the face.

GVHD occurs not only when there is a mismatch of classical MHC class I or class II molecules, but also in the context of disparities in minor histocompatibility antigens; such minor differences are likely to be present in all donor–recipient pairs other than identical twins, even when HLA-matched. Mature CD4 T cells in the graft that are activated by allogeneic molecules produce a 'cytokine storm' that recruits other T cells, macrophages, and natural killer (NK) cells to create the inflammation characteristic of GVHD. Although B cells may also be present in GVHD inflammation, they do not play a significant role in causing or sustaining GVHD.

GVHD is arbitrarily called 'acute' if it occurs less than 100 days after the transplant, and 'chronic' if it develops after 100 days. Chronic GVHD differs from acute GVHD in other respects and is a more severe and intractable problem. The presence of alloreactive T cells in the donor bone marrow is usually detected in routine laboratory testing by the mixed lymphocyte reaction (MLR) (see Fig. 15.3), in which lymphocytes from the potential donor are mixed with irradiated lymphocytes from the potential recipient. If the donor lymphocytes contain alloreactive T cells, these will respond by cell division. Although the MLR is routinely used for the selection of donors it does not accurately quantify alloreactive T cells. Although the limiting-dilution assay more precisely counts the frequency of alloreactive T cells, it is too cumbersome for routine clinical use.

The case of John W. Wells: a curative therapy becomes a problem.

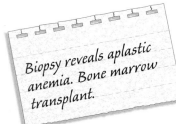

7-year-old boy with severe anemia. Order bone marrow biopsy.

Biopsy reveals aplastic anemia. Bone marrow transplant.

John was healthy until he was 7 years old, when his mother noticed that he had become very pale. She also noticed small hemorrhages (petechiae) on the skin of his arms and legs and took John to the pediatrician. Apart from the pallor and skin petechiae, a physical examination showed nothing unusual. The pediatrician ordered blood tests, which revealed that John was indeed very anemic. His hemoglobin was 7 g dl^{-1} (normal 10–15 g dl^{-1}) and platelet count was 20,000 per μl^{-1} (normal 150,000–250,000 μl^{-1}). His white blood cell count was also lower than normal. The pediatrician sent John to a hematology consultant for a bone marrow biopsy.

The biopsy showed that John's bone marrow had very few cells and that red cell, platelet, and white cell precursors were almost completely absent. Aplastic anemia (bone marrow failure) of unknown cause was diagnosed. Aplastic anemia is ultimately fatal but can be cured by a successful bone marrow transplant. Fortunately, John had an HLA-identical 11-year-old brother who could be the bone marrow donor. John was admitted to the Children's Hospital and given a course of busulfan to eradicate his own lymphocytes. He was then given 2×10^8 nucleated bone marrow cells per kg body weight obtained from his brother's iliac crests.

John did well for 3 weeks after the bone marrow transplant and was then sent home to recover. However, on the 24th day after the transplant he was readmitted to hospital with a skin rash and watery diarrhea consistent with acute GVHD. On admission he had a patchy red rash on palms and soles, scalp, and neck. He had no fever and was not jaundiced. His lungs were clear and the heartbeat normal. The liver and spleen were not enlarged.

John was treated with corticosteroids and the immunosuppressive drug tacrolimus (FK506). His skin rash faded, but the intestinal symptoms did not abate and the diarrhea became more profuse. He was given a test dose of antithymocyte serum but developed a fever and shaking chills, so this therapy had to be discontinued. John developed diffuse bleeding from the colon and a colonoscopy revealed diffuse GVHD throughout the large bowel. After 6 weeks in which the watery diarrhea did not improve, John was given an anti-CD3 monoclonal antibody by intravenous injection, to try and eliminate the T cells, as well as octreotide (an inhibitor of intestinal vasoactive peptide) and another immunosuppresssive drug, mycophenolate mofetil. This therapy provided no relief of his symptoms, which were finally controlled by injection of a monoclonal antibody to CD2 (another cell-surface marker of T cells) every 3 days. This brought about a 90% decrease in the volume of his stool and the intestinal bleeding stopped. Treatment with anti-CD2 antibody was continued weekly for 2 months and his symptoms completely disappeared. John was sent home, with continuing treatment with low doses of corticosteroid, after 6 months of hospitalization for GVHD.

GVHD developing; try tacrolimus.

Patient not responding. Try monoclonal antibodies.

Graft-versus-host disease (GVHD).

Graft-versus-host disease was first described over 30 years ago by Billingham, Brent, and Medawar, who gave allogeneic lymphocytes to newborn mice. The mice became runted (their growth was retarded), lymphoid tissue was destroyed, and they developed diarrhea and necrosis of the liver. GVHD was first recognized in human infants with severe combined immunodeficiency disease who inadvertently received allogeneic lymphocytes in a blood transfusion (Fig. 47.2). For a recipient to develop GVHD, the graft must contain immunocompetent cells, the recipient must express major or minor histocompatibility molecules that are lacking in the graft donor, and the recipient must be incapable of rejecting the graft.

The first clinical manifestation of GVHD is a bright red rash that characteristically involves the palms and soles. The rash usually begins on the face and neck and progresses to involve the trunk and limbs, particularly the palms and soles. The rash may itch a great deal and its onset may be accompanied by fever. After the skin manifestations appear (Fig. 47.3), the gastrointestinal tract becomes involved (Fig. 47.4) and profuse watery diarrhea is produced.

Fig. 47.2 GVHD on trunk and limbs of affected infant.

Fig. 47.3 GVHD in the skin. Panel a: early GVHD in the skin. Lymphocytes are emerging from blood vessels (lower arrow) and adhering to the basal layer of the epidermis (upper arrow). Panel b: the basal cells of the epidermis begin to swell and vacuolate. Their nuclei become condensed (dark staining) as these cells die (arrow). Panel c: advanced destruction of the skin by GVHD, with sloughing of the epidermis (arrow). Photographs kindly provided by Dr Robert Sackstein.

Fig. 47.4 GVHD in the colon. Inflammatory cells have invaded the crypts of the intestine and destroyed the normal architecture (arrow). Photograph kindly provided by Mark Shlomchik.

Liver function tests may become abnormal and reveal destruction of hepatic tissue (Fig. 47.5). Eventually, other tissues such as the lungs and bone marrow become sites of GVHD inflammation. The only satisfactory therapy at present for GVHD is elimination of the T cells that initiate the reaction, either by immunosuppressive drugs or, as in John's case, by anti-T-cell monoclonal antibodies.

Some bone marrow transplant centers now routinely deplete donor bone marrow of T cells before transplantation, to try and avoid GVHD. But this treatment is still somewhat controversial and is not universally followed. Untreated bone marrow seems to provide a 'graft-versus-leukemia' reaction, which makes relapse less likely in the case of bone marrow transplants for the treatment of leukemia.

Fig. 47.5 Liver damage as a result of GVHD. Inflammatory cells have invaded the liver and destroyed the hepatic ducts (arrows). Photographs kindly provided by Dr Robert Sackstein.

Questions.

1 Bone marrow is often depleted of T cells before transplantation in order to try and avoid GVHD. However, in the treatment of leukemia by bone marrow transplantation, T cells in the graft can have a beneficial effect. How do you explain this?

2 CD4 T cells in the graft that recognize foreign histocompatibility molecules become activated and produce the cytokine interferon (IFN)-γ. This helps sustain and increase GVHD. Why?

3 John was given a monoclonal antibody to CD3 but it did not help him. When he was given an antibody to CD2, however, he showed sustained improvement. How do you explain this difference?

4 Why are the skin and intestinal tract the major sites of GVHD?

Answers

Case 1

Answer 1

First you find out that Nicholas has had all his routine immunizations. He received DPT (diphtheria, pertussis, and tetanus antigens) and oral live poliovirus vaccine at ages 3, 4, and 5 months, and a booster of both before entering kindergarten. He was also given MMR (mumps, measles, and rubella live vaccines) at 9 months of age. At the same time, he was given Hib vaccine (the conjugated capsular polysaccharide of *Haemophilus influenzae*, type b; Fig. A1.1). His growth and development have been normal. He suffered a middle ear infection (otitis media) at age 24 months. Other than that he has had no other illnesses, except for a common cold each winter. You feel comfortable that he is protected against infection with *Haemophilus influenzae* from the Hib vaccine. However, your concern about the possibility of pneumococcal infection leads you to advise the surgeon to immunize Nicholas against pneumococcal capsular poysaccharides by giving him conjugated pneumococcal polysaccharide vaccine. You also advise prophylactic antibiotics, to be taken at a low dose daily but at higher doses when Nicholas has any dental work done, or any invasive surgical procedure.

Answer 2

The typhoid vaccine was given subcutaneously and a response was mounted in a regional lymph node. The sheep red blood cells were given intravenously, and, in the absence of a spleen, failed to enter any secondary lymphoid tissue where an immune response could occur.

Answer 3

The defect is inherited as an autosomal recessive. The parents are normal but each carries this recessive gene. Furthermore, they are consanguineous, a setting in which autosomal recessive disease is encountered more frequently than in outbred people. Chance would predict that one in four (that is, two) of their eight children would be affected. Each pregnancy provides a one in four chance of the fetus' inheriting the abnormal gene from both parents. As it turned out, this happened in five of Mrs Vanderveer's eight pregnancies. Since Betsy married a normal man, all her children are heterozygous for the defect, like their maternal grandparents, and have normal spleens.

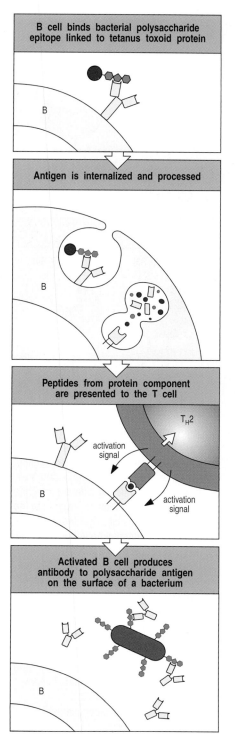

Fig. A1.1 *Haemophilus influenzae* b vaccine is a conjugate of bacterial polysaccharide with the tetanus toxoid protein, which enhances the immune response by allowing a polysaccharide-specific B cell to recruit T-cell help. The B cell recognizes and binds the polysaccharide, internalizes and degrades the toxoid protein to which it is attached, and displays peptides derived from it on surface MHC class II molecules. Helper T cells generated in response to earlier vaccination against the toxoid recognize the complex on the B-cell surface and activate the B cell to produce antibody against the polysaccharide. This antibody can then protect against infection with *H. influenzae* type b.

Case 2

Answer 1

There has been random inactivation of the X chromosomes in Randy's mother's neutrophils. Therefore 50% of her neutrophils have a normal X chromosome and 50% have an X chromosome bearing the CGD defect. The cells bearing the CGD defect X chromosome have not been selected against. That half does not reduce NBT, whereas the half bearing the normal X chromosome does.

Answer 2

Streptococcus pneumoniae, the pneumococcus, does not produce catalase which converts hydrogen peroxide (H_2O_2) into water and oxygen and is thus far less resistant to intracellular killing than microorganisms that are catalase producers.

Answer 3

Because of persistent antigenic stimulation he is making more immunoglobulins (antibodies) than a normal person. In fact, all chronic infections, such as malaria, result in hypergammaglobulinemia.

Answer 4

In leukocyte adhesion deficiency (see Case 3), there is a defect in the mobility of the leukocytes so that their emigration from the circulation to sites of infection is impeded. The microbicidal capacity of phagocytes in leukocyte adhesion deficiency is unimpaired. Children with leukocyte adhesion deficiencies therefore never develop inflammatory lesions, which are caused by activated phagocytes, and are equally susceptible to pneumococcal and other bacterial infections (see Question 2). In contrast, the leukocytes in chronic granulomatous disease have normal mobility and can reach sites of infection in a normal manner but cannot efficiently destroy bacterial infections when they reach them. Hence the development of abscesses and granulomas.

Case 3

Answer 1

It is autosomal recessive. The parents are both healthy; they have had an affected male and an affected female child. These facts lead to the conclusion that the leukocyte adhesion deficiency is inherited as an autosomal recessive trait. In fact the gene encoding CD18 has been mapped to the long arm of chromosome 21 at the position 21q22. An examination of the leukocytes of the parents would reveal decreased expression of CD18 but sufficient for normal function.

Answer 2

Monoclonal antibodies to CD18 can induce a mild phenotype of leukocyte adhesion deficiency. Such monoclonal antibodies have been used to prevent graft rejection in recipients of kidney grafts and, when administered before bone marrow transplantation, can prevent graft-versus-host disease.

Answer 3

T cells express the β_1 integrin VLA-4 (Fig. A3.3), and its interaction with VCAM-1 seems to be sufficient to enable T cells to home and function normally. VLA-4 is not expressed on neutrophils and macrophages, which are much more dependent on β_2 integrins for their adhesion to other cells. B cells

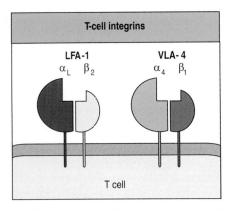

T-cell integrins

LFA-1
α_L β_2

VLA-4
α_4 β_1

T cell

Macrophage integrins

Mac-1 (CR3)
α_M β_2

gp150.95 (CR4)
α_X β_2

macrophage

Fig. A3.3 T cells and phagocytes express different integrins. Integrins are heterodimeric proteins containing a β chain, which defines the class of integrin, and an α chain, which defines the different integrins within a class. The α chain is larger than the β chain and contains binding sites for divalent cations that may be important in signaling. LFA-1, a β_2 integrin, and VLA-4, a β_1 integrin, are expressed on T cells and are important in the migration and activation of these cells. Macrophages and neutrophils express all three members of the β_2 integrin family, LFA-1, Mac-1 (also known as CR3), and gp150.95 (also known as CR4). Like LFA-1, Mac-1/CR3 binds the immunoglobulin superfamily molecules ICAM-1, ICAM-2, and ICAM-3, but in addition it is a complement receptor (for the fragment iC3b). gp150.95/CR4 also binds complement and stimulates phagocytosis.

also home normally in leukocyte adhesion deficiency and this is probably due to the integrin $\alpha_4{:}\beta_7$, which is not defective in this disease.

Answer 4

Luisa's brother had delayed separation of the umbilical cord and Luisa developed an infection at the site of umbilical cord separation. The role that neutrophils and macrophages play in wound healing is not well understood. Nevertheless, it is apparent from cases with leukocyte adhesion deficiency that the movement of white blood cells into wounds is vital to normal tissue repair.

Answer 5

Fucose is a defining determinant of the sialyl-Lewis^x element, which is the ligand whereby white blood cells bind to selectins (see Fig. 3.1). Consequently, infants who cannot make this determinant have very high white blood cell counts because the leukocytes cannot roll on the endothelium to begin the process of leukocyte emigration from the bloodstream.

Answer 6

After the transplant her leukocytes were found to express CD18.

Answer 7

This was done to destroy her abnormal cells and create 'space' for the transplanted cells. Such preparation is not performed in the case of severe combined immunodeficiency because the lymphoid compartment is already devoid of T cells and thus 'space' is available for the transplanted cells without any treatment.

Answer 8

PHA and Con A are so-called nonspecific T-cell mitogens. The mitogenic response requires cell–cell interactions that depend upon the interaction of LFA-1 with ICAM-1, as well as VLA-4 with VCAM-1. In leukocyte adhesion deficiency, one of these interactions is missing.

Case 4

Answer 1

Histamine release on complement activation is caused by C3a (the small cleavage fragment of C3) and the main chemokine is C5a (the small cleavage fragment of C5). These are both generated by the C3/C5 convertase, which in the classical pathway is formed from C4b and C2a. In HANE, C4b and C2a are both generated free in plasma. C4b is rapidly inactivated if it does not bind immediately to a cell surface; for that reason, and because the concentrations of C4b and C2a are relatively low, no C3/C5 convertase is formed, C3 and C5 are not cleaved and C3a and C5a are not generated.

The edema in HANE is caused not by the potent inflammatory mediators of the late events in complement activation but by C2b generated during the early events, and by bradykinin generated through the uninhibited activation of the kinin system.

Answer 2

The only other complement component that should be decreased is C2, which is also cleaved by C1. C1 plays no part in the alternative pathway of complement activation, so complement activation by the alternative pathway is not affected. The terminal components are not affected either. The unregulated activation of the early complement components does not lead to the formation of the C3/C5 convertase (see Question 1), so the terminal components are not abnormally activated. The depletion of the early components of the classical pathway does not affect the response to the normal activation of complement by bound antibody because the amplification of the response through the alternative pathway compensates for the deficiency in C4 and C2.

Answer 3

This is not hard to explain; as we have already remarked, the alternative pathway of complement activation is intact and thus, although the classical pathway is affected by deficiencies in C2 and C4, these are compensated for by the potent amplification step from the alternative pathway.

Answer 4

Stanozolol is a well-known anabolic androgen that has been used illegally by Olympic competitors. For unknown reasons, anabolic androgens suppress the symptoms of HANE, and that is why stanozolol was prescribed to Richard. Patients, especially females, do not like to take these compounds because they cause weight gain, acne, and sometimes amenorrhea. Preparations of purified C1INH are now available and intravenous injection of C1INH prepared from human donors is safe and very effective in halting the symptoms of the disease. C1INH is not available for therapeutic use in the United States.

Answer 5

In practice, you would administer epinephrine immediately in any case, because most such emergencies are due to anaphylactic reactions and epinephrine is a harmless drug. If the laryngeal edema is anaphylactic, it will respond to the epinephrine. If it is due to hereditary angioneurotic edema, it will not. Anaphylactic edema is also likely to be accompanied by urticaria and itching, and the patient may have been exposed to a known allergen. Most patients know if they are allergic or have a hereditary disease, and they should be asked if they have had a similar problem before.

Answer 6

HANE does not skip generations: it is therefore likely that its effects are dominant. It clearly affects both males and females, so it cannot be sex-linked. If the gene has a dominant phenotype, and Richard's two children are normal, then it follows they cannot have inherited the defective gene from their father, and their children cannot inherit the disease from them.

Richard has inherited his abnormal C1INH gene from his mother. Since he has a normal C1INH gene from his father, you might expect that he would have 50% of the normal level of C1INH. However, the tests performed by his immunologist revealed 16% of the normal level. In general, functional C1INH tests in HANE patients reveals between 5% and 30% of normal activity. How could this be explained? There are two possibilities: decreased synthesis (that is, less than 50% synthesis from only one gene); or increased consumption of C1INH as a result of increased C1 activation. Both explanations have been shown to be correct. Patients with HANE synthesize about 37–40% of the normal amount of C1INH, and C1INH catabolism is 50% greater than in normal controls.

Case 5

Answer 1

As he lives longer his adaptive immunity against these common bacteria becomes better and he has come to rely less on innate immune mechanisms for protection against infection. Bacteria coated with antibodies can be phagocytosed independently of complement via the Fc receptors on phagocytes.

Answer 2

A similar result. The factor B would be rapidly destroyed and its rate of synthesis could turn out to be normal. The overproduction of C3b in the absence of factor I leads to an increased binding of factor B to C3b and its subsequent cleavage by factor D. Thus, factor B is being consumed excessively as a result of the deficiency in factor I. The lack of factor B, like the C3 deficiency, is secondary to the basic defect in factor I.

Answer 3

His serum levels of C3 and factor B rose to normal. C3b disappeared from his serum. The effect lasted for about 10 days.

Answer 4

Deficiency in factor H. Because factor H is needed for the cleavage of C3b by factor I in the blood, factor H deficiency should result in clinical symptoms identical with those of factor I deficiency. In fact this is true; several families with factor H deficiency have been studied and they show symptoms indistinguishable from factor I deficiency.

Answer 5

He is producing large amounts of C3b, which binds to complement receptor 1 (CR1) on red blood cells and leads to their agglutination by anti-C3 (see Case 6).

Case 6

Answer 1

The CH_{50} is the quantity of complement required for 50% lysis of 5×10^8 optimally sensitized sheep red blood cells in 1 hour at 37°C. (The sheep red blood cells are usually sensitized with a rabbit antibody against the sheep cells.) The complement titer is expressed as the number of CH_{50} contained in 1 ml of undiluted serum from the patient. This is determined by plotting $\log y/(1-y)$ (where y is the percentage lysis) against the log of the amount of serum. The plot is linear near $y/1-y = 1$ (50% lysis) (Fig. A6.1). Because this assay measures the functional integrity of all the complement components comprising the classical pathway, it is the single best screening test when a general diagnosis of complete complement deficiency is being considered.

Answer 2

The bacteria are temporarily vulnerable to killing by complement when they divide. At this time the bacterial membrane is exposed and is vulnerable to attack by the membrane-attack complex. The association between genetic deficiencies of membrane-attack complex proteins and neisserial infections illustrates that an important aspect of host defense against these infections is the killing of extracellular bacteria by complement-mediated lysis.

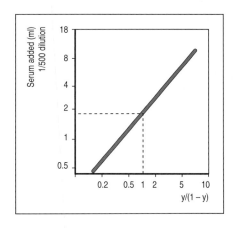

Fig. A6.1 Determination of CH$_{50}$. This is the quantity of complement required for 50% lysis of 5×10^8 optimally sensitized red blood cells in 1 hour at 37°C. Complement titers are expressed as the number of CH$_{50}$ contained in 1 ml of undiluted serum. This is usually determined by plotting log10 $y/(1-y)$ (where y is percentage lysis) against the log of the amount of serum. This plot is linear near $y/(1-y) = 1$ (50% lysis). In the example shown, the serum contains 250 CH$_{50}$ ml^{-1}.

Bound C43b binds to the receptor CR1 on erythrocyte surfaces

CR1

immune complex

In the spleen and liver, phagocytic cells remove the immune complexes

FcR

phagocyte

Fig. A6.3 Erythrocyte complement receptor CR1 helps to clear immune complexes from the circulation. Immune complexes bind to CR1 on erythrocytes, which transport them to the liver and spleen, where they are removed by macrophages expressing receptors for both Fc and bound complement components.

Neisseria that escape killing enter a variety of cell types and establish an intracellular infection.

Answer 3

In humans, clearance of immune complexes from the blood is largely effected by their attachment to complement receptor 1 (CR1) on the surface of red blood cells. One of the ligands for CR1 is C4b (the other is C3b). All the C1 components are required for the formation of C4b, and if C4b is part of an immune complex, the binding of the complex by CR1 is facilitated (Fig. A6.3). Immune complexes bound to the red-cell surface are transported to the liver and spleen, where C4b (or C3b) is converted by factor I to C4bi (or C3bi); this facilitates uptake of the immune complexes by phagocytes via complement receptor CR3 and others, and their destruction. In the absence of C4b, immune complexes are less efficiently attached to red cells and are therefore less efficiently cleared from the blood.

Case 7

Answer 1

Anakinra is a recombinant protein that competitively blocks the binding of IL-1β to its receptor, IL-1R, and would thus be expected to mitigate the effects of the excess IL-1β produced in the cryopyrinopathies. Pretreatment with anakinra before exposure to cold has been shown to prevent the development of symptoms and the elevation of acute-phase reactants in patients with familial cold autoinflammatory syndrome, and it also improves symptoms and corrects biochemical abnormalities in Muckle–Wells syndrome. Children with NOMID/CINCA have shown similarly promising results, with resolution of fever, rash, uveitis (inflammation of the middle layer of the eyeball, the uvea, comprising the choroid, iris and cilary body), and the relief of excess pressure exerted by the CSF.

Answer 2

Yes. It has recently been found that patients with this disease have an increased secretion of IL-1β, and have shown remarkable improvement when treated with anakinra. Unlike the periodic fever syndromes discussed above, however, the causative gene mutation has not yet been identified.

Answer 3

Colchicine inhibits the assembly of microtubules in cells by binding to β-tubulin subunits. In the context of autoinflammatory disease it is thought to act mainly by inhibiting the neutrophil response. It inhibits microtubule-dependent processes in the cell, including the division of neutrophil precursors and the secretion of pro-inflammatory mediators by mature neutrophils. In addition, colchicine has been shown to modulate the production of chemokines and pro-inflammatory prostanoids (prostaglandins and leukotrienes) by neutrophils, and it is also thought to inhibit neutrophil adhesion to the endothelium, a necessary step in the migration of neutrophils out of the blood and into tissues. Interestingly, it has also recently been shown to block IL-1β processing and secretion. Together, all of these actions are anti-inflammatory.

Answer 4

Caspase-1 is required to process the precursor forms of IL-1 and IL-18 proteolytically to produce mature active cytokines that can be secreted. It is now known that the cytokine IL-33 also requires processing by caspase-1. IL-33 is an IL-1-like cytokine that signals through the IL-1 receptor-related protein ST 2. IL-33, via ST 2, activates the transcription factor NFκB and mitogen-activated protein (MAP) kinases, and drives the production of the T_H2-associated cytokines IL-4, IL-5, and IL-13. In addition to IL-33, one might expect that blood levels of cytokines that are induced by the action of IL-1β, such as IL-6, would be increased in response to caspase-1 activation.

Case 8

Answer 1

Several immunodeficiencies result in susceptibility to pneumococcal infection. These include defects of innate immunity, including congenital asplenia (see Case 1) and defects within the complement pathways (see Cases 2–4). Other defects in the NFκB activation pathway that lies downstream of TLRs and other cell-surface receptors include NEMO deficiency (see Case 9) and mutations in IκB that prevent its degradation and release of NFκB. In addition, defects of adaptive immunity that result in impaired antibody production, such as X-linked agammaglobulinemia (see Case 10) and common variable immunodeficiency (see Case 13), result in increased susceptibility to Gram-positive bacteria, such as pneumococci and staphylococci.

Answer 2

These infections are usually associated with fever. IRAK4 deficiency, however, leads to an early block in the TLR/IL-1R signaling pathways and barely detectable or no TLR-induced production of pro-inflammatory cytokines. The virtual absence of pro-inflammatory cytokines and the inability of IRAK4-deficient patients to respond to what little IL-1 might be produced results in an impaired febrile response (see Case 7). Thus, a history of little or no fever associated with recurrent pyogenic infections supports a possible diagnosis of IRAK4 deficiency.

Answer 3

So far, there are too few IRAK4-deficient patients for us to be sure. More cases of IRAK4 deficiency need to be identified and followed throughout their lives before conclusions can be drawn. The 18 patients identified so far all display increased susceptibility to invasive infections with pyogenic bacteria in childhood. As they mature into adolescence, however, susceptibility to such infections becomes variable, and many no longer show significantly increased susceptibility.

Answer 4

The antibody response to polysaccharides seems to require initial signaling via TLR-2 (which recognizes lipoteichoic acid of Gram-positive bacterial cell walls and lipoproteins of Gram-negative bacteria) and TLR-4 (which recognizes LPS) in dendritic cells. This may be because stimulation of dendritic cells via these TLRs is essential for their ability to induce the differentiation of T cells (especially T_H1) that 'help' B cells to make the antipolysaccharide antibodies. The response to Pneumovax, for example, seems to depend on the presence of such TLR ligands in the vaccine. It has been shown that depletion of endotoxin from the vaccine renders it unable to elicit an antibody response in mice.

Answer 5

The answer to this important question is not clear. The production of type I interferons in response to the ligation of TLRs 7, 8, or 9 is markedly diminished in IRAK4-deficient patients, and interferon synthesis is variably affected in response to TLR-3 ligation by long double-stranded RNA. The apparent integrity of the antiviral defenses in these patients is therefore surprising. One explanation could be that humans can make relatively intact adaptive immune response to viruses as a result of responses by cytotoxic T cells, which kill the infected cells, and through antiviral antibodies produced by B cells. There is probably redundancy between the adaptive and innate immune responses, with (innate system) NK cells cytotoxic for virus-infected cells being activated by T-cell-derived IL-2 and IFN-γ. Other intracellular antiviral immune responses are produced via activation of the RNA-dependent protein kinase pathway and via activation of the cytoplasmic protein RIG1 by double-stranded RNA, which leads to the synthesis of type I interferons.

Answer 6

Experience in managing these patients is limited. However, they are maintained in good health into their adolescence by a combination of prophylactic antibiotics and regular infusions of intravenous immunoglobulin.

Case 9

Answer 1

One important defense against mycobacteria is the activation of macrophages to synthesize cyokines such as IL-18 and TNF-α after recognition of microorganisms via Toll receptors. These cytokines amplify immune responses, activate macrophages' intracellular killing activity and induce nitric oxide, which is important in the destruction of the intracellular mycobacteria. In patients with NEMO mutations, these cytokines are not made in such quantity because Toll receptors signal via activation of NFκB by the IKK complex, which contains NEMO. NF-α and IL-18 themselves also signal through the NFκB pathway, so their effect is severely reduced. This in turn reduces the synthesis of the important cytokine interferon-γ (IFN-γ) by T cells, which is normally induced by IL-12 acting synergistically with IL-18. Thus, NEMO deficiency also results in decreased IFN-γ production. Because IFN-γ also activates macrophages' ability to kill intracellular bacteria, a lack of NEMO leads to an inability of macrophages to kill mycobacteria that have been taken up by phagocytosis and have become resident in their endosomes. The persistently infected macrophages continue to activate T cells, leading to granuloma formation. Patients with *NEMO* mutations thus have some similarities to children with mutations in IL-12, the IL-12 receptor, and the IFN-γ receptor in terms of their susceptibility to mycobacteria (see Case 30).

Answer 2

The killer activity of natural killer (NK) cells is deficient in most patients with *NEMO* mutations. Patients deficient in NK cells are known to suffer from recurrent CMV and herpes virus infections. This suggests that activation of NK cells via their invariant activating receptors in response to viruses is dependent on intact IKK and NFκB activation.

Answer 3

Isotype class switching can be induced by various members of the TNF family (to which CD40 ligand belongs), which are expressed on the surface of activated dendritic cells and engage receptors on B cells. This engagement activates NFκB in a NEMO-independent fashion by selective activation of IKKα dimers by the kinase NIK which results in the processing of the auto-inhibited NκB subunit p100 into its active form p52. This small amount of

B-cell activation might also explain the presence of lymph nodes in patients with *NEMO* mutations.

Answer 4

IKK is the major means of phosphorylating IκB, but other kinases can also perform this function, thereby circumventing the block to some extent. When IκB cannot be phosphorylated at all because of a mutation at the phosphorylation site, the pathway is completely blocked and a more severe immunodeficiency results.

Case 10

Answer 1

In every somatic cell of a female, one of the two X chromosomes is inactivated. Which of the X chromosomes is inactivated is a random process so that each is normally active in 50% of the cells on average. However, if the normal X chromosome is inactivated in a pre-B cell of Bill's mother or grandmother, that cell has no normal *btk* product and cannot mature. All of their B cells therefore have the normal X chromosome active (Fig. A10.1). This makes it appear that in the B cells of Bill's mother and grandmother the inactivation of the X chromosome has been nonrandom. If we have a marker that allows us

Fig. A10.1 Arrested B-cell development in X-linked agammaglobulinemia is responsible for apparently nonrandom X inactivation in the B cells of female carriers. In normal individuals, B-cell development proceeds through a stage in which the pre-B-cell receptor consisting of μ:λ5:Vpre-B transduces a signal via Btk, triggering further B-cell development. In males with XLA, no signal can be transduced and, although the pre-B-cell receptor is expressed, the B cells develop no further. In female mammals, including humans, one of the two X chromosomes in each cell is permanently inactivated early in development. Because the choice of which chromosome to inactivate is random, half of the pre-B cells in a carrier female express a wild-type *btk*, and half express the defective gene. None of the B cells that express *btk* from the defective chromosome can develop into mature B cells. Therefore, in the carrier, mature B cells always have the nondefective X chromosome active. This is in sharp contrast to all other cell types, which express the nondefective chromosome in only half of the population. Apparently nonrandom X chromosome inactivation in a particular cell lineage is a clear indication that the product of the X-linked gene is required for the development of cells of that lineage. It is also sometimes possible to identify the stage at which the gene product is required, by detecting the point in development at which X-chromosome inactivation develops bias. Using this kind of analysis, one can identify carriers of traits like XLA without needing to know the nature of the gene.

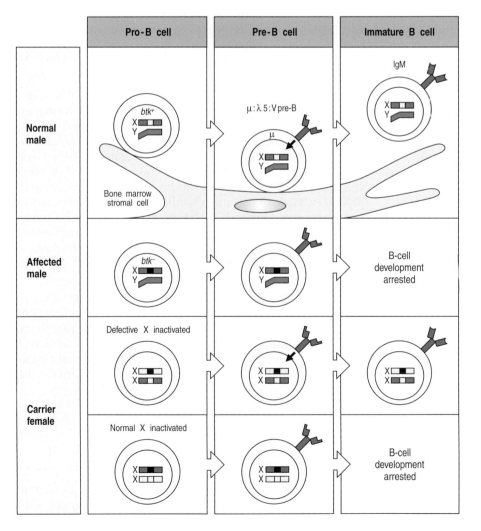

to distinguish between the two X chromosomes of Bill's aunts we could determine whether their B cells exhibit random or nonrandom X inactivation. In fact it turns out that one of Bill's aunts is a carrier and the other is not.

Answer 2

Maternal IgG crossed the placenta into Bill's circulation during fetal life. He was protected passively by the maternal IgG for 10 months.

Answer 3

Live polio vaccines are made from viruses with a disabling mutation in the gene that allows the virus to enter the motor nerve cells and cause paralysis. When a normal infant is given live, attenuated poliovirus orally, the poliovirus establishes a harmless infection in the gut. Within 2 weeks the infant makes IgG and IgA antibodies that neutralize the poliovirus and prevent the infection from spreading, so that as the infected gut cells die the infection is terminated. When male infants with X-linked agammaglobulinemia are given this same vaccine, they are incapable of making any antibodies and the infection can persist. They continue to excrete the poliovirus from their gut. After a time there may be a mutation in some of the viruses that causes them to reacquire the ability to enter nerve cells (neurotropism). These so-called revertant viruses disseminate through the bloodstream and infect the neurons in the spinal cord, thus causing paralytic poliomyelitis. Another example of a virus that may disseminate from the gastrointestinal tract in males with X-linked agammaglobulinemia are the echoviruses. These are cleared readily by normal individuals, but without the benefit of either IgA or IgG antibodies against the infecting serotype the virus can disseminate to the central nervous system and cause meningoencephalitis.

Answer 4

Certain plant lectins, such as phytohemagglutinin and concanavalin A, cause virtually all T cells to divide and are therefore known as nonspecific mitogens. Antigens to which the host has previously been exposed also cause T cells to divide *in vitro*. After 72 hours, exposure either to nonspecific mitogens or to specific antigens, ^{3}H-thymidine (tritiated thymidine) was added to Bill's T-cell cultures. Tritiated thymidine becomes incorporated into the DNA of dividing cells. The stimulation indices (the number of tritium counts in the stimulated cultures divided by the number of counts in similar cultures not exposed to mitogen or antigen) were normal for both mitogens and antigens. Bill's T cells responded to tetanus and diphtheria toxoids because he had been immunized with these inactivated bacterial toxins before the diagnosis was established.

Answer 5

Hereditary deficiency of complement component C3. The fixation of C3 to the bacterial surface, either by the classical or by the alternative pathways of complement activation, leads to its cleavage into a succession of fragments, two of which (C3b and iC3b) bind to complement receptors on the surface of phagocytic cells and enhance phagocytosis. iC3b binds the most potent complement receptor (CR3), and is the most important opsonizing agent for the ingestion and phagocytosis of encapsulated bacteria (see Case 16).

Answer 6

Tonsils are 80–90% B cells.

Answer 7

The rate at which IgG is catabolized depends on its concentration. In normal individuals, IgG has a half-life of approximately 21 days. In males with X-linked agammaglobulinemia, because the concentration is lower, IgG has

Bill weighs 75 kg.

The vascular volume is 8% of body weight or 0.08 liters kg^{-1}.

Bill's blood volume is therefore 75 × 0.08 = 6 liters.

Bill's hematocrit (the portion of the blood composed of cells) is 45%.

Therefore his plasma is 55% of his blood volume or 6 × 0.55 = 3.3 liters (3300 ml).

Bill injects himself with 10 g gamma globulin or 10,000 mg. He thereby raises his IgG level by 10,000/3300 or roughly 3 mg ml^{-1} or 300 mg dl^{-1}.

Half the IgG equilibrates into the extravascular pool, so that Bill has really raised his plasma level by only 150 mg dl^{-1}.

Fig. A10.7 IgG was radiolabeled with ^{125}I and then injected intravenously. After 10 minutes to allow for complete mixing of the radioactive dose in the blood, a plasma sample was obtained and the radioactivity was assayed. The amount of radioactivity at that time point was considered to be 100%. Plasma samples were obtained subsequently at frequent intervals and the percentage of residual radioactivity was determined and plotted on semi-log graph paper. The blue line shows the rate of disappearance of IgG in a normal person and the red line the rate in a male with X-linked agammaglobulinemia. From inspection of the curve it can be determined how long it takes for the radioactivity to decrease by 50%, say from 50% to 25%.

a half-life of approximately 28 days. Overall, Bill's optimal level of 600 mg dl^{-1} decreases to about 450 mg dl^{-1} after a week because of catabolism, as well as minor losses in saliva, tears, the gut and other secretions.

IgG, like all plasma proteins, distributes into the extravascular space: half the body IgG is in the blood and the other half is in the extravascular space. A dose of gamma globulin administered intravenously would equilibrate with the extravascular fluid in 24 hours. The dose of 10 g is arrived at as shown in Fig. A10.7.

Answer 8
Signaling via the pre-B-cell receptor is critical for B-cell development. Therefore, a deficiency of any one of the nonredundant components of pre-B-cell receptor signaling can cause a failure of B-cell development. Defects transmitted as an autosomal recessive have been described in the surrogate light chain Vpre-B, Igα, Igβ, and the B-cell linker protein BLNK.

Case 11

Answer 1
There is no DNA switch region 5′ to the Cδ gene. A single transcript of VDJCμCδ is alternatively spliced to yield the μ or δ heavy chain (see Fig. 11.4). On the other hand, there are DNA switch regions 5′ to all the other heavy-chain C genes, and isotype switching must occur before functional transcripts

of these genes are made. Isotype switching requires recombinase enzymes, which are activated in response to signals from T cells.

Answer 2

These experiments tell us that the activation of macrophages against this opportunistic microorganism requires binding of CD40 on the macrophage surface to the CD40 ligand expressed on activated T cells.

Answer 3

X-linked hyper IgM patients are deficient in antibody responses to T-cell dependent antigens but can make IgM antibodies to antigens that can stimulate a B-cell response without T-cell help. The blood group antigens are sugar groups that are also found on bacteria in the gut and can activate B cells in the absence of T-cell help. The secreted bacterial toxins, on the other hand, are proteins that cannot elicit a B-cell response in the absence of T cells. Without CD40 ligand, Dennis' T cells would have been unable to activate his B cells to respond to these protein antigens. He would also have been unable to make antibodies to *Streptococcus pyogenes*, because the antigenic component of the bacterial capsule is a protein.

Answer 4

The polysaccharide capsules of these pyogenic bacteria are resistant to destruction by phagocytes unless they are opsonized. IgM is not, in itself, an opsonin because there are no receptors on phagocytic cells for the Fc portion of IgM. IgM can promote the phagocytosis of bacteria only by activating complement, leading to the deposition of C3 fragments on the bacterial surface. The C3 is recognized by complement receptors on the phagocytic cells. IgG, however, is more efficient than IgM in promoting the phagocytosis of most bacteria. In addition, there is a range of Fc receptors for IgG isotypes on phagocytes, and IgG1, IgG2, and IgG3 antibodies all promote complement activation on bacterial surfaces. This means that bacteria coated with IgG stimulate phagocytosis through two different classes of receptors, Fc and C3. This results in much more efficient phagocytosis than stimulation through a single class of receptor.

Answer 5

Both newborns and people taking immunosuppressive drugs such as cyclosporin A exhibit increased susceptibility to both pyogenic and opportunistic infections.

A second reason for the increased susceptibility of newborns to some pyogenic infections is the immaturity of many of their B cells. Newborns are normally protected until their lymphocytes mature by preexisting maternal IgG.

Cyclosporin A also inhibits transcription of the *IL-2* gene, thereby preventing the expansion of T-cell clones activated by antigen. This means that all T-cell mediated immune responses, including cytotoxic T-cell responses, are suppressed by cyclosporin A.

Answer 6

In a word, no! In the case of X-linked agammaglobulinemia we saw that the gene defect in a tyrosine kinase, Btk, was critical for the normal maturation of B cells. Thus pro-B cells with the X chromosome carrying the *btk* mutation did not mature; only the pro-B cells with the normal X chromosome matured. This appeared as nonrandom X-chromosome inactivation in the B cells of the carriers. The CD40 ligand is not needed for the normal maturation of T cells. Its expression is induced in mature T cells by activating them. Dennis' mother and sister exhibit random X-chromosome inactivation in their T cells and this test is of no help in determining the carrier state in the X-linked hyper IgM immunodeficiency syndrome. The defect in Dennis' *CD40L* gene may be either a deletion or a missense mutation, and would need to be analyzed to find a genetic marker that could then be looked for in his mother and sister.

Case 12

Answer 1

Although the most notable features of X-linked hyper IgM syndrome are elevated IgM and lack of other immunoglobulin isotypes, the defect is actually of T cells, not B cells. Lack of expression of CD40 ligand by T cells not only affects B-cell function. It also results in failure to activate monocytes/macrophages and dendritic cells via CD40. Interleukin-12 is not synthesized. In the absence of CD40 signaling, pulmonary macrophages are inefficient in killing *Pneumocystis carinii*, leading to pneumonia, and liver macrophages may be similarly deficient in killing *Cryptosporidium*, leading to chronic inflammation of the bile ducts (cholangitis). Lack of dendritic cell activation via CD40 may impair their ability in turn to elicit robust T-cell responses. This may contribute to the susceptibility to opportunistic infections in X-linked hyper IgM syndrome. In AID deficiency, the defect is solely in the B cells, and only affects antibody production, resulting in a phenotype similar to other B-cell deficiencies such as X-linked agammaglobulinemia. There is susceptibility to pyogenic bacteria, in particular, leading to frequent bacterial infections of the ears, lungs and sinuses, but no increased risk of opportunistic infections.

Answer 2

As the syndrome in this patient resembles that seen in CD40 ligand deficiency, we might hypothesize that she has a defect in CD40, the receptor for CD40 ligand. In fact, several cases of hyper IgM syndrome secondary to defects in CD40 have now been reported. As might be expected, these patients are clinically indistinguishable from those with CD40 ligand deficiency, except in the pattern of inheritance.

Answer 3

In this case it is likely that the defect is in another of the proteins required in the B cell for class switching. One of the genes involved in class switching is that for uracil-DNA glycosylase (*UNG*), which removes uracil from uridine. Recently, a defect in *UNG* has been described in a patient with hyper IgM syndrome.

Answer 4

B cells from patients deficient in CD40 ligand undergo normal class switch recombination in response to CD40 ligation and cytokine action, whereas B cells from AID patients do not. Measurement of IgE synthesis after stimulation of cultured blood lymphocytes with anti-CD40 antibody plus IL-4 is an excellent indicator of class switching because blood lymphocytes normally contain very few or no B cells that have already switched to IgE. B cells from patients deficient in CD40 ligand make normal amounts of IgE in response to anti-CD40 plus IL-4. In contrast, B cells from patients with AID deficiency, or CD40 deficiency or defects in genes downstream of CD40 fail to secrete IgE in response to anti-CD40 and IL-4.

Answer 5

Enlargement of lymph nodes in children is usually due to expansion in the number and size of germinal centers. The interaction of CD40 with its ligand is required for germinal center formation, and patients with CD40 ligand deficiency or CD40 deficiency have no germinal centers. In contrast, AID is not required for germinal center formation. Mitotic stimuli result in normal cell proliferative responses in these patients, and infections result in enlarged lymph nodes, as observed in Daisy.

Case 13

Answer 1

TACI, like other members of the TNFR family, such as TNFR-I and Fas, might require ligand-induced trimerization for signaling. The TRAFs that associate with TACI cytoplasmically have been shown to have a higher affinity for trimeric receptors. Therefore, in the case of mutations in the cytoplasmic domain of TACI, the recruitment of mutant and normal TACI subunits into a trimeric complex will compromise the binding of downstream signaling molecules and thus interfere with signaling. In this case, the mutant TACI will be acting as a dominant negative.

Answer 2

There is considerable variation in the severity of clinical symptoms in both related and unrelated individuals with the same *TACI* mutations. As seen in this case, in the same family, the *TACI* mutation could be associated with CVID and IgA deficiency. This suggests that the penetrance of the *TACI* mutation is determined by the particular genotype of the individual. In addition, environmental modifiers could also be involved. Indeed, some family members of patients with CVID carry the same *TACI* mutation but have few or no symptoms. Furthermore, some of the common mutations in *TACI* found in CVID have been identified in a small percentage of 'normal' subjects on blood screening.

Answer 3

A deficiency in CD19, which is a component of the B-cell co-receptor complex, mutations in the co-stimulatory protein ICOS, and possibly a mutation in BAFF-R, have been found.

Case 14

Answer 1

His mother had nonrandom inactivation of the X chromosome in her T cells, thereby demonstrating that she carried an X-linked gene required for the normal maturation of T cells.

Answer 2

From what we know of their functions, defects in the receptors for IL-2, IL-4, IL-9, IL-15, and IL-21 do not appear relevant to the early block in T-cell development seen in X-linked SCID, because all of them activate mature lymphocytes or other effector cells. However, the receptor for IL-7 is thought to be important for pre-T-cell growth in humans and mice, and in mice (which are deficient in both B and T cells when they lack the γ_c chain) for pre-B-cell growth as well. Mice with a defect in the IL-7 receptor alone suffer from blocks in T- and B-cell development that resemble those seen in mice lacking the γ_c chain. A loss of IL-7 receptor function is therefore likely to be the most important loss of function responsible for X-linked SCID.

Answer 3
In Martin's case, the advantage of choosing his mother as donor lies in the facility of proving the establishment of chimerism. The transplanted T cells, responding to PHA, would have an XX karyotype in this male infant. If the father had been used as a donor this valuable marker would not be available to make it easy to distinguish between the transplanted cells and Martin's own cells. Conversely, half the maternal T-cell precursors will not mature because they carry the genetic defect. This would not be a problem if Martin's father had been used as the donor.

Answer 4
To prevent graft-versus-host disease. The mother's bone marrow donation almost certainly contained mature T cells capable of reacting with the paternal HLA antigens inherited by Martin from his father. Recognition of the paternal HLA antigens in Martin by the T-cell antigen receptors of the maternal T cells would incite graft-versus-host disease. Martin was also treated with a short course of the immunosuppressant drugs cyclosporin and prednisone after receiving the transplant, to suppress any graft-versus-host disease that might have arisen.

Answer 5
The *P. carinii* organisms are present in lung fluid and in the pulmonary macrophages. They do not incite an inflammatory response until the infant has T cells bearing a T-cell antigen receptor for *P. carinii* antigens. After a successful transplant, the infant is rendered chimeric. The transplanted T cells recognize *P. carinii* antigens and incite an inflammatory response, which makes the pneumonia more severe.

Answer 6
They die of progressive BCG infection. This ordinarily harmless attenuated bacillus is a pathogen in individuals with compromised cell-mediated immunity.

Answer 7
They developed a progressive vaccinia infection, which spread contiguously in the skin from the site of inoculation—so-called vaccinia gangrenosa. It proved invariably fatal in these infants. No live vaccines of any kind should be given to an immunodeficient child or adult.

Answer 8
Gene therapy. The γ_c gene is introduced by retroviral transfer into bone marrow stem cells of patients with X-linked SCID and the cells are injected into the blood. This therapy has resulted in full immune reconstitution in patients with X-linked SCID. Unfortunately, some of these patients developed leukemia, so that further gene therapy trials have been halted for the moment.

Case 15

Answer 1
The thymus gland can shrink (involute) as the result of many kinds of stress, particularly infection.

Answer 2

The brief, mild graft-versus host reaction was probably due to incompatibility at one or more minor histocompatibility loci.

Answer 3

The T cells of a patient with SCID are incapable of responding to a mitogenic stimulus. As Roberta's cells could not respond to John's it was not necessary to add mitomycin to the mixed lymphocyte reaction.

Answer 4

NK cells, like most other cells in the body, probably have sufficient 5′ nucleotidase to prevent excessive dATP accumulation.

Case 16

Answer 1

The few T cells that were able to rearrange their T-cell receptor genes must have had a T_H2 phenotype and secreted large amounts of interleukin-4 (IL-4) and interleukin-5 (IL-5).

IL-4 is required for switching to IgE synthesis and IL-5 for the recruitment of eosinophils. The few B cells in the patient (which were below the limit of detection) must have been induced to switch immunoglobulin class to IgE.

Answer 2

The few clones of T cells that were able to mature were activated, as shown by their surface expression of MHC class II molecules. The activated clones would have expanded within the lymph node.

Answer 3

T-cell DNA fragments containing the T-cell receptor genes were subjected to gel electrophoresis and then blotted with a J_β cDNA probe that will react with all DNA bands that contain $V_\beta(D)_\beta-J_\beta$ rearrangements as well as with J_β gene segments that are not rearranged. The DNA from a normal T-cell population would give a multiplicity of different $V_\beta(D)_\beta-J_\beta$ bands, whereas Ricardo's T cells yield very few bands of rearranged DNA.

Answer 4

DNA containing RAG genes cloned from Ricardo's cells and placed under the control of a cytomegalovirus promoter was inserted into a fibroblast cell line, which was then able to express the RAG proteins. A similar fibroblast line containing inserted normal RAG genes was used as a control. Substrate DNA containing the heptamers and nonamers in the appropriate arrangement was then added to the fibroblast culture. After incubation for several hours, the percentage of the substrate DNA that had undergone recombination was measured.

Answer 5

Both his parents are normal, but he had an affected brother and an affected sister. This indicates Mendelian autosomal recessive inheritance of the defect. If the defect had been X-linked recessive, a female would not have been affected; if it were dominant, one of the parents would have to have been affected.

Answer 6

The T cells that are present are activated, as shown by their expression of

MHC class II molecules, and express homing receptors for the skin. In the skin, the activated T cells secrete chemokines that attract other inflammatory cells, such as monocytes and eosinophils into the skin. The perivascular inflammation in the skin causes the blood vessels to dilate, and this appears as a bright red rash.

Case 17

Answer 1
The intrathymic maturation of CD8 T cells depends on the expression of MHC class I molecules on thymic epithelial cells. Conversely, maturation as CD4 T cells requires interactions with the MHC class II molecules also present on thymic epithelium. As Tatiana and Alexander do not lack MHC class II molecules, their CD4 T cells are normal and they have normal humoral immunity.

Answer 2
The maturation of CD8 T cells bearing γ:δ chains occurs after these cells emigrate from the thymus and is independent of MHC class I expression, whereas the maturation of α:β CD8 T cells occurs in the thymus and is dependent on the expression of MHC class I molecules.

Answer 3
No. Delayed-type hypersensitivity reactions are provoked by antigen-specific CD4 T cells (see Case 9).

Answer 4
The factors that help B cells to mature and secrete immunoglobulins are derived from activated CD4 T cells. These cells were normal in Tatiana and Alexander. Factors that suppress B-cell responses are secreted by CD8 T cells, of which the children had very few. They were therefore not very efficient at terminating B cell-mediated humoral immune reactions and tended to over-produce antibody. In Case 18, we saw the opposite phenomenon in patients with MHC class II deficiency, who have a deficiency of CD4 T cells and very low levels of serum immunoglobulins.

Answer 5
The patients with TAP1 deficiency clinically resembled those with TAP2 deficiency.

Case 18

Answer 1
The maturation of CD4 T cells in the thymus depends on the interaction of thymocytes with MHC class II molecules on thymic epithelial cells. When the MHC class II genes are deleted genetically in mice, the mice also exhibit a deficiency of CD4 T lymphocytes.

Answer 2
The polyclonal expansion of B lymphocytes and their maturation to immunoglobulin-secreting plasma cells requires helper cytokines from CD4 T cells, such as interleukin-4. Helen's hypogammaglobulinemia is thus a consequence of her deficiency of CD4 T lymphocytes.

Answer 3

Helen's T cells, although decreased in number, are normal and are not affected by the defect. They are capable of normal responses to nonspecific mitogens and to an allogeneic stimulus in which the antigen is presented by the MHC molecules on the surface of the (nondefective) allogeneic cells and thus does not require to be processed and presented by the defective cells. However, the failure of her lymphocytes to respond to tetanus toxin *in vitro* resulted from the fact that, in this situation, there were no cells that could present antigen on MHC class II molecules to the CD4 T cells.

Answer 4

Yes. Helen's T cells would be capable of recognizing the foreign MHC molecules on the grafted skin cells and would reject the graft.

Case 19

Answer 1

The IgG could also be shown to belong to a single subclass of IgG, that is IgG1, IgG2, IgG3, or IgG4. Furthermore, it would be possible to show that a single variable-region gene was rearranged to form this IgG.

Answer 2

The proliferation of malignant plasma cells in the bone marrow crowded out the red blood cell and white blood cell precursors. This creates a limitation on space in the bone marrow.

Answer 3

Although her serum IgG concentration is quite elevated, almost all the IgG is secreted by the myeloma cells and is monoclonal. In fact, she has very little normal polyclonal IgG and has been effectively rendered agammaglobulinemic by her disease. In addition, her white blood cell count is decreased and she has too few neutrophils ($<1000\ \mu l^{-1}$) to ingest bacteria in the bloodstream and lungs effectively.

Answer 4

We learned in the case of X-linked agammaglobulinemia that the turnover or fractional catabolic rate of IgG is concentration-dependent. In the young man with X-linked agammaglobulinemia, the administered IgG lasted longer than it would in a normal person. In Mrs Archer, the very elevated level of serum IgG would result in rapid catabolism of the IgG administered.

Answer 5

No. M-components appear in the blood as people age. About 10% of healthy individuals in the ninth decade of life have M-components. This is called benign monoclonal gammopathy. Without bone lesions and the presence of many malignant plasma cells in the bone marrow, the diagnosis of multiple myeloma cannot be made. Some people have IgM M-components in their blood. This is due to another malignancy of plasma cells called Waldenström's macroglobulinemia, which differs in many ways from multiple myeloma and is a more benign disease.

Answer 6

The best explanation is that the malignant transformation of a plasma cell occurred as class switching was taking place. The progeny of this cell can make two immunoglobulin classes, for example IgM and IgG. It is not known whether some cells make IgM and some IgG, or all of the cells make both, although it seems most likely that each cell makes only one isotype.

Case 20

Answer 1

IL-13 has activities that overlap with IL-4. The two molecules, which are both produced by activated CD4$^+$ or CD8$^+$ T cells, are structurally very similar, and probably arose by gene duplication. Both cytokines promote IgE synthesis, B-cell proliferation, and MHC class II gene expression. There are, however, some important differences between them. IL-13 seems to be produced at similar levels by helper T cells of the T$_H$0, T$_H$1, and T$_H$2 types and is produced very rapidly after T-cell activation, whereas IL-4 is primarily the product of T$_H$2 cells and appears later in the response. IL-13 does not have the T$_H$2-promoting activity of IL-4 and, unlike IL-4, does not induce T-cell proliferation. IL-13 promotes both eosinophil survival and recruitment.

Answer 2

IL-4 induces isotype switching to IgE and IgG4 in human B cells. In addition, IL-4 promotes T-cell differentiation to the T$_H$2 phenotype. T$_H$2 cells produce more IL-4, further enhancing humoral immune responses. In addition, T$_H$2 cells make IL-5, IL-6, IL-10, and IL-13, which promote B-cell expansion and immunoglobulin synthesis.

Answer 3

The demonstration that the tumor cells all express the same T-cell receptor. Proliferation of T cells can arise either as the result of clonal malignant transformation (leukemia/lymphoma) or from polyclonal expansion driven either by antigen or as the result of dysregulated proliferation. Antigen-driven T cells are polyclonally expanded. As a result, they express a heterogeneous population of T-cell receptors of various specificities at their cell surface. These receptors are constructed during T-cell development by random splicing of genomic V, D, and J segments. As neoplasms are clonal, T-cell leukemias and lymphomas express only a single T-cell receptor (Fig. A20.3).

Answer 4

Corticosteroids are known to inhibit the transcription of a number of cytokines associated with allergic states, including IL-4, IL-5, and GM-CSF. Corticosteroids function by interacting with cytosolic receptors, inducing a conformational change in these receptors that enables the steroid:receptor complex to act as a transcription factor and bind to specific sequences known as glucocorticoid response elements (GREs) in the promoters of certain genes in the nucleus (Fig. A20.4). Among the genes containing GREs in their promoters is the anti-inflammatory gene family IκBα. Normally, the IκB protein binds to NFκB transcription factors in the cytosol, preventing their translocation to the nucleus where they promote the transcription of a wide variety of pro-inflammatory genes (including that for GM-CSF). After the corticosteroid-induced upregulation of IκB, NFκB-regulated gene transcription is suppressed and the production of a number of pro-inflammatory cytokines is reduced.

Answer 5

Although CD4 T$_H$2 lymphocytes represent a major source of IL-4, other cells have been reported to produce this cytokine. In mice, a unique population of

Fig. A20.3 The unique rearrangement events in each T cell can be used to identify tumors of T cells. Tumors are the outgrowth of a single transformed cell. Thus, each cell in a tumor has an identical pattern of rearranged T-cell receptor genes. This figure shows the gel electrophoresis of DNA fragments containing the T-cell receptor β-chain constant regions; the DNA is obtained either from the placenta (lane P), a tissue in which the T-cell receptor genes are not rearranged, or from peripheral blood lymphocytes from two patients suffering from T-cell tumors (lanes T$_1$ and T$_2$). Bands corresponding to the unrearranged C$_\beta$1 and C$_\beta$2 genes can be seen in all lanes. Additional bands corresponding to specific rearrangements (arrowed) can be seen in each of the tumor samples, indicating that a large proportion of the cells in the sample carry an identical gene rearrangement. Note that these are the only discrete additional bands that can be seen in these samples; no bands deriving from rearranged genes in the normal lymphocytes also present in the patient's serum can be seen, as no one rearranged band is present at sufficient concentration to be detected in this assay. Photograph courtesy of T. Diss.

| Steroid receptors are found in the cytoplasm complexed with a heat-shock protein Hsp90 | Steroids cross the cell membrane and bind to the steroid receptor complex, releasing Hsp90 | The steroid receptor can now cross the nuclear membrane | In the nucleus the steroid receptor binds to specific gene regulatory sequences and activates transcription |

Fig. A20.4 Mechanisms of steroid action. Corticosteroids are lipid-soluble molecules that enter cells by diffusing across the plasma membrane and bind to their receptors in the cytosol. The binding of corticosteroid to the receptor displaces a dimer of a heat-shock protein named Hsp90, exposing the DNA-binding region of the receptor. The receptor:steroid complex then enters the nucleus and binds to specific DNA sequences in the promoter regions of steroid-responsive genes. Corticosteroids exert their effects by modulating the transcription of a wide variety of genes.

thymocytes bearing the cell-surface antigen NK1.1 may represent a major source of IL-4 early on in immune responses. NK1.1$^+$ T cells express an unusually restricted pattern of α:β T-cell receptors (V$_\alpha$24, J$_\alpha$Q7:V$_\beta$11) and respond to antigen-presenting cells bearing the MHC class I-like molecule CD1. Unlike CD4 T$_H$2 cells, they release large amounts of IL-4 within 30–90 minutes of primary stimulation. In addition to these thymus-derived cells, mast cells and basophils have also been reported to produce IL-4.

Transgenic mice overexpressing the NK1.1-type T-cell receptor have a marked tendency towards the production of IL-4 and IgE, suggesting a potential role for this population of cells in allergy. However, mice in which the β_2-microglobulin gene has been knocked out, and that therefore cannot express CD1 and are deficient in NK1.1$^+$ thymocytes, develop normal IgE and IL-4 responses to allergic stimuli and parasitic infection. This indicates that NK1.1$^+$ T cells are not essential for such responses.

Case 21

Answer 1
Mycobacteria, particularly atypical mycobacteria, are ubiquitous in the environment and any infection is normally contained by T-cell action. AIDS, however, is characterized by a marked reduction in the number of CD4 T$_H$1 cells. Hence the production of IFN-γ is compromised and patients with AIDS have difficulty in activating their macrophages and clearing mycobacteria.

Answer 2
The delayed-type hypersensitivity reaction is provoked by a few T cells that are specific for tuberculin. After binding antigen, these T cells secrete chemokines that nonspecifically recruit macrophages and other inflammatory cells. The secretion and action of these chemokines is not dependent on IFN-γ. It is therefore not surprising that Clarissa and her cousins developed positive tuberculin skin tests.

Answer 3

Granulomas form where there is local persistence of antigen, antigen-specific T cells, and activated macrophages. As these children could not activate their macrophages it is not surprising that they did not form granulomas. Granulomas can be beneficial in that they wall off and prevent the spread of microorganisms. These children could not do that and we find the mycobacteria spreading via the blood stream—a highly unusual finding in mycobacterial disease.

Answer 4

Infection with pyogenic bacteria such as pneumococci is controlled and terminated by antibody and complement. In fact Clarissa had a very high level of IgG (1750 mg dl^{-1}), probably as a result of increased IL-6 production induced by the chronic mycobacterial infection. Most viral infections, such as chickenpox, are terminated by cytotoxic CD8 cells. Activation of these cells is not dependent on IFN-γ. However, there is evidence from mice that a lack of the IFN-γ receptor increases susceptibility to certain viruses, including vaccinia virus and lymphocytic choriomeningitis virus. Salmonella take up residence as an intracellular infection of macrophages and become inaccessible to antibody and complement; they can be destroyed in this site only when macrophages are activated by IFN-γ.

Case 22

Answer 1

This strongly suggests that her mother did not have a mutation at the *WAS* locus on her X chromosome, but rather that Mrs Sloper was a new mutant. If her mother had been a carrier, there would have been only a 1 in 32 (2^5) chance of the mother's not having an affected child among five boys.

Answer 2

Signals that direct the maturation of hematopoietic stem cells into the various lineages are transmitted by their contact with stromal cells in the bone marrow. Presumably this interaction, like T-cell–B-cell interaction, requires cytoskeletal reorientation, and thus will be impaired in cells containing an active affected chromosome. The stem cells bearing an active normal X chromosome thus have a survival advantage.

Answer 3

Polysaccharide-specific B cells would be present in normal numbers. The failure of response to polysaccharide antigens in WAS results from the failure of T-cell–B-cell collaboration, not from a block in B-cell maturation.

Answer 4

You might try to give antibody against the B-cell cell-surface protein CD40 along with the immunogen. Ligation of CD40 by the CD40 ligand borne by activated T cells is a signal for a resting B cell to start dividing and to undergo isotype switching. The antibody should act like the CD40 ligand and induce isotype switching in B cells (see Case 11).

Answer 5

Theophylline is one of the methyl xanthines (like caffeine) that inhibits the phosphodiesterase responsible for the dephosphorylation of cyclic AMP. Like the β_2 agonists used in bronchial asthma (see Case 33), it raises the intracellular cyclic AMP level and this induces the relaxation of smooth muscle around the airways.

Case 23

Answer 1
In retrospect, the family history and the patient's past medical history provided clues that he might be affected by XLP. The patient himself had low IgG and recurrent otitis. His uncle and grandfather were affected by some of the manifestations of XLP, including recurrent lymphoma and thrombocytopenia.

Answer 2
In some patients, family history as well as a classic clinical presentation (fulminant infectious mononucleosis) provide a strong basis for a clinical diagnosis. Often, however, there is no clear family history and the severe lymphoproliferation cannot be easily distinguished from other forms of malignant lymphohistiocytosis with hemophagocytosis. The molecular defect responsible for many cases of XLP has been identified as a mutation or deletion of the gene encoding SAP (*SH2D1A*). In difficult cases, a molecular diagnosis is desirable, and PCR analysis of the four exons encoding SAP can be performed.

Answer 3
In a girl, mutation of an X-chromosome gene such as *SH2D1A* would be very unlikely to account for her problem. An autosomal disorder is more probable. If the parents were related, suspicion of an autosomal defect would be heightened. Mutations in molecules that partake in pathways that involve SAP are good candidates. These include signaling pathways initiated by the engagement of SLAM-family receptors.

Answer 4
Destruction of Alexander's bone marrow by radiation or chemotherapy (with busulfan and cytoxan) followed by administration of HLA-matched marrow from a normal sibling or unrelated donor would replace all lymphoid precursors with normal SAP-expressing cells of donor origin. This approach has been used successfully in the treatment of XLP.

Answer 5
SAP has been shown to have a crucial role in CD4 T-cell function; it is essential for late B-cell help and the development of long-term humoral immunity. SAP knockout mice and patients with XLP generate a normal acute antiviral humoral response but show near-complete absence of virus-specific long-lived plasma cells and memory B cells.

Case 24

Answer 1
Fas and FasL are homotrimeric signaling complexes (see Fig. 24.2). If one element of the trimer is mutant, the trimer is rendered ineffective and cannot deliver the signal to downstream elements of the pathway that ultimately cause cell death. This type of effect is called a dominant-negative effect.

Answer 2
Some family members, even though they show no clinical evidence of ALPS, will show impaired lymphocyte apoptosis *in vitro*. It is clear that environmental and/or other genetic factors play a role in the full expression of the ALPS phenotypes as in other genetically inherited diseases. This is called variable expressivity.

| In a normal cell, cytochrome *c* is present only in mitochondria | When programmed cell death is induced, the mitochondria swell and leak, releasing cytochrome *c*, which binds to Apaf-1 | The Apaf-1/cytochrome *c* complex activates caspase, which cleaves I-CAD, releasing CAD to enter the nucleus and cleave DNA | Bcl-2 binds to mitochondrial membranes, blocking the swelling and so blocking the process that leads to cell death |

Fig. A24.3 Bcl-2 inhibits the processes that lead to programmed cell death. In normal cells cytochrome *c* is confined to the mitochondria (first panel). However, during apoptosis the mitochondria swell, allowing the cytochrome *c* to leak out into the cytosol (second panel). There it interacts with the protein Apaf-1, forming a cytochrome *c*:Apaf-1 complex that can activate caspases. An activated caspase cleaves I-CAD (see Fig. 24.2), which leads to DNA fragmentation (third panel). Bcl-2 interacts with the mitochondrial outer membrane and blocks the mitochondrial swelling that leads to cytochrome *c* release (last panel).

Answer 3

Vaccinia (the virus used for smallpox vaccination) expresses a protein, Crm A, which inhibits caspases. Herpes simplex has two genes, Us5 and Us3, which encode genes that also inhibit caspases. On the other hand, Epstein–Barr virus, which causes acute infectious mononucleosis (see Case 27), produces a protein that resembles Bcl-2, which prevents apoptosis and renders infected cells resistant to killing by cytotoxic T cells (Fig. A24.3)

Answer 4

Yes, it would be. The lethality in mice of knocking out caspase-8 points to the importance of this enzyme in fetal tissue remodeling, but there are known interspecies differences. Moreover, a point mutation in human caspase-8, at the site where it interacts with the Fas complex, might interfere with its function in Fas-induced apoptosis but not in other, Fas-independent, processes. Thus a missense mutation in caspase-8 could conceivably cause ALPS.

Case 25

Answer 1

In Foxp3-deficient mice, infusion of relatively small numbers of T_{reg} cells controls the disease symptoms. So it is likely that, in humans also, a small number of natural T_{reg} cells is sufficient for immune regulation, and that Billy can make sufficient T_{reg} cells from the bone marrow stem cells that derive from his sister.

Answer 2

The 'conditioning' leading up to a bone marrow transplant comprises treatment with cytotoxic drugs (see Cases 3 and 18) that kill all rapidly dividing cells, including the CD4 effector T cells responsible for the uncontrolled inflammatory response in IPEX. As these cells are destroyed, the autoimmune response that they have produced will be dampened. Immunosuppressant drugs have the same effect in reducing the activation and proliferation of T cells, and this is why they are used to treat IPEX and other autoimmune disorders.

Answer 3

Because IPEX patients are unable to downregulate the immune activation triggered by infections, their disease frequently flares up on exposure to pathogens and even after vaccination. IVIG is useful in forestalling infections or ameliorating their impact when they occur. Vaccination is contraindicated in IPEX patients because of the risk of disease flare-up.

Answer 4

Venous infusion of immunocompetent purified naive CD4 T cells into mice with severe immunodeficiency (for example, *scid* mice or Rag-deficient mice (see Case 16)) results in colitis. Infusion of CD4 CD25 T_{reg} cells into these recipients can both prevent and reverse the colitis. This suggests that CD4 CD25 T_{reg} cells may have therapeutic potential in human autoimmune diseases.

Answer 5

Patients with IL-2Rα deficiency present a clinical picture with similarities to IPEX. IL-2 signaling is essential for the maintenance of T_{reg} cells. Therefore T_{reg} activity is deficient in the absence of IL-2, IL-2Rα (CD25) and the transcription factor STAT5, which is important for transducing the IL-2 signal. This may explain the similarity of the symptoms of IL-2Rα deficiency to those of IPEX.

Case 26

Answer 1

Superantigens, but not conventional antigens, can activate naive T cells. Superantigens will thus induce the proliferation of lymphocytes from newborns and from the thymus, as previous exposure to the antigen and expansion of the number of antigen-reactive cells is not required. Superantigens do not require processing by accessory cells and are thus able to induce the proliferation of purified T cells in the presence of paraformaldehyde-treated monocytes, which lack the capacity to process antigen. Direct binding of a labeled protein to cells positive for MHC class II, or its co-precipitation with MHC class II molecules, confirms it as a superantigen.

Answer 2

During the evolution of an adaptive immune response to conventional antigen, a cascade of events must occur over a relatively long period of time. The antigen has to be internalized, processed, and presented as peptide:MHC complexes by antigen-presenting cells. The complexes are recognized only by those T cells bearing a T-cell receptor specific for the antigen-derived peptides—a fraction of a percent (<0.1%) of the entire T-cell pool. These few antigen-specific T cells must then proliferate and bystander cells be recruited before an effective response can be mounted. In contrast, superantigen-induced immune activation is independent of antigen processing, thus bypassing the first step, and immediately activates a sizeable fraction of T cells. A very small number of superantigen molecules is sufficient to activate a T cell with the appropriate V_β region in its receptor (<10 molecules per T cell). Activation results in a massive secretion of T-cell cytokines, which include IL-2, IFN-γ, TNF-α, and LT. In addition, superantigens can directly activate monocytes and dendritic cells by cross-linking their surface MHC class II molecules. Cross-linking is effected by superantigens bound to T-cell receptor β chains and/or because a number of superantigens, including

TSST-1, have two distinct binding sites for MHC class II molecules. Cross-linking of MHC class II molecules causes a rapid and massive release of cytokines such as IL-1, TNF-α, IL-6, CXCL8 (formerly known as IL-8), and IL-12. This is associated with the upregulation of B7 co-stimulatory molecules on these cells, which, together with cytokine action, further amplifies T-cell activation by superantigen. Thus, minute amounts of superantigen are sufficient to rapidly activate a large number of T cells and monocytes/macrophages, resulting in an amplificatory loop and in a massive outpouring of cytokines, which leads to the rapid appearance of clinical symptoms.

Answer 3
Liver injury may occur as a result of decreased organ perfusion during hypotension. However, immunologic mechanisms may also contribute to injury. Hepatocytes express Fas, a cell-surface molecule crucial for the induction of apoptosis (programmed cell death). T-cell activation by super-antigens and the massive release of cytokines results in the upregulation of the natural ligand for Fas—FasL—on the surface of circulating lymphocytes. Cross-linking of Fas on hepatocytes by FasL on circulating lymphocytes results in the triggering of apoptosis in hepatocytes. In addition, circulating cytokines such as TNF-α are also capable of triggering cell death and can result in liver injury.

Answer 4
Protection against toxic shock is conferred by antibodies against the super-antigen, which neutralize it before it can cause disease. In order to stimulate an antibody response, the superantigen must be recognized, internalized, and processed by superantigen-specific B cells which then present the antigenic peptides to antigen-specific T cells. These are activated to become helper T cells that can in turn stimulate the production of superantigen-specific antibodies on reexposure to the superantigen. Antibodies to other antigens that cross-react with the superantigen may also confer protection.

In humans, there is evidence that during and following TSST-associated illness, $V_\beta2$ T cells become anergic and thus cannot provide help to superantigen-specific B cells. Patients with TSS therefore fail to develop TSST-1-specific antibody. So Claire is, unfortunately, likely to be at risk of another episode of TSS. Hopefully, she will eventually develop anti-TSST-1 antibodies.

Case 27

Answer 1
Cell-mediated cytotoxic activity against EBV-infected cells is the main method of controlling EBV replication. Although EBV-specific antibodies are produced in a normal infected host, they do not seem to have a major role in controlling the virus. Patients who lack the ability to produce specific antibody, but who have intact T-cell cytotoxic responses, are in most instances able to fight the infection effectively.

Answer 2
Bone marrow transplantation is most often performed in patients whose immune systems have either been destroyed with high doses of chemotherapy or in patients with primary immune deficiency. Even when donor and recipient are matched for the major histocompatibility antigens, some degree of minor incompatibility is inevitable. A major complication that can arise after a bone marrow transplant is graft-versus-host disease (GVHD; see Case 47).

Dendritic cells express high levels of B7 and can activate naive CD8 T cells

CD8 T cell

CD28

B7

virus-infected dendritic cell

Activated CD8 T cell makes IL-2, driving its own proliferation and differentiation

IL-2

virus-infected dendritic cell

Fig. A27.3 Naive CD8 T cells can be activated directly by potent antigen-presenting cells. Naive CD8 T cells encountering peptide:MHC class I complexes on the surface of dendritic cells that express high levels of co-stimulatory molecules (top panel) are activated via separate signals (1 and 2) to produce IL-2 (bottom panel) and proliferate in response to it. They eventually differentiate into armed CD8 cytotoxic T cells that are then able to attack any cell-type infected with the same virus. Interaction of CD40L on helper T cells and CD40 on dendritic cells strongly enhances the capacity of dendritic cells to activate CD8 cytotoxic T cells, by causing an increased expression of co-stimulatory molecules and secretion of cytokines, e.g., IL-12.

In GVHD, mature T lymphocytes in the bone marrow graft are activated and begin to attack the host's tissues. To prevent this, bone marrow grafts are often treated to remove mature T cells. If a bone marrow donor has been infected with EBV, they will have a small number of transformed B cells carrying the virus (about one in a million B cells). In the donor, these B cells are being 'held in check' by cytotoxic cells; there is an equilibrium between cell division and death of EBV-infected B cells. If mature T cells are removed from a marrow graft, but B cells are left, then the transformed B cells will escape from surveillance by cytotoxic T cells and might begin to proliferate at a high rate. Removing both mature B and T cells from donor bone marrow results in a lower rate of EBV lymphoproliferative disease after transplantation. An alternative explanation is that if a transplant recipient is infected with EBV the destruction of their immune system before transplantation removes enough cytotoxic cells to tip the balance in favor of the donor transformed B cells.

Treatment of EBV-related lymphoproliferative disease after transplantation has demonstrated the potential utility of adoptive immunotherapy. EBV-specific cytotoxic T cells are used directly from donor blood or can be expanded *in vitro* by culture with EBV-transformed donor cells. The cytotoxic T cells are then transfused into the transplant recipient with lymphoproliferative disease and can kill the dividing donor B cells. This treatment has achieved remission of EBV lymphoproliferative disease in up to 90% of patients in various studies so far. The major complication of this treatment is acute or chronic GVHD.

Answer 3
Many adults will have circulating in their blood activated and armed cytotoxic T cells specific for EBV (Fig. A27.3). When a blood sample is cultured and then infected with EBV, the B cells will become infected and the activated cytotoxic T cells will then immediately destroy the infected cells. Transformed B cells can be prepared from these individuals either by removing T cells from the blood sample before culture or by adding inhibitors of T-cell activation such as cyclosporin A. It is exceedingly unlikely that a fetus will have been infected by EBV, and so their blood sample is unlikely to contain any activated EBV-specific cytotoxic T cells.

Answer 4
EBV activates B cells polyclonally, that is without respect to the antigen specificities of the infected cells. A significant percentage (5–10%) of circulating B lymphocytes bear antigen receptors of low affinity for several cross-reacting carbohydrate, nucleotide, or glycoprotein antigens. If infected with EBV and activated, these cells will begin to secrete polyspecific IgM antibodies. Many of these antibodies will bind relatively nonspecifically to erythrocytes of other species such as horse, ox, cow, or sheep and are thus called heterophile ('other-loving') antibodies. They are found in approximately 90% of patients with EBV.

Answer 5
These patients have no B cells, the cellular host of EBV.

Case 28

Answer 1
The goal of treatment for symptomatic cryoglobulinemia associated with hepatitic C virus (HCV) infection is to eradicate the HCV or at least reduce the

level of viral infection. There are no specific inhibitors of HCV replication currently available; however, interferon-α (IFN-α) is effective in treating HCV infection in some patients. Although the mechanism of action remains uncertain, there is evidence that in patients with chronic HCV, IFN-α clears the virus by preventing new infection of liver cells rather than by inhibiting viral replication within cells. There is also some evidence that IFN-α may tip the balance between the T_H1 and the T_H2 immune response in favor of a T_H1 response. Although IFN-α has been shown to improve many of the features of HCV-induced cryoglobulinemia, cessation of therapy often, unfortunately, results in a relapse of the disease. The combination of ribavirin with IFN-α improves both biochemical and virologic responses in chronic HCV infection and has been beneficial in symptomatic cryoglobulinemia. Ribavirin is a guanosine nucleoside analog that is active against many RNA and DNA viruses.

For severe systemic illness such as the intestinal vasculitis, plasmapheresis is an option to rapidly lower circulating titers of autoantibodies or immune complexes. Plasmapheresis involves withdrawal of blood from the patient's vein, separation of the plasma from blood cells, either by centrifigation or membrane separation, and reinfusion of the patient's blood cells, which are resuspended in donor plasma or another replacement solution (such as albumin, normal saline, or lactated Ringer's solution).

Answer 2

Rituximab is a chimeric anti-CD20 monoclonal antibody consisting of human IgG1κ constant regions and murine variable regions. The CD20 antigen is expressed constitutively on B-lineage cells. Rituximab bound to CD20-positive cells leads to very efficient depletion of circulating B lymphocytes by a range of mechanisms. These include complement fixation, which results in formation of the membrane-attack complex, and antibody-dependent cell-mediated cytotoxicity (ADCC). The initial clinical application of rituximab was in the treatment of patients with B-cell lymphoma refractory to standard therapy; however, its use has now been extended to treat a range of autoimmune diseases, including chronic idiopathic thrombocytopenic purpura (see Fig. 40.1), autoimmune hemolytic anemia (see Case 39), and cold agglutinin disease (see Case 39).

In Billy's case, rituximab might help his immune complex-mediated disease in a number of ways. The most straightforward interpretation is that rituximab causes B-cell depletion, leading to decreased antibody production and a decrease in the level of circulating immune complexes. However, B cells are important not only in generating antibody responses but also in their role as antigen-presenting cells and cytokine-producing cells. Removal of B cells by rituximab may have additional benefits in symptomatic cryoglobulinemia.

A major concern about the use of rituximab is the risk of infections. In patients with HCV-related cryoglobulinemia it is possible that reduction in anti-HCV antibody levels after B-cell depletion might allow uncontrolled replication of the HCV. Further studies need to be carried out to determine whether this theoretical risk is important in patients.

Case 29

Answer 1

Antibodies against other streptococcal antigens, especially streptococcal DNase B, can be tested for.

Answer 2

Hereditary angioneurotic edema is caused by a deficiency of the inhibitor of C1, the first component in the classical pathway of complement activation. This inhibitor (C1INH) regulates the activity of the complement pathway and prevents its inappropriate activation. Thus the erythema marginatum is probably caused by complement activation.

Answer 3

The arthritis is caused by immune complexes formed in the joint being taken up by phagocytes, which then release prostacyclin and prostaglandin E_2 (PGE_2). These mediators increase blood flow in small blood vessels as well as increasing vascular permeability, thereby enhancing edema and swelling, which also causes pain. Aspirin acetylates and inhibits the enzyme cyclooxygenase and thereby inhibits the conversion of arachidonic acid to prostaglandin G_2 (PGG_2). This is a critical step in the formation of all prostaglandins.

Answer 4

As activated antigen-specific T cells are necessary to form granulomas, it is likely that the T cells in the Aschoff nodules have specific antigen receptors for streptococcal antigens or for peptides that cross-react with streptococcal antigens.

Answer 5

IgG antibodies bound to the streptococci fix complement on the bacterial surface. Complement components are recognized by complement receptors on phagocytes, thus facilitating the uptake and destruction of the bacteria. After C3b is bound to the bacterial surface, its conversion to C3bi, which binds to various complement receptors on phagocytes, requires factor H and factor I. If factor H is bound to M protein it can no longer act as a cofactor for factor I, and thus conversion of C3b to C3bi and bacterial uptake is inefficient.

Case 30

Answer 1

Absence of delayed-type hypersensitivity to a wide range of antigens unrelated to *M. leprae* is called anergy. This should not be confused with T-cell or B-cell anergy, although it might operate by similar mechanisms. In tuberculoid leprosy, there is a strong delayed-type hypersensitivity to *M. leprae* and no anergy. The existence of anergy in the lepromatous form of leprosy but not in the tuberculoid form is most probably due to the presence of regulatory CD8 T cells in lepromatous leprosy. The CD8 T cells secrete the cytokines IL-10 and LT and thereby suppress antigen presentation by macrophages. These cytokines not only influence the T_H1 versus T_H2 phenotype, as discussed above, but can suppress T-cell responses to other unrelated antigens. IL-10 and LT suppress not only the *M. leprae*-specific T cells, but also neighboring T cells, leading to global hyporesponsiveness, which was manifest in Ursula's case as anergy to candida and mumps antigens. However, Ursula's case is somewhat atypical; in many patients with lepromatous leprosy the unresponsiveness is confined to *M. leprae* and responses are made to other antigens. Other pathogens use the IL-10 pathway to produce anergy; the Epstein–Barr virus, for example, produces a protein, VIL-10, that is homologous to IL-10. Measles virus induces anergy by binding to CD46 on monocytes and inhibiting their production of IL-12.

Answer 2

The immune response in patients with lepromatous leprosy is skewed towards the T_H2 phenotype, leading to a disseminated infection. Because tuberculoid leprosy involves a T_H1 response and significantly reduced symptoms, we would like to switch the response to the T_H1 phenotype. Cytokines with the potential to inhibit T_H2 and induce a T_H1 response are IL-2, IFN-γ, and IL-12. Local injection of IFN-γ has been shown to lead to partial reversal of anergy and reduction of lesions. IFN-γ has also been shown to be effective in the treatment of other similar diseases, such as leishmaniasis. In the visceral form of leishmaniasis, the T-cell response is also skewed to the T_H2 phenotype. This is in contrast to the cutaneous form of leishmaniasis, which is accompanied by a T_H1 response. IL-12 might also be beneficial, as it can induce T_H1 cells and does not activate T_H2 cells.

Answer 3

In lepromatous leprosy, a humoral immune response driven by T_H2 cells predominates, with vigorous antibody production, leading to hypergammaglobulinemia as observed in Ursula. The cytokines produced in T_H2 responses lead to enhanced immunoglobulin production. IL-4 induces isotype switching to IgE and increased production of IgG4 and IgE. IL-10 stimulates the production of IgG1 and IgG3, whereas IL-5 stimulates immunoglobulin production globally. Therefore, it is not uncommon to find hypergammaglobulinemia in patients who are producing a vigorous T_H2 response to an antigen.

Answer 4

Ursula's T_H2 response to *M. leprae* leads to increased production of IL-4 and IL-10. When she encounters a new antigen, her immune system will be awash with IL-4, triggering a T_H2 response to that antigen. This T_H2 response with its associated IL-4 production leads to an IgE response. Because asthma and other atopic diseases are T_H2-driven diseases involving IgE production, Ursula has a higher risk of developing asthma.

Case 31

Answer 1

The lymph node would exhibit marked, if not exuberant, follicular hyperplasia, indicating an ongoing B-cell response to the virus. In SCID and XLA the lymph nodes are very small. In SCID the lymph node would contain no or very, very few lymphoid cells. In XLA the lymph node would have no follicles, no germinal centers, and no B cells or plasma cells. T cells would be present but not in an organized array (Fig. A31.1).

Answer 2

In infants, the HIV infection typically runs a more rapid course, and infants often die before they reach 1 year old. Mr Thomas, however, had been infected for many years before symptoms of immunodeficiency started to appear. This difference is probably due to the fact that infants are immunologically immature and naive, whereas an infected adult has a functionally mature immune system and decades of acquired adaptive immunity to many different antigens. This has consequences for both the response to HIV itself and the susceptibility to other infections. We have seen that T cells in newborn infants are not fully 'turned on.' For example, in hyper IgM the CD40 ligand is not readily expressed when T cells of newborns are activated (see Case 11). Their T cells do not synthesize interferon-γ in normal amounts, nor are their cytotoxic T lymphocytes readily activated. This functional immaturity is

Fig. A31.1 Lymph node sections from patients with severe combined immunodeficiency (SCID), X-linked agammaglobulinemia (XLA), and AIDS.

probably why young infants have difficulty confining and walling off infections, particularly those that require adaptive immunity mediated by T cells. Tuberculosis offers a clear example of this. It is a fast-spreading, highly lethal infection in young infants, whereas in older children and immunologically normal adults, the infection is usually confined to the lung, or more rarely to other organs.

HIV infection in infants occurs before they have had an opportunity to develop any adaptive immunity to common infections, and this means they are prone to infections not seen in adult AIDS. An adult will already have antibodies to the common pyogenic bacteria and will not be particularly susceptible to pyogenic infections, whereas these are frequently observed in affected infants. Adults will also have been exposed to common viruses, such as Epstein–Barr Virus (EBV). This is normally encountered early in life and contained as a latent infection (see Case 3). Virtually all adults have been infected with EBV by the end of the second or third decade of life, and primary EBV infection is therefore not a threat to HIV-infected adults. For an HIV-infected infant, however, a first encounter with this virus causes bizarre manifestations such as parotitis (inflammation of the parotid gland, like mumps) and a form of pneumonia characterized by pulmonary lymphoid hyperplasia.

Answer 3
The answer to this question is not known precisely. The immune response to the virus is illustrated in Fig. A31.3. Antibody to HIV seems to have a minor

Fig. A31.3 The immune response to HIV. Infectious virus is present at relatively low levels in the peripheral blood of infected individuals during a prolonged asymptomatic phase but is replicated persistently in lymphoid tissues. During this period, CD4 T-cell counts gradually decline, although antibodies and CD8 cytotoxic T cells directed against the virus remain at high levels. Two different antibody responses are shown in the figure, one to the envelope protein of HIV, env, and one to the core protein p24. Eventually, the levels of antibody and HIV-specific cytotoxic T lymphocytes (CTLs) also fall, and progressively more infectious HIV appears in the peripheral blood.

role in resisting the progress of infection. HIV-specific cytotoxic CD8 cells arise as virus levels decline from the peak associated with primary infection, and seem to have a more important role in containing the infection. Rare individuals with mutations in the co-receptors for HIV (CCR5 and CXCR4) are resistant to HIV infection.

Answer 4

The virus burden in these individuals is very low and, in some cases, the only detectable viruses carry mutations in genes such as *nef* or *tat*, which are vital to HIV replication in the infected host. However, viruses able to replicate in culture can be isolated from most of these so-called 'long-term non-progressors.' These patients seem to be able to contain replication-competent virus, most probably by continuing to maintain a successful cytotoxic CD8 T-cell response.

Answer 5

They inhibit the reverse transcriptase of HIV. Zidovudine (AZT; 3'-azido,2',3'-dideoxythymidine) is a nucleoside analogue that is phosphorlated inside the cell and used as a substrate by the reverse transcriptase of HIV (Fig. A31.5). HIV reverse transcriptase synthesizes a DNA complement of the viral RNA, at the start of a new round of virus replication in a newly infected cell. The incorporation of zidovudine blocks further extension of this DNA strand and thereby stops replication of the virus. Two other nucleoside analogues, ddI (dideoxyinosine) and ddC (dideoxycytosine) inhibit HIV replication by a similar mechanism and are also used to treat HIV infection. Lamivudine (also known as 3TC) is an enantiomer of a dideoxy analog of cytidine. Efavirenz (also called Sustiva) is a non-nucleoside inhibitor of HIV reverse transcriptase. Unfortunately, mutation allows the virus to acquire resistance to all these drugs. Replication of HIV (and other known retroviruses) is error-prone, and the virus mutates as it replicates in the infected host. Resistance to zidovudine requires multiple mutations but can arise in only a few months. Combining HIV protease inhibitors with zidovudine or other nucleoside analogs has dramatically improved the survival of HIV-infected patients.

Answer 6

HIV produces an aspartyl protease, structurally related to pepsin and renin, which has two aspartic-acid residues in the active site. This protease is required to splice the HIV Gag proteins after they are synthesized and before the packaging of the virus into its coat and its budding from the cell surface. Inhibitors of this HIV protease have been designed and are very effective in halting HIV replication. Should the HIV in Mr Thomas become resistant to the reverse transcriptase inhibitors, adding a protease inhibitor to his drug regimen might keep the virus in check for some time longer.

Answer 7

The precise answer to this question is unknown. It is clear that the cells producing the virus are killed, either by cytotoxic T lymphocytes or by direct cytotoxic effects of the virus. It is also possible that the death of noninfected 'bystander' cells contributes to CD4 T-cell depletion. HIV is known to have a cytotoxic effect on CD4 T cells in culture. The viral capsular gp120 binds and cross-links the CD4 molecule, which depresses T-cell function and may induce apoptosis, even when the CD4 cells are not themselves infected by HIV. If, as seems likely, the early killing of HIV-infected CD4 cells by cytotoxic T cells serves to contain the virus and prevent greater CD4 T-cell depletion in the next round of HIV infection and replication, a declining ability to mount cytotoxic responses, especially to new viral variants, could be very important. Patients infected with HIV show impairment of T-cell function, especially

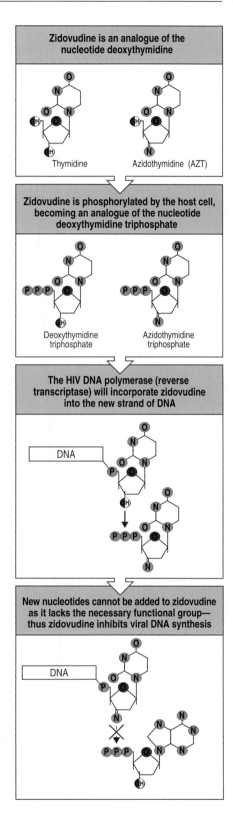

Fig. A31.5 The mechanism of action of the antiretroviral drug zidovudine (AZT).

memory cell responses, even during the asymptomatic phase. They are hyper-gammaglobulinemic, and humoral responses seem favored at the expense of cell-mediated immunity. This immunoregulatory bias impedes inflammatory responses, and possibly also cytotoxic responses to HIV-infected cells.

Answer 8

The CD4 T-cell count is, statistically speaking, the best indicator of the time-course of progression to AIDS. Other factors such as lifestyle and the incidence of intercurrent infections do not seem statistically significant.

Answer 9

Tumor necrosis factor-α (TNF-α). It causes loss of appetite (anorexia) and increased expenditure of body heat (thermogenesis).

Case 32

Answer 1

John's hoarseness resulted from angioedema of the vocal cords. His wheezing was due to forced expiration of air through bronchi that had become constricted. In this case, constriction resulted from the release by activated mast cells of histamine and leukotrienes that caused the smooth muscles of the bronchial tubes to constrict.

Answer 2

John's parents were instructed to avoid feeding him any food containing peanuts and to read labels of packaged foods scrupulously to avoid anything containing peanuts. They were advised to inquire in restaurants about food containing peanuts. Because green peas, also a legume, contain an antigen that cross-reacts with peanuts and might also incite an anaphylactic reaction, peas were withdrawn from John's diet. A Medi-Alert bracelet, indicating his anaphylactic reaction to peanuts, was ordered for John. The parents were also given an Epi-Pen syringe pre-filled with epinephrine to keep at home or while travelling, in case John developed another anaphylactic reaction.

Answer 3

Epinephrine acts at β_2-adrenergic receptors in smooth muscle surrounding blood vessels and bronchi. It has opposing effects on the two types of muscle. It contracts that surrounding the small blood vessels, thereby constricting them, stopping vascular leakage and raising the blood pressure. It relaxes that of the bronchi, making breathing easier.

Answer 4

Albuterol, which is also a β_2-adrenergic agent. It can be inhaled to alleviate bronchial constriction more effectively.

Answer 5

Histamine and tryptase are released by activated mast cells; high levels in the blood indicate the massive release from the mast cells that occurs during an anaphylactic reaction.

Answer 6

Immediately after a systemic anaphylactic reaction the patient is unresponsive in a skin test owing to the massive depletion of mast-cell granules and failure of the blood vessels to respond to mediators. This is called tachyphylaxis and lasts for 72–96 hours after the anaphylactic reaction. For this reason, John had to come back to the allergy clinic a few days later for his tests.

Answer 7
Increased incidence of peanut allergy has been related to the increasing topical use of peanut oil based creams to treat dry skin in infants. Although peanut oil is not the culpable allergen, it is often contaminated by allergenic peanut proteins.

Answer 8
It was recently shown that depletion of IgE by the administration of a humanized mouse anti-human IgE antibody that binds circulating IgE, but not mast-cell-bound IgE, results in protection from peanut anaphylaxis. This therapy works because it results in the eventual depletion of mast-cell-bound IgE, which is in equilibrium with serum IgE. It is safe because the anti-IgE antibody does not trigger mast-cell activation.

Case 33

Answer 1
During inspiration, the negative pressure on the airways causes their diameter to increase, allowing an inflow of air. During expiration, the positive expiratory pressure tends to narrow the airways. This narrowing is exaggerated when the airway is inflamed and bronchial smooth muscle is constricted, as in asthma. This causes air to be trapped in the lungs with an increase in residual lung volume at the end of expiration. Breathing at high residual lung volume means more work for the muscles and increased expenditure of energy; this results in the sensation of tightness in the chest. The high residual lung volume is also the cause of the hyperinflated chest observed on the chest radiograph. The peribronchial inflammation in asthma causes bronchial marking around the airways.

Answer 2
Chronic allergic asthma is not simply due to constriction of the smooth muscles that surround the airway. It is largely due to the inflammatory reaction in the airway, which consists of cellular infiltration, increased secretion of mucus, and swelling of the bronchial tissues. This explains the failure of bronchodilators, which dilate smooth muscles, to maintain an open airway and their failure to reverse completely the decreased air flow during Frank's acute attacks. Steroids are therefore given to combat the inflammatory reaction of the late-phase response.

Answer 3
Analysis of the genetic linkage of atopy to various polymorphic DNA markers suggests that this trait maps to chromosome 5q11 in the area containing the genes for IL-4, IL-5, and IL-9, cytokines that are involved in regulating the IgE response. Because IL-4 is required for the induction of B cells to undergo isotype switching to IgE, it is possible that the underlying cause of the high IgE levels in atopic individuals is the excessive production of IL-4.

Answer 4
Allergic individuals have a tendency to respond to allergens with an immune response skewed to producing T_H2 cells rather than T_H1. The cells produce the interleukins IL-4 and IL-13, cytokines that induce IgE production in

humans. T_H2 cells also make IL-5, which is essential for eosinophil maturation. Furthermore, activated T cells and bronchial epithelial cells secrete CCL11 (formerly known as eotaxin), which attracts eosinophils in the airways. Production of IL-4 and IL-5 by T_H2 cells responding to allergens in atopic individuals explains the frequent association of IgE antibody response and eosinophilia in these patients.

Answer 5

IgE-mediated hypersensitivity to an allergen is tested for by injecting a small amount of the allergen intradermally. In allergic individuals, this is followed within 10–20 minutes by a wheal-and-flare reaction at the site of injection (see Fig. 33.7), which subsides within an hour. The wheal-and-flare reaction is due mainly to the release of histamine by mast cells in the skin. This increases the permeability of blood vessels and leakage of their contents into the tissues, resulting in the swollen wheal; dilation of the fine blood vessels around the area produces the diffuse red 'flare' seen around the wheal. This reaction is almost completely inhibited by antagonists to the histamine type 1 receptor, the major histamine receptor expressed in the skin.

Answer 6

The recurrence of the redness and swelling at the site of previous immediate allergic reactions represents the late-phase response characterized by a cellular infiltrate.

Answer 7

Because Frank has made IgE antibodies against numerous allergens, it is very likely that should he become exposed to a rabbit, he would mount an IgE antibody response to rabbit allergens, which would probably cause allergic symptoms. A negative skin test would probably simply reflect the fact that Frank has not yet been exposed to rabbit allergens. Therefore Frank should be advised not to get a rabbit, regardless of the result of the skin test.

Answer 8

Repeated administration of relatively high doses of allergen by subcutaneous injection is thought to favor antigen presentation by antigen-presenting cells that produce IL-12. This results in the induction of inflammatory CD4 T cells (T_H1 cells) rather than T_H2 cells. The presence of T_H1 cells tends to lead to an IgG antibody response rather than an IgE response because the T_H1 cells produce interferon-γ, which prevents further isotype switching to IgE. The IgG antibody competes for antigen with IgE. Furthermore, IgG bound to allergen inhibits mast-cell activation (via FcϵRI) and B-cell activation (via surface Ig) by allergen because of inhibitory signals delivered subsequent to binding of Fcγ receptors on these cells. This is thought to be one mechanism damping down the allergic response. Another is no further boosting of IgE production because IL-4 and IL-13 are not secreted. Existing IgE levels themselves may not fall by much, because IFN-γ does not affect B cells that have already switched to IgE production.

Answer 9

Most human allergy is caused by a limited number of inhaled protein allergens that elicit a T_H2 response in genetically predisposed individuals. These allergens are relatively small, highly soluble protein molecules that are presented to the immune system by the mucosal route at very low doses. It has been estimated that the maximum exposure to ragweed pollen allergens is less than 1 μg per year. It seems that transmucosal presentation of very low doses of allergens favors the activation of IL-4-producing T_H2 cells and is particularly efficient at inducing IgE responses. The dominant antigen-presenting cell type in the respiratory mucosa expresses high levels of co-stimulatory B7.2 molecules. Expression of B7.2 on antigen-presenting cells is thought to favor the development of T_H2 cells. In contrast, injection of antigen subcutaneously in large doses, as occurs on vaccination, results in antigen uptake in the local lymph nodes by a variety of antigen-presenting cells and favors the development of T_H1 cells, which inhibit antibody switching to IgE.

Case 34

Answer 1
Corticosteroids bind to steroid receptors in inflammatory cells such as T cells and eosinophils. The steroid:receptor complex is translocated into the nucleus, where it can control gene expression, including the expression of cytokine genes, by binding to control elements in the DNA. In addition, corticosteroids increase the synthesis of the inhibitor of the transcription factor NFκB, which controls the expression of multiple cytokine genes. One effect is to inhibit the synthesis of cytokines and the release of preformed mediators and arachidonic acid metabolites. Although topical steroids are very effective, excessive or prolonged use of powerful steroids can lead to local skin atrophy.

Answer 2
The immunosuppressant cyclosporin A acts primarily on T cells and interferes with the transcription of cytokine genes. The drug binds to an intracellular protein, cyclophilin, and this complex in turn inhibits calcineurin, which normally dephosphorylates nuclear factor of activated T cells (NFAT), a major cytokine gene transcription factor. FK506, or tacrolimus, is another immunosuppressant with a spectrum of activity similar to that of cyclosporin. Tacrolimus binds to the cytoplasmic protein FK506-binding protein, and this complex also inhibits calcineurin. Tacrolimus has a smaller molecular size and higher potency than cyclosporin A and, perhaps because of these features, it seems to be effective as a topical formulation.

Answer 3
Patients with AD have defective cell-mediated immunity, which is required for the control of herpesvirus infections. Cell-mediated immunity involves T_H1 CD4 cells and CD8 cytotoxic cells; patients with AD have selective activation of T_H2 rather than T_H1 cells, as shown by their reduced delayed-type hypersensitivity skin reactions. They also have decreased numbers and function of CD8 cytotoxic T cells. Furthermore, monocytes from patients with AD secrete increased amounts of IL-10 and prostaglandin E_2 (PGE$_2$). Both IL-10 and PGE$_2$ inhibit the production of the T_H1 cytokine IFN-γ, and IL-10 also inhibits T-cell-mediated reactions.

Answer 4
Scratching causes tissue damage that stimulates the keratinocytes to secrete cytokines and chemokines (IL-1, IL-6, CXCL8, GM-CSF, and TNF-α). IL-1 and TNF-α induce the expression of adhesion molecules such as E-selectin, ICAM-1, and VCAM-1 on endothelial cells, which attract lymphocytes, macrophages, and eosinophils into the skin. These infiltrating cells secrete cytokines and inflammatory mediators that perpetuate keratinocyte activation and cutaneous inflammation.

Answer 5
The skin of more than 90% of patients with AD is colonized by *Staphylococcus aureus*. Recent studies suggest that *S. aureus* can exacerbate or maintain skin

inflammation in AD by secreting a group of toxins known as superantigens, which cause polyclonal stimulation of T cells and macrophages (see Case 26). T cells from patients with AD preferentially express T-cell receptor β chains $V_\beta3$, 8, and 12, which can be stimulated by staphylococcal superantigens, resulting in T-cell proliferation and increased IL-5 production. Staphylococcal superantigens can also induce expression of the skin homing receptor (CLA) in T cells, which is mediated by IL-12. In addition, nearly half of AD patients produce IgE directed against staphylococcal superantigens, particularly SEA, SEB, and toxic shock syndrome toxin-1 (TSST-1). Basophils from AD patients who produce antitoxin IgE release histamine on exposure to the relevant toxin. These findings suggest that local production of staphylococcal exotoxins at the skin surface could cause IgE-mediated histamine release and thereby trigger the itch–scratch cycle that exacerbates the eczema.

Answer 6

Inhaled allergens can enter capillaries in the lungs and thence be carried to the skin in the bloodstream. Binding of allergen to Langerhans cells and mast cells bearing surface IgE in the skin results in cytokine and mediator release, which causes edema and inflammation. In patients with AD and asthma, inhalation challenge with dust-mite allergen results in both bronchial and skin reactions in a subset of patients. Skin reactions occurred 1.5–17 hours after the inhalation challenge with mite allergen. Sensitization directly through the skin can also occur. A mouse model of AD suggests that sensitization directly through the skin can even result in allergen-induced asthma. In this model, patch application of allergen to the shaved skin of a normal mouse results in an eczematous dermatitis and subsequent allergen-specific airway hypersensitivity such that a single first-time encounter with the allergen by inhalation causes airway hyperresponsiveness typical of the asthmatic state.

Case 35

Answer 1

The activation of complement-generated C3a, which releases histamine from mast cells and causes hives. The swelling around the mouth and eyelids is a form of angioedema. There is a more complete discussion of the role of the complement and the kinin systems in the pathogenesis of angioedema in Case 37.

Answer 2

Gregory almost certainly had developed vasculitis in the small blood vessels of his brain and this compromised oxygen delivery to his brain.

Answer 3

He had red cells and albumin in his urine, which indicated an inflammation of the small blood vessels in his kidney glomeruli. He also developed purpura in his feet and ankles. Purpura (which is the Latin word for purple) indicates hemorrhage from small blood vessels in the skin that are inflamed and have become plugged with clots. A skin biopsy of one lesion showed the deposition of IgG and C3 around the small blood vessels, suggesting that an immune reaction was taking place.

Answer 4

Massive follicular hyperplasia, polyclonal B-cell activation, and many mature plasma cells in the medulla. The massive B-cell activation in the lymph nodes leads to an overflow of plasma cells from the medulla of the nodes into the efferent lymph. It is otherwise very, very rare to find plasma cells in the blood, as were found in Gregory's blood. They find their way to the bloodstream via the thoracic duct. The enlargement of the spleen was almost certainly due to hyperplasia of the white pulp. Some plasma cells probably enter the blood from the hyperplastic follicles in the spleen.

Answer 5
The acute-phase reaction is caused by interleukin (IL)-1 and to a greater extent by IL-6, which are released from monocytes that have been activated by the uptake of immune complexes. The acute-phase response consists of dramatic changes in protein synthesis by the liver. The synthesis of albumin drops sharply, as does the synthesis of transferrin. The synthesis of fibrinogen, C-reactive protein, amyloid A, and several glycoproteins is rapidly upregulated. The precise advantage to the host of the acute-phase reaction is not well understood, but it is presumably a part of innate immunity, which aids host resistance to pathogens before the adaptive immune system becomes engaged.

Answer 6
His serum C1q level was decreased. This almost always indicates complement consumption by immune complexes via the classical pathway. (In hereditary angioneurotic edema (see Case 4) the C1q level is normal; in this disease the activation of complement is not caused by immune complexes.) The level of C3 in Gregory's serum was also lowered: a further indication of complement consumption (see Case 43).

Answer 7
No! The skin test is positive when there are IgE antibodies bound to the mast cells in the skin. Gregory did not have IgE antibodies to penicillin, as confirmed by the negative RAST test. Serum sickness is caused by complement-fixing IgG antibodies.

Case 36

Answer 1
The 33-mer peptide has a predominance of proline residues and this structure is crucial to its resistance to digestion by gastrointestinal enzymes. A bacterial propyl endopeptidase could potentially catalyze the breakdown of this peptide, thereby preventing it from interacting with TTG. Removing the 33-mer and its antigenic epitopes from the subepithelial space would eliminate the interaction of HLA DQ2-bound peptides with T cells and would halt the inflammatory process.

Answer 2
Testing is recommended in asymptomatic children who are in high-risk groups (type 1 diabetes mellitus, autoimmune thyroiditis, Down syndrome, Turner syndrome, Williams syndrome, IgA deficiency, and first-degree relatives of CD patients) starting at age 3 years, as long as they have not been put on a gluten-free diet. The reasoning behind testing asymptomatic children is that celiac disease can be clinically difficult to detect but there is reason to believe that ongoing inflammation in the gut has deleterious effects and should be diagnosed and treated as early as possible.

An initial negative screen for celiac disease does not preclude its emergence later in some people. Repeat screening for type 1 diabetics and patients with

Down syndrome and first-degree relatives with celiac disease should be performed periodically. The optimal frequency of such testing has not yet been clarified. An alternative to interval testing is to HLA type the high-risk individuals. In the absence of HLA DQ2 or DQ8, celiac disease is extremely rare.

Answer 3

A level of 220 ppm gluten had been arbitrarily designated as 'gluten free', but there is emerging evidence that even very small amounts of gluten are toxic to people with celiac disease and a new limit of 20 ppm is being considered. The amount of gluten in foods is not the only challenge to patients; unclear labeling also poses a major obstacle to maintaining a gluten-free diet.

Answer 4

Oats do not contain gluten, but they can be cross-contaminated with wheat as a result of storage in silos that hold both types of grains. Patients newly diagnosed with celiac disease should not eat oats, but they can be reintroduced in small amounts once the disease is in remission.

Case 37

Answer 1

Pentadecacatechol can be transferred from the initial point of contact to other areas of the skin by the fingernails after scratching the itchy lesion at the primary site of hapten introduction. This is why it is essential to cut the fingernails short and thoroughly wash off the skin and scalp to remove the chemical and prevent further spread.

Answer 2

The half-life of some of the proteins haptenated by pentadecacatechol can be quite long. CD4 memory T cells will continue to be activated as long as the haptenated peptides are being generated. In Paul's case this went beyond the third week following contact with poison ivy.

Answer 3

Once an individual is sensitized, the reaction often becomes worse with each exposure, as each reexposure not only produces the hypersensitivity reaction but generates more effector and memory T cells. Memory T cells that mediate delayed hypersensitivity reactions, such as contact sensitivity to poison ivy and the tuberculin test, can persist for most of the life of the individual.

Answer 4

You could perform a patch test. In this test a patch of material impregnated with the hapten is applied to the skin under seal for 48 hours. The area is then examined for redness, swelling, and vesicle formation. Alternatively, peripheral blood mononuclear cells can be incubated with the hapten and T-cell proliferation assessed 6–9 days later.

Answer 5

The risk of Brian's developing poison ivy sensitivity is at least as high as that for a normal child. This is because antibody plays no discernible role in the genesis of delayed hypersensitivity reactions and T-cell function is normal in X-linked agammaglobulinemia. In fact, clinical observations suggest that boys with X-linked agammaglobulinemia may develop more severe forms of poison ivy. It has been suggested that, in the absence of antibody, more hapten is available for conjugation with self proteins and that, in the absence of antigen presentation by B cells, the T-cell response is skewed more towards $T_H 1$ cells.

Answer 6

The artifically induced tuberculin reaction is a good model of a delayed hypersensitivity reaction. This skin test detects infection with the bacterium *Mycobacterium tuberculosis*, or previous immunization against tuberculosis with the attenuated live vaccine BCG. Small amounts of tuberculin, a protein derived from *M. tuberculosis*, are injected subcutaneously; a day or two later, a sensitized person develops a small, red, raised area of skin at the site of injection. In countries where BCG is administered routinely to babies, the tuberculin test can be used to test for T-cell function. This is because antigen-specific memory T cells are long-lived, and the sensitivity to tuberculin will persist throughout life. In the USA, children are not immunized with BCG. However, they all receive a full course of diphtheria and tetanus vaccines, which in each case contain purified protein toxoids as the antigen. Contact sensitivity to these two antigens can be used to test T-cell function. Alternatively, antigen derived from the yeast-like fungus *Candida albicans*, which is a normal inhabitant of the body flora, can be used to induce a delayed hypersensitivity reaction in the skin.

Answer 7

Some of the commoner environmental causes of delayed hypersensitivity reactions are insect bites or stings, which introduce insect venom proteins under the skin, and skin contact with chemicals in the leaves of some plants, or with metals such as nickel, beryllium, and chromium (Fig. A37.7). Nickel sensitivity is quite common and often occurs at the site of contact with nickel-containing jewelry. Contact sensitivity to beryllium has been well documented in factory workers engaged in manufacturing fluorescent light bulbs. Celiac disease is a type of delayed sensitivity reaction seen in people who are allergic to the protein gliadin, a constituent of wheat grains and flour. Patients with celiac disease thus have to avoid all food products containing wheat flour.

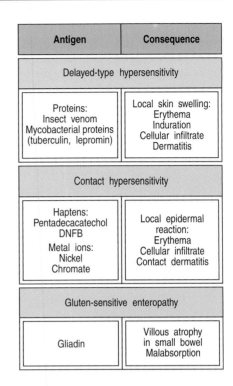

Antigen	Consequence
Delayed-type hypersensitivity	
Proteins: Insect venom Mycobacterial proteins (tuberculin, lepromin)	Local skin swelling: Erythema Induration Cellular infiltrate Dermatitis
Contact hypersensitivity	
Haptens: Pentadecacatechol DNFB Metal ions: Nickel Chromate	Local epidermal reaction: Erythema Cellular infiltrate Contact dermatitis
Gluten-sensitive enteropathy	
Gliadin	Villous atrophy in small bowel Malabsorption

Fig. A37.7 Some type IV hypersensitivity reactions. Depending on the source of antigen and its route of introduction, these clinical conditions have different names and consequences.

Case 38

Answer 1

Graves' disease is caused by an autoantibody against the receptor for TSH in the thyroid gland. The autoantibody acts as an agonist (like TSH) and stimulates the thyroid to become hyperactive (Fig. A38.1). In contrast, Robert's thyroid gland was being gradually destroyed by the infiltration of cytotoxic T cells specific for a thyroid antigen; this autoimmune disease is known as Hashimoto's disease. He could possibly also have developed an anti-TSH antibody that antagonized TSH. In fact, patients with Graves' disease can sometimes go on to develop Hashimoto's disease, and both stimulatory and blocking autoantibodies can be present in both diseases.

Answer 2

The loss of adrenal cortical hormones as a result of the autoimmune destruction of his adrenal cortex caused Robert's pituitary gland to secrete greatly increased amounts of ACTH. ACTH is composed of 39 amino acids, the amino-terminal 14 amino acids of which can be cleaved off by trypsin-like enzymes. This 14 amino-acid peptide is called melanocortin and stimulates melanocytes in the skin to produce the brown pigment melanin. The receptor for melanocortin also binds intact ACTH, albeit at a lower affinity. The increased

Fig. A38.1 Feedback regulation of thyroid hormone production is disrupted in Graves' disease. Graves' disease is caused by autoantibodies specific for the receptor for thyroid-stimulating hormone (TSH). Normally, thyroid hormones are produced in response to TSH and limit their own production by inhibiting the production of TSH by the pituitary (left panels). In Graves' disease, the autoantibodies are agonists for the TSH receptor and therefore stimulate the production of thyroid hormones (right panels). The thyroid hormones inhibit TSH production in the normal way but do not affect production of the autoantibody; the excessive thyroid hormone production induced in this way causes hyperthyroidism.

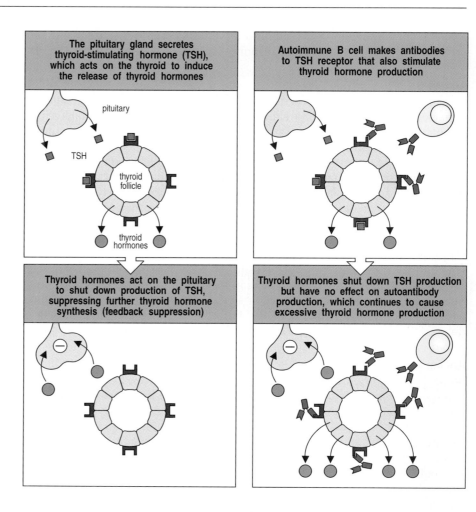

amounts of ACTH, and probably of melanocortin, were what led to the increased pigmentation of Robert's scrotum and the tissue around his nipples.

Answer 3

They had twice the normal number of CD4+ and CD8+ effector/memory cells in their lymph nodes. This apparently resulted from a lack of negative selection of autoreactive cells in the thymus.

Answer 4

This experiment implies, but does not prove, that *aire* acts in the thymus, and that the expression of *aire* in peripheral organs is of no importance. The lymphocytes transferred from the normal mice had left the thymus with self-reactive clones deleted, whereas this had not occurred in the *aire*-deficient lymphocytes. To show definitively that the thymus is responsible for the disease, the investigators transplanted either knockout or control thymuses into *nude* mice (which do not have a thymus). Only those mice that received a knockout thymus developed autoimmune disease.

Case 39

Answer 1

A plasma exchange (plasmapheresis). In this procedure the patient's blood is repeatedly removed 300 to 500 ml at a time and the blood is centrifuged (in this case at 37°C, at which temperature the IgM antibodies would be eluted from the red blood cells). The cells are resuspended in normal plasma and infused back into the patient. This is a relatively efficient procedure for removing IgM antibodies because 70% of IgM is in the plasma compartment and only 30% of IgM is in the extravascular compartment. In contrast (as we learned in Case 10), only 50% of IgG is in the vascular compartment and 50% is in the extravascular space.

Answer 2
They usually do not cause anemia because they do not fix complement (C1q) and there are no Fc receptors for these immunoglobulin classes on cells of the macrophage lineage, which bear only Fcγ receptors. Fcε receptors are expressed on mast cells and B cells, and their engagement would not result in the destruction of red blood cells.

Answer 3
The internal thiol ester that binds covalently to the IgM autoantibody is situated in the C3d and C4d fragments. The complement components C3b and C4b that are bound to the autoantibody may be digested by Factor I to C3c + C3d and C4c + C4d. C3c and C4c would be released from the antigen:antibody complex and thus antibodies against them would not agglutinate the red blood cells.

Answer 4
There is almost certainly no T-cell antigen receptor for this carbohydrate antigen, so that help to move the B cells into follicles would not occur and specific B cells would undergo apoptosis in the T-cell zone (Fig. A39.4). Alternatively, the B cells might be rendered anergic by soluble antigen or undergo programmed cell death through interaction of Fas and Fas ligand.

Although the precise mechanism of autoimmunization awaits elucidation, the molecular characterization of the autoantigen I and of the host cell adherence receptors for *Mycoplasma pneumoniae* have indicated collectively that the disorder has its origin in the interaction of the infective agent with carbohydrate attachment sites on host cells. The mycoplasma adheres to the ciliated bronchial epithelium, and also to red cells and a variety of other cells via ligands that consist of long carbohydrate chains of I antigen type that are capped with sialic acid (see Fig. 39.5). When bound to the carbohydrate chain that contains the I-antigen sequence, the mycoplasma may act as a carrier and the I antigen may act as a hapten.

Case 40

Answer 1
It would be likely to last 1–2 weeks. The infant has the disease because maternal IgG antibodies to the acetylcholine receptor have crossed the placenta from the maternal circulation to the fetal circulation. The infant is not synthesizing these autoantibodies; he or she has acquired the disease passively by transfer of the antibodies (Fig. A40.1). The maternal IgG antibodies bind to the acetylcholine receptors in the baby, and the complex of the receptor with bound IgG antibodies is internalized into the cell and degraded. Within 10–15 days all the maternal IgG antibodies to the acetylcholine receptor are adsorbed from the babies' blood and the symptoms abate.

Answer 2
Azathioprine is an immunosuppressive agent. It is converted in the liver to 6-mercaptopurine, which inhibits DNA synthesis. Thus the growth of rapidly

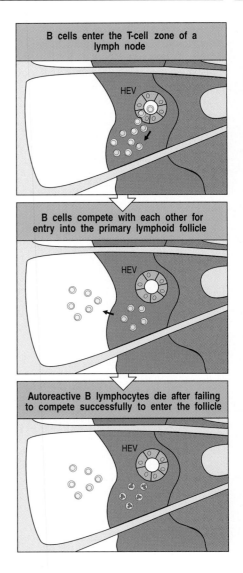

Fig. A39.4 Autoreactive B cells do not compete effectively to enter primary lymphoid follicles in peripheral lymphoid tissue. In the top panel B cells are seen entering the T-cell zone of a lymph node through high endothelial venules (HEVs). Those with reactivity to foreign antigens are shown in yellow and autoreactive cells are gray. The autoreactive cells fail to compete with B cells specific for foreign antigens for exit from the T-cell zone and entry into primary follicles (center panel). This is because B cells reactive to foreign antigens receive signals from antigen-specific T cells which promote their activation and survival. In contrast, the autoreactive B cells fail to receive survival signals and undergo apoptosis in the T-cell zone (bottom panel).

Fig. A40.1 Antibody-mediated autoimmune disease can appear in the infants of affected mothers as a consequence of transplacental transfer of immunoglobulin G.

Patient with myasthenia gravis makes antibodies against acetylcholine receptor	Transfer of antibodies across placenta into the fetus	Newborn infant has symptoms of myasthenia gravis

dividing cells, such as B cells and T cells, is inhibited and the immune response is suppressed. Unfortunately the effects of azathioprine are not specific. It suppresses not only the formation of antibodies to the acetylcholine receptor but also all other immune responses. Patients taking azathioprine become susceptible to infections. If used for very prolonged periods it is associated with the development of lymphomas. The reasons for this are not well understood.

Answer 3

Fig. A40.3 summarizes ways in which infectious diseases can break self tolerance and induce, or worsen, an autoimmune disease. An infectious agent may expose hidden antigens, or may increase expression of MHC molecules and co-stimulators on tissue cells so as to induce an autoimmune response. B cells already primed to make an autoantibody may receive help from nearby T cells activated by an infection, especially if pathogens become attached to self molecules. Pathogens may induce responses that cross-react with self molecules. Bacterial and viral superantigens can overcome clonal anergy and break tolerance to self antigens. At present, relatively little is known about the induction of human autoimmune disease, and there are only a few examples in which the evidence for any one of these mechanisms is strong.

Mechanism	Disruption of cell or tissue barrier	Infection of antigen-presenting cell	Binding of pathogen to self protein	Molecular mimicry	Superantigen
Effect	Release of sequestered self antigen; activation of non-tolerized cells	Induction of co-stimulatory activity on antigen-presenting cells	Pathogen acts as carrier to allow anti-self response	Production of cross-reactive antibodies or T cells	Polyclonal activation of autoreactive T cells
Example	Sympathetic ophthalmia	Effect of adjuvants in induction of EAE	? Interstitial nephritis	Rheumatic fever ? Diabetes ? Multiple sclerosis	? Rheumatoid arthritis

Fig. A40.3 Ways in which infectious agents can break self tolerance. Because some antigens are sequestered from the circulation, either behind a tissue barrier or within the cell, it is possible that an infection that breaks cell and tissue barriers may expose hidden antigens (first column). A second possibility is that the local inflammation in response to an infectious agent may trigger the expression of MHC molecules and co-stimulators on tissue cells, inducing an autoimmune response (second column). In some cases, infectious agents may bind to self proteins. Because the infectious agent induces a helper T-cell response, any B cell that recognizes the self protein will also receive help. Such responses should be self-limiting once the infectious agent is eliminated, because at this point, the T-cell help will no longer be provided (third column). Infectious agents may induce either T-cell or B-cell responses that can cross-react with self antigens. This is termed molecular mimicry (fourth column). T-cell polyclonal activation by a bacterial superantigen could overcome clonal anergy, allowing an autoimmune process to begin (fifth column). There is little evidence for most of these mechanisms in human autoimmune disease. EAE, experimental autoimmune encephalomyelitis.

Fig A41.4 The structure and metabolism of the cytotoxic drug cyclophosphamide. Cyclophosphamide is administered as a stable 'pro-drug' which is transformed enzymatically in the body to phosphoramide mustard, a powerful and unstable DNA alkylating agent.

Case 41

Answer 1
The monomeric IgG in the gamma globulin administered intravenously binds to high-affinity Fc receptors on macrophages and causes the release of immunosuppressive cytokines such as transforming growth factor-β(TGF-β), interleukin (IL)-10 and the IL-1 receptor antagonist. In patients with pemphigus vulgaris, clinical improvement after intravenously administered gamma globulin is noted in about 6 weeks and decreased antibody titers are found after 6 months.

Answer 2
The reasons for this difference are not understood. We know that the IgG1 antibody reacts with a different epitope of desmoglein-3 than does the IgG4 antibody, but this in itself does not provide an explanation. IgG1 fixes complement, and thus we must infer that the proteinases generated by complement activation do not digest desmoglein-3.

Answer 3
Isotype switching is required to make the IgG4 antibodies in addition to or in lieu of the IgG1 antibodies. Switching to IgG4 is stimulated by the cytokine IL-4 released by activated T_H2 cells. If we take T cells from asymptomatic individuals who make IgG1 antibodies and T cells from patients who make IgG4 antibodies and stimulate them with desmoglein-3 we might find that the patients' T cells make more IL-4 (and thus undergo more isotype switching) than those of the asymptomatic individuals. Higher levels of IL-4 production in patients' T cells are indeed found in such an experiment.

Answer 4
Cyclophosphamide (Fig. A41.4) is an alkylating agent that interferes with DNA synthesis and therefore stops cell division. Although it has many bad side effects such as anemia, thrombocytopenia, and hair loss, it is effective in halting lymphocyte cell division, and thus suppresses immune reactions.

Answer 5
Pneumocystis carinii is an opportunistic pathogen that is a frequent cause of pneumonia in immunosuppressed patients. In this case the high-dose corticosteroid treatment suppressed T-cell functions and trafficking, causing increased susceptibility to this potentially serious infectious agent.

Case 42

Answer 1
Rheumatoid factors can be found in the serum of patients with other immune-complex diseases. We encountered an example in the case of mixed essential cryoglobulinemia (Case 28). Patients with hypergammaglobulinemia and

chronic infection may also have circulating immune complexes and rheumatoid factor.

Answer 2

Mast cells express this receptor and use it to take up immunoglobulin complexes. This causes the immediate release from mast cells of preformed TNF-α and IL-1. Mast cells are the only cells with such cytokines already preformed.

Answer 3

They upregulate the expression of the integrin CD11:CD18 (LFA-1) on the leukocytes, and this increase in integrin expression promotes the binding of the leukocytes to the blood-vessel wall and their emigration from the blood vessels. A recombinant soluble IL-1 receptor antagonist called anakinra (Kineret) has been used successfully in the treatment of rheumatoid arthritis.

Answer 4

This monoclonal antibody is a chimeric human–mouse immunoglobulin against which patients may develop antibodies. These antibodies would render the therapy useless and might even cause anaphylactic reactions. Better anti-TNF monoclonal antibodies are those that are completely humanized and therefore do not elicit an immune response. As TNF-α is normally important in containing infections, however, with all classes of anti-TNF agents there is the risk of a preexisting infection, such as tuberculosis, becoming reactivated when it is inhibited. Patients about to receive infliximab or another anti-TNF agent should have a tuberculin skin test to ensure they are free of tuberculosis. The activation of tuberculosis in the absence of TNF-α suggests that this cytokine is critical in activating macrophages to contain the latent infection.

Case 43

Answer 1

The serum level of complement proteins C3 and C4 is lowered in SLE by the large number of immune complexes that bind to them, triggering their cleavage. The depletion of these proteins is therefore proportional to the severity of the disease. Successful immunosuppressive therapy is reflected in a rise in the serum level of C3 and C4. Measurement of either C3 or C4 is sufficient: it is not necessary to measure both, and C3 is most usually measured.

Answer 2

The objective of these tests was to establish whether Nicole had autoimmmune hemolytic anemia, which occurs in SLE when there are antibodies against erythrocytes. Nicole did not have hemolytic anemia (see Case 39).

Answer 3

Because ultraviolet light provokes the onset of SLE and causes relapses.

Answer 4

She had not developed glomerulonephritis. If she had, her urine would have contained protein and red blood cells.

Answer 5

Because of the constant stimulation of their B cells by autoantigens, patients with SLE have a greatly expanded B-cell population and consequently an

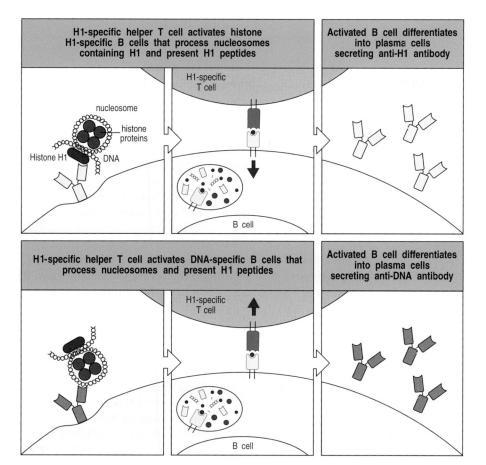

Fig. A43.6 Autoantibodies to various components of a complex antigen can be stimulated by an autoreactive helper T cell of a single specificity. In SLE, patients often produce autoantibodies to all of the components of a nucleosome, or of some other complex antigen. The most likely explanation is that all the autoreactive B cells have been activated by a single clone of autoreactive T cells specific for a peptide of one of the proteins in the complex. A B cell binding to any component of the complex through its surface immunoglobulin can internalize the complex, degrade it, and return peptides derived from the relevant protein to the cell surface bound to class II MHC molecules, where they stimulate helper T cells. These, in turn, activate the B cells. The figure illustrates this scheme for a T cell specific for the H1 histone protein of the DNA:protein complex comprising the nucleosome, and two B cells specific for the histone protein in one case, and double-stranded DNA in the other.

increased number of plasma cells secreting immunoglobulin. A lymph node biopsy from Nicole would have exhibited follicular hyperplasia in the cortex and increased numbers of plasma cells in the medulla.

Answer 6

In the first place, a large multimolecular complex such as a nucleosome carries many separate epitopes, each of which can stimulate antibody production by a B cell specific for that epitope. Any of these antibodies can precipitate the nucleosome particle to form an immune complex. Such potentially autoreactive B cells probably exist normally in the circulation but, provided that T-cell tolerance is intact, they are never activated because this requires T cells to be reactive against the same autoantigen. SLE is probably caused by a failure of T-cell tolerance. T cells for each of the components of the complex antigen will not be needed to induce antibodies against its individual components. As Fig. A43.6 shows, a T cell that is specific for one protein component of a nucleosome could activate B cells specific for both protein and DNA components.

Case 44

Answer 1

The oligoclonality of the immunoglobulins in the cerebrospinal fluid reflects the activation of a limited number of B-cell clones that have gained entry into the central nervous system after the breakdown of the blood–brain barrier. Only those B cells that recognize antigen via their surface immunoglobulin receptor and receive a stimulatory signal from an activated T cell will proceed to synthesize and secrete immunoglobulins.

Answer 2
Corticosteroids and cyclophosphamide (a powerful cytotoxic drug) inhibit T-cell proliferation and thus interfere with the secretion of cytokines that drive the inflammation and further T-cell activation. The mechanism of action of IFN-β is not known. More recently, a monoclonal antibody, natalizumab (Tysabri), that targets the α_4 integrin subunit has been re-approved in the United States for a restricted subset of patients with MS, after being withdrawn from the market in 2005 after three patients developed progressive multifocal leukoencephalopathy due to the JC virus. This drug is aimed at blocking the movement of leukocytes from the blood into sites of inflammation.

Answer 3
The patients got markedly worse. IFN-γ upregulates the expression of MHC class II molecules and thus enhances antigen presentation. In addition, it drives the differentiation of T_H1 cells, which are involved in the pathogenesis of MS.

Answer 4
Proteins eaten as part of food have long been known not to elicit routine immune responses. The reason seems to be that there are antigen-specific mechanisms in the gut for suppressing peripheral immune responses to antigens delivered by mouth. One is that when T cells in gut-associated lymphoid tissues are presented with orally delivered protein antigens in the absence of an infection, a lack of co-stimulatory signals induces the T cells to become anergic. Another involves the development of regulatory T cells (see Case 7), which can actively suppress antigen-specific responses after rechallenge with antigen. Such cells produce cytokines, including interleukin-4 (IL-4), IL-10, and TGF-β, which inhibit the development of T_H1 responses and are associated with low levels of antibody and virtually absent inflammatory T-cell responses. However, attempts to treat MS in humans by feeding the MBP antigen have proved unsuccessful.

Answer 5
EAE cannot be induced in mice lacking CD28. CD28 on T cells is the receptor for the B7 co-stimulatory molecules, which are essential for the activation of naive antigen-specific T cells, including T cells that recognize MBP. In contrast, mice in which the cell-surface protein CTLA-4, another receptor for B7 molecules, has been knocked out, develop EAE more readily than their normal littermates. This is because CTLA-4 binds B7 molecules about 20 times more strongly than does CD28, and normally delivers an inhibitory signal to the activated T cell. In CTLA-4 knockout mice, this inhibitory signal is missing and so the T cells are more readily activated.

Case 45

Answer 1
The red blood cells, coated with Rh antibody, adhere tightly to the Fc receptors of macrophages in the red pulp of the spleen, Kupffer cells of the liver, and elsewhere. It turns out that most IgG antibody to the Rh antigen is of the IgG3 or IgG1 subclass, the IgG subclasses that bind most tightly to the high-affinity Fcγ receptor (CD64) (Fig. A45.1). The macrophages destroy the antibody-coated red cells, as they would antibody-coated bacteria (Fig. A45.2).

Free immunoglobulin does not bind to Fc receptors	Aggregation of immunoglobulin on bacterial surface allows binding to Fc receptors	Conformational change in immunoglobulin on binding to antigen may expose Fc-binding site
bacterium Fc receptors macrophage		
No destruction of bacterium	Activation of macrophage leading to destruction of bacterium	

Fig. A45.1 Bound antibody is distinguishable from free immunoglobulin by its state of aggregation and/or by conformational change. Free immunoglobulin molecules cannot bind Fc receptors. Antigen-bound immunoglobulin, however, can bind because several antibody molecules bound to the same surface bind to Fc receptors with high avidity. Some studies also suggest that antigen binding and aggregation induce a conformational change in the Fc portion of the immunoglobulin molecule, increasing its affinity for the Fc receptor. Both effects probably contribute to discrimination by Fc receptors between free and bound antibody.

Answer 2

If the mother and father are ABO identical, the fetus has a high likelihood of also being ABO compatible with its mother. If fetal blood enters the maternal circulation, it is likely to last much longer if the mother has no ABO alloantibodies to the fetal cells. For example, if the fetal red blood cells were type B, they would be quickly hemolyzed if the mother were red blood cell type A; she would have anti-B alloantibodies. Rapid destruction of the fetal red blood cells by hemolysis would impede alloimmunization of the mother.

Answer 3

Rh-negative cells were used because they cannot be hemolyzed by the IgG Rh antibodies that have crossed the placenta.

Answer 4

You shouldn't, because the amount given (300 µg of IgG) is insufficient to cause the fetus harm. This amount of antibody raises the maternal titer in the indirect Coombs test to <1:4, a titer of Rh antibody that cannot cause significant hemolysis in the fetus. If the maternal titer is >1:4 she has been alloimmunized by her fetus.

Answer 5

She has IgM anti-Rh antibodies; they agglutinate Rh-positive cells in saline, unlike IgG anti-Rh antibodies. As IgM antibodies do not cross the placenta you would have no immediate concern about her pregnancy. However, she should be tested repeatedly to be sure that she is not developing a positive Coombs test, signifying the presence of IgG antibodies.

Case 46

Answer 1

These infections are normally controlled by macrophages activated by T cells. Christopher's cell-mediated immunity was suppressed by azathioprine, prednisolone, prednisone, and cyclosporin A.

Answer 2

It prevents the synthesis of interleukin-2 and other cytokines that are essential to the clonal expansion of T cells and also the activation of their effector functions. Cyclosporin A acts by inhibiting the activation of the cytokine genes by the transcriptional activator NFAT. NFAT is a cytoplasmic phospho-

Fig. A45.2 A rosette of Rh-positive (Rh⁺) red cells. These are coated with anti-Rh antibodies, which adhere to the Fc receptors on a macrophage (central cell). The macrophage is pitting the red blood cells and destroying them. Photograph courtesy of J. Jandl.

protein that is activated when it is dephosphorylated by calcineurin. Dephosphorylation enables NFAT to translocate to the nucleus, where it activates the synthesis of cytokines. Cyclosporin A inhibits the dephosphorylation of NFAT by calcineurin.

Answer 3

The graft became infiltrated with lymphocytes that caused a general inflammatory response, as well as a specific attack on the cells in the graft by cytotoxic CD8 T cells. The graft became swollen with edema fluid and inflammatory cells. The renal capsule was stretched by the edema fluid and cellular infiltrate, and this induced pain.

The donor was HLA-A2,11 and HLA-B7,35. If Christopher received another graft with the markers HLA-A11, or HLA-B7 or 35 he would reject such a graft more rapidly because he would have been sensitized to these HLA markers. This is second-set rejection.

Answer 4

To avoid hyperacute rejection, for which there is no therapy. If these events had occurred, a valuable kidney graft would have been wasted. On the other hand, HLA mismatching would lead to problems with first-set rejection, which could be managed with immunosuppressive therapy.

Answer 5

Anti-ICAM interferes with the interaction of LFA-1 on lymphocytes and other leukocytes with ICAM-1 expressed on endothelial cells. Thus it would impede the emigration of white cells from the circulation into the graft (see Case 3). When Christopher's kidney graft was threatened with rejection, it became important to get rid of the T cells mustering to attack it. Antibody to CD3 would remove T cells from the circulation and lower the total body T cells as it apparently did in this case (only 15% of Christopher's peripheral blood lymphocytes (PBL) reacted with anti-CD3, compared with a normal value of 75–80%).

Answer 6

Juvenile onset diabetes is strongly associated with certain MHC class II alleles (see Fig. 46.3). It is likely that Christopher and his father's diabetic relatives share one or more of these alleles.

Answer 7

HLA-DR2 is least frequently encountered in people with IDDM and is therefore associated with a lower risk.

Answer 8

Not necessarily. The fact that HLA-DR3 and HLA-DR4 are associated with IDDM does not necessarily mean that they are involved directly in the mechanism of susceptibility. The HLA-DR3 and HLA-DR4 genes could simply be very close to some other gene that is therefore usually inherited with them and is the real reason for the IDDM susceptibility. This problem, which is known as linkage disequilibrium, applies to any genetically defined disease association. In fact, the inheritance of susceptibility to IDDM is now known to be more closely associated with the inheritance of specific alleles of HLA-DQ genes, which are adjacent to the HLA-DR genes in the MHC (see Fig. 18.2). Molecules belonging to the HLA-DQ type are difficult to type by using antibodies, and this association has only become clear since it became possible to look directly at the DNA encoding them. Most of the information we have on susceptibility to IDDM therefore comes from patients who were typed only for HLA-DR alleles.

Answer 9

HLA-DR and HLA-DQ are class II MHC molecules that present peptide antigens to CD4 T cells. You might therefore expect that specific alleles of these molecules are particularly efficient at presenting peptides derived from pancreatic β cells and this could explain why T cells become activated against these cells in IDDM. But it is, in fact, not at all obvious how this would work, and the mechanism of disease susceptibility in IDDM is an outstanding issue in research immunology. The problem is that pancreatic β cells themselves cannot present their peptides to CD4 T cells because they do not express MHC class II molecules (and even if they did, MHC class II molecules do not present antigens generated endogenously). The direct target of the autoimmune attack by T cells on pancreatic β cells is therefore not known, and neither is the connection between that target and the MHC class II molecule or molecules associated with IDDM.

The mechanism of protection by HLA alleles associated with low risk is also unclear. One obvious possibility is that these HLA molecules cannot bind antigenic pancreatic β-cell peptides. However, this cannot be the explanation, because the protective alleles actually prevent the development of IDDM in individuals with strong susceptibility alleles. This has been demonstrated conclusively in NOD mice, into which genes for protective MHC molecules can be introduced, and the resulting transgenic animals can be shown to be protected from developing IDDM. The explanation for these dominant protective effects is again unknown.

Answer 10

Perhaps not surprisingly, the proportion of NOD mice developing diabetes has been found to vary depending on how carefully the animal houses are kept free from infection. Perhaps more surprisingly, high rates of infection are associated with lower rates of IDDM. This is another unexplained but important phenomenon: there are other autoimmune diseases in which decreased susceptibility is associated with infectious disease. For example, systemic lupus erythematosus, which we discuss in Case 43, is never seen in black Africans in Africa where parasitic disease is endemic. When African blacks migrate to temperate countries where there is no parasitic disease, however, the incidence of lupus quickly approaches that seen in the native Caucasian population. In other autoimmune diseases, by contrast, infection clearly increases susceptibility (see Case 40).

Case 47

Answer 1

The engrafted T cells appear to recognize allogeneic antigens on the recipient's hematopoietic cells and thus will attack the leukemic cells. One such antigen, HB-1, which is a B-cell lineage marker, is expressed by acute lymphoblastic leukemia cells, which are B-lineage cells, and by B lymphocytes transformed by Epstein–Barr virus (EBV).

Answer 2

IFN-γ induces the expression of MHC molecules on cells and this makes GVHD worse, as it provides more targets for the donor T cells.

Answer 3

It is not entirely clear why the anti-CD2 was more therapeutically efficacious compared with the anti-CD3. One reason might be that anti-CD3 antibodies rapidly cause the disappearance of CD3 molecules from the surface of T cells, and so the T cells can be no longer recognized by the antibodies, whereas anti-CD2 antibody does not have this effect on CD2.

Answer 4

One reason could be that the skin and intestine express a higher level of MHC molecules than other tissues. The intestinal tract is also likely to be damaged by the preparative cytotoxic treatments given to destroy the recipient's bone marrow. The damage induces the production of cytokines: as well as inducing MHC molecules, these can also drive GVHD and make the tissue susceptible to immunological attack.

Figure Acknowledgments

Case 1

Fig. 1.2 bottom panel from *The Journal of Experimental Medicine* 1972, **135**:200–219. Copyright 1972 The Rockefeller University Press.

Case 4

Fig. 4.5 from Williams, K.: Hereditary angioneurotic edema. *Clinical Immunology* 1975, **4**:174–188. Copyright © 1975, with permission from Elsevier.

Case 6

Fig. 6.2 from *Blut* 1990, **60**:309–318. © Springer-Verlag.

Case 7

Fig. 7.1 from *The Immune System*, 2nd edition, by Peter Parham. © 2005 by Peter Parham. Used by permission of Garland Science.

Fig. 7.3 from Notarangelo, L. et al.: Primary immunodeficiency diseases: an update from the International Union of Immunological Societies Primary Immunodeficiency Diseases Classification Committee Meeting in Budapest, 2005. *Journal of Allergy and Clinical Immunology* 2006, **117**:883–896. Copyright © 2006, with permission from the American Academy of Allergy, Asthma, and Immunology.

Fig. 7.4 from Boschan, C.: Neonatal-onset multisystem inflammatory disease (NOMID) due to a novel S331R mutation of the CIAS1 gene and response to interleukin-1 receptor antagonist treatment. *American Journal of Medical Genetics Part A* 2006, **140A**:883–886. Copyright © 2006 John Wiley & Sons, Inc. Reprinted with permission of Wiley-Liss, Inc., a subsidiary of John Wiley & Sons, Inc.

Fig. 7.5 from Petrilli, V., Papin, S., Tschopp, J.: The inflammasome. *Current Biology* 2005, **15**:R581. Copyright © 2005, with permission from Elsevier.

Case 8

Fig. 8.3 from McDonald, D.R., Brown, D., Bonilla, F.A., Geha, R.: Interleukin receptor-associated kinase 4 deficiency impairs toll-like receptor dependent innate anti-viral immune responses. *Journal of Allergy and Clinical Immunology* 2006, **118**(6):1357–1362. Copyright © 2006, with permission from the American Academy of Allergy, Asthma, and Immunology.

Case 13

Figs. 13.1 and **13.2** from Castigli, E., Geha, R.S.: Molecular basis of common variable immunodeficiency. *Journal of Allergy and Clinical Immunology* 2006, **117**:740–747. Copyright © 2006, with permission from the American Academy of Allergy, Asthma, and Immunology.

Fig.13.3 from Cunningham-Rundles, C., Bodian, C.: Common variable immunodeficiency: Clinical and immunological features of 248 patients. *Clinical Immunology* **92**:34–48. Copyright © 1999, with permission from Elsevier.

Fig.13.4 reprinted by permission from Macmillan Publishers Ltd: *Nature Genetics* 2005, **37**:829–834, © 2005.

Case 14

Fig. 14.2 bottom left panel from *Diagnostic Immunopathology* (2nd edition), Eds. R.B. Colvin, A.K. Bahn, and R.T. McCluskey. New York, Raven Press 1994, 246–247. © 1995, Raven Press.

Index